KEATON RYON

# Abyssal Sapience

### Deep Time
### Hidden Minds
### Unveiled Origins

Keaton Ryon

# KEATON RYON

Published by Teumessian House
La Jolla, California
teumessianhouse.com

First edition

Printed in the United States.

ISBN: 979-8-9942315-7-9

Some material in this book reflects speculative interpretation and should not be taken as established scientific fact.

Cover concept and design by Keaton Ryon.

KEATON RYON

# DEDICATION

To the hidden part of us that asks the first question. To the bonds
that shape us, even when their names elude us. To all that vanishes
into the deep and returns, unbidden, in memory.

KEATON RYON

# CONTENTS

KEATON RYON

KEATON RYON

## ACKNOWLEDGMENTS

To all the scholars, seekers, and wanderers who dared to ask questions without easy answers, the author is grateful. Some pursued explanations through the rigor of science. Others searched in the hidden folds of imagination. All were moved by the same impulse, to reach into the unknown and bring something back into the light. This work stands on the shoulders of that restless curiosity.

# PROLOGUE

**The Weight of Time: On the Memory of Matter**

---

*"Deep time is measured in units that humble the human instant."*

Robert Macfarlane, Underland: A Deep Time Journey

---

T ime.

We use this word... all the time. Its ubiquity is so enmeshed in our daily existence that we seldom think of it beyond its function as a guidepost or dangling lure. A temporal carrot on an apocryphal stick.

"What time is it?"
"When will they arrive?"
"We're already late!"

The moment we anticipate always gives way to the next. We sense this progression more than we openly acknowledge it. But humanity's perception of the ticking of time "passing" is an illusion. Faith and science rarely see eye to eye. Yet on this point, they seem to agree. The interconnected lattice of individual experiences we refer to as time is merely a helpful construct. It is a way for biological brains to make sense of the physical world as we move through it.

The ancient Greeks often spoke of Chronos versus Kairos.[1] Chronos refers to a measured, sequential interval, representing a linear progression of time. In contrast, Kairos embodies the powerful presence of a timeless moment that exists outside the usual flow and order of time.

In Judaism, time tends to be seen as cyclical, a repeating pattern that connects our perceived present to a series of divine emanations.[2] Islam also presents this broader perspective. "...a day with your Lord is like a thousand years of those which you count." [Qur'an, Surah Al-Hajj 22:47]

Eastern philosophies, such as Buddhism and certain schools of Hindu thought, teach that one must transcend temporal illusions to achieve true freedom.

Turning to the realm of scientific inquiry and revelation, we have many luminaries to consider. But in the author's humble opinion, the most foundational thought remains Einstein's theory of relativity and his model of the *Block Universe*, first envisioned by his former teacher, Hermann Minkowski. Einstein summarized this unwieldy and profound concept with the following statement.

> *"The distinction between past, present, and future is only a stubbornly persistent illusion."*[3]
>
> *- Albert Einstein*

This statement is easy to dismiss. But the idea is not just foundational to human understanding. It is woven into the structure of creation itself.

Consider the following.

* **Relativistic Time Dilation**: This is a popular concept in science fiction. The faster an observer travels through space, the more slowly their experience of temporal passage unfolds. This is not a perceptual illusion, but a *genuine* phenomenon. It is a measurable deceleration of the chronological flow of time relative to those who remain stationary or move at lower speeds.

Science fiction fans may recognize this concept from Joe Haldeman's seminal masterpiece, *The Forever War*, in which soldiers travel at near-light speed, caught in a conflict that spans the galaxy. While only weeks or months pass for them aboard their ships, decades or even centuries elapse on Earth.[4]

And of course, there's Rod Serling's unmistakable influence on the same idea in the classic film *Planet of the Apes*.[5] No spoilers, but most readers likely recall how that one ends.

* **Relativistic Simultaneity**: What is "now"? Is our measure of reality what the clock shows, or is it tied to a position in space? If it's 3 p.m. Tuesday in Melbourne, Florida, then it's 5 a.m. in Melbourne, Australia. We grasp this practically, but it reflects the less intuitive implications of this phenomenon.

What time is it on a planet orbiting a star in the Andromeda galaxy 2.5 million light-years away? It's tempting to answer "Who cares?" but consider how entirely relative our perception of chronology is to our position in the cosmos.

* **Gravitational Time Dilation**: The stronger the gravitational field, the more slowly events unfold within its sphere of influence. The most compelling dramatic exposition of this model the author has encountered is Christopher Nolan's script for the film *Interstellar*.

In this fictional account, space explorers (deliberately unnamed here to avoid spoilers) visit a planet orbiting close to a black hole. Its extreme gravity warps time so inordinately that every hour on the surface equates to seven years back on Earth. This creates a fantastically taut and compelling narrative tension. The crew must act quickly or risk returning to find that everyone they once knew and loved has long since aged and vanished.[6]

It's a powerful scenario, made all the more effective by how faithfully it translates real physics into human consequence.

A real-world example of gravitational time dilation can be found in the atomic clocks aboard GPS satellites. Orbiting about 20,200 kilometers (12,550 miles) above Earth's surface, these clocks tick faster than those on the ground.[7]

This discrepancy must be constantly corrected for. Otherwise, these satellites would drift disastrously off course.

Another striking example of relativistic time dilation is observed in muon decay. Similar to the soldiers in *The Forever War*, these tiny particles travel at near-light speed. They decay rapidly into other particles, typically within about 2 microseconds.

For reference, a microsecond is one millionth of a second. Muons are created high in Earth's atmosphere when cosmic rays collide with atomic nuclei, tens of kilometers above the surface. At rest, they shouldn't survive long enough to reach the surface from their birthplace high in the atmosphere. Yet because they travel so quickly, their subjective temporal flow slows drastically, allowing them to survive far longer than expected from a stationary perspective. This relativistic extension of their lifespan enables them to reach ground-based detectors, providing direct and measurable evidence of temporal dilation at near-light speed.[8]

So, what are the implications? They might seem heady or esoteric at first. They are anything but.

Minkowski's formulation of space-time, later expanded and given physical meaning by Einstein's theory of relativity, conceives reality as a four-dimensional structure in which time exists alongside space.[9] Within this framework, which is referred to as the Block Universe or eternalism, the past, present, and future coexist as a single, continuous whole. We do not move through time. Rather, consciousness partitions it, slicing through the continuum like a knife through a freshly baked loaf of bread.

Time, then, is not a river flowing past us but a dimension within which awareness perceives its own unfolding. What we experience as change is the mind encountering each cross-section of that greater whole. It is this succession of moments that together form the illusion of time's passage, a rhythm within an eternal heartbeat.

It is not necessary to fully comprehend these theories right away. Even the author can occasionally struggle to visualize them. Still, they are useful, not as definitive answers, but as new lenses through which to view deep time. They encourage the exploration of ideas that require us to take flight from the nest of our shared, familiar perspective as humans living on this planet.

Extending the bread analogy, each slice of the loaf represents a single flash of awareness within the continuum. We imagine ourselves living in an individual slice, then the next, and so on. However, under the Block Universe model, the entire loaf exists simultaneously.

Our minds encounter one slice after another, but every slice is always present. Nothing passes away. Nothing materializes from a formless void. In Einstein's view, all points in the continuum coexist. Nothing begins. Nothing ends. All is everlastingly present.

Although frequently framed in scientific terms, this idea resonates deeply with spiritual traditions across cultures and centuries. One of the most powerful expressions of this comes in God's response to Moses when the patriarch asks for the Divine name.

*"And God said unto Moses, I AM THAT I AM."*

*(Exodus 3:13-15, KJV)*

One need not subscribe to any specific faith to recognize the symmetry between Einstein's Block Universe and this declaration of perpetual being. It reflects the same fundamental truth: eternalism, which is the opposite of presentism.

God is described as eternal, always existing, and beyond any linear sequence of events. This mirrors the scientific notion that all individual points exist simultaneously.

We can overlay this spiritual understanding onto logic-based frameworks. Modern physics, too, gestures toward continuity rather than annihilation. The quantum no-hiding theorem demonstrates that information is never truly lost, but rather transformed. The first law of thermodynamics declares the same of energy. It cannot be created or destroyed. It can only be converted from one form to another. Taken together, these principles echo the ancient intuition that nothing in existence truly vanishes. All things observable or unobservable persist in some form. They do not begin or end.[10]

They have always been.

A popular expression of this truth is the phrase, *"We are all made of stardust."* While it might sound like a wistful bumper sticker or novelty T-shirt slogan, it is, in fact, a scientifically accurate reflection of our reality.

All the elements that compose the visible and invisible universe, including human beings, were forged in the hearts of ancient stars.

When a star dies, these elements are expelled in a violent explosion and scattered across the cosmos by supernovae.

But even those elements didn't originate in the heart of a star. They coalesced in an even earlier age, forming the stars themselves.

So where were they before that?

Whether you lean toward faith or science, the answer remains the same. They always were.

The main thrust of this book is not eternity, time dilation, or cosmological models. Nor is it an attempt to comprehend the whole enchilada.

Instead, the author seeks to explore a topic that intersects with these ideas, a subject both adjacent to and emergent from them. This prologue is meant to unmoor the reader's mind, attitude, and assumptions from the comfort of simpler paradigms. We carry these tidy shorthands like a quietly ticking timepiece in our breast pocket.

Let's leave those behind, for now. Together we will confront the invisible violence of time itself.

This inexorable force of the cosmos swallows all. Civilization. Identity. Memory. Self and other. Other and self. Eventually, it spits them back, scrambled, estranged, and hauntingly familiar.

Not unlike the final, macabre revelation beheld by marooned astronaut George Taylor from *Planet of the Apes*, raging at the buried past on a ruined shore.

So here we stand, at the threshold of perception.

There's no need to criticize those who came before for their failings or blind spots. We can instead turn our gaze toward places that are yet to be explored.

And they abound, dear reader.

In their unfathomed depths, something blossoms, strange and proliferating.

Inside this eternal moment is an infinite expanse of possibility. It is an echoing remainder of what is unglimpsed, unknown, and perhaps unknowable.

Together, we will assail this sealed tomb of occluded enigmas and forgotten manifolds.

Behind it lie not mere revelations, but an epistemological rupture. It is one that can no longer be denied or ignored.

Descend now into the abyss, where our perception falters, and worlds exist far beyond ordinary comprehension.

Shake off your temporal cloak. We undertake this quest together.

The author is but one soul, pounding at the chamber door of eternities past and epochs yet to come. But together, we can turn the tide and crack open this forgotten mystery, buried deep in our consciousness and deeper still in the farthest, unknown reaches of this planet.

It is the author's hope and firm belief that in breaching this barrier, we will together encounter the greatest of our grandfathers and grandmothers.

In doing so, we will complete a circuit. It is one we have been pursuing for so long that we have forgotten what our fingertips anticipate.

The singularity does not exist in the past, the future, or even the present.

It resides outside of time.

And this is where we must encounter it.

KEATON RYON

# CHAPTER ONE

### Unknown Unknowns: The Presence of What is Missing

*"... there are known knowns; there are things we know we know. We also know there are known unknowns; that is to say, we know there are some things we do not know. But there are also unknown unknowns - the ones we don't know we don't know."*

*— Donald Rumsfeld, U.S. Department of Defense press briefing, February 12, 2002*

When pushing far enough into obfuscation, you occasionally stumble upon a startling clarity that might be impossible to find elsewhere. Was there ever a more clear-cut case than Donald Henry Rumsfeld's now-famous statement above? Throughout his long career, Rumsfeld wore many titles. Most notably, he served as both the youngest and, decades later, the oldest Secretary of Defense in U.S. history. First under Gerald Ford, then under George W. Bush. While his legacy includes many memorable actions and remarks, this one stands apart for its peculiar blend of logic and absurdity. It proved so compelling that famed documentarian Errol Morris built an entire film around it (*The Unknown Known*, 2013), using the quote as the centerpiece. It's a fascinating watch. The author encourages readers to seek it out, regardless of political affiliation, ideology, or lack thereof.

Setting aside fever dreams of hegemony and calculated ambiguity, let's instead focus on the concept of "unknown unknowns." It's a phrase that, despite its origins in political doublespeak, perfectly frames the core thrust of this chapter. Born of weaponized vagueness, it distills a clear, pointed model of what lies beyond our ability to see, touch, or measure. Nowhere does the concept of "unknown unknowns" feel more apt than when we turn our focus to the boundless and mysterious waters at our feet.

As we stand at the ocean's edge, we gaze over this implacable behemoth with reverence and awe, mindful of its unimaginable power. Its skin ripples under wind and sun. Beneath, tsunamis build, currents churn, and a multitude of lives flourish, from whales and dolphins to plankton and krill. Instinct tells us this is a different universe from our own, close enough to touch yet fundamentally unknowable. We are separated from it by deep evolutionary divides. Life that evolved in water followed a path so divergent from ours that we can't even agree whether most of its denizens possess language, use tools, or fit into any of the familiar frameworks we insist on applying.

Too often, we plant ourselves on a throne of ascendancy, crowning humanity as the sole gatekeeper of "intelligence." But perhaps those who truly guard that gate remain utterly unconcerned with, and possibly unaware of, our inflated self-regard.

The author does not intend to diminish humankind's contributions to Earth's history. But a healthy dose of perspective is never a bad thing. It is comforting, of course, to regard one's own species as preeminent. Yet can we truly order an incomplete list? Can we categorize that which we have yet to conceive?

What feeds the human soul more, the satiation of dominance or the thrill of confronting the unknown?

Human eyes are drawn to that which reflects the light. It's how we are made. This form of beauty is undeniable. But the light only penetrates so far into the darkness. What remains unseen is unfathomable.

Faith and science both offer innumerable examples of this truth. The author will not recite them all here. The faith of any tradition rests on the assurance of things not seen. Science, by contrast, employs the scientific method, a disciplined process of accumulating knowledge through careful observation and rigorous skepticism. It remains the gold standard of inquiry precisely because cognitive assumptions can so easily distort interpretation.

And what if there was a place on Earth, a colossal one, that has never truly been observed? Never measured. Never comprehended. Never experienced.

Our Homeboy of Perpetual Scientific Potable Quotables, Carl Sagan, once remarked: *"The absence of evidence is not evidence of absence."*[11] Indeed, some of us are not blinded by a lack of vision, but by a failure to ask the right questions. Or perhaps by a lack of willingness to approach the unfamiliar at all, much less breach it.

This book is being written in what we casually refer to as the "early 21st century." In this era, humanity has begun to dream in earnest about life beyond its cradle.

Our focus shifts from time to time, but at present it rests squarely on our orbital neighbor and so-called angry red cousin, Mars. Also known throughout the ages as Ares, Angaraka, Ma'adim, and Huo Hsing, it has sparked legend, fear, and curiosity in equal measure.

We theorize and postulate on its history and provenance, pouring ever greater portions of our stored treasure into funding mechanical missionaries that land on its surface to gather data. We measure air and soil, hunting for any sign of life or even the potential for it, that might bolster dreams of future migration.

To date, humanity has launched just over 50 spacecraft across the 140-million-mile gap, orbiting and probing to capture our neighbor's vital signs and hidden truths.[12] We have learned so much.

We still know so little.

And yet...

There is a place far more mysterious than Mars, a realm we know far less about. When you stand on a beach, this environment begins roughly 1,000 meters away, or less than a mile from shore. It is the deep ocean, separated from us not so much by distance as by the categorical absence of light and the crushing weight of immense pressure.

These two factors, combined, push this frontier even further beyond our collective imagination than Mars. We can *see* Mars. But if there were a planet where no natural light could penetrate, and where the atmosphere would flatten even the sturdiest materials we could invent, would we dare send even a single mission?

The answer lies in the stark difference between the number of expeditions funded to explore this enigmatic realm here on Earth versus similar environments found in our solar system. "Follow the money" is a reliable maxim. So, let's Woodward and Bernstein this for a moment.

NASA's budget for fiscal year 2024 totaled about $27 billion. Nearly half of that (around $13 billion) supported spaceflight programs, including human missions, low-Earth-orbit operations, and deep-space exploration.[13] This figure does not include the considerable sums invested by private aerospace firms such as SpaceX and Blue Origin, whose financial disclosures remain limited. Even so, industry estimates suggest that SpaceX alone generated over $13 billion in 2024 revenue, with roughly a third of that tied directly to launch and orbital operations.[14] Blue Origin's annual spending, while less transparent, likely exceeded $1 billion through NASA contracts and engine production.[15] Together, these figures suggest that roughly $18–20 billion was funneled into sending humans and machines beyond Earth's atmosphere in a single year. This scale of investment rivals that of some national space programs.

Consider Venus. It is a suffocating planet cloaked in toxic clouds, hot enough to melt lead, and pressurized like the brutal depths of Earth's ocean trenches. Yet humanity has launched forty-six missions toward Venus, roughly half of which succeeded in reaching orbit, returning data, or touching the surface.[16] By contrast, humanity has sent only a handful of expeditions to Challenger Deep, the deepest known point on our own planet.[17]

We have sent probes to Saturn's moon Titan, dropping instruments into its methane-soaked skies. We even dream of cracking open Europa's icy crust to reach the alien ocean hidden beneath its frozen skin. All this before fully plumbing the darkest waters of our own planet.

That's not just a technological choice. It's a psychological one.

We are drawn upward and outward toward the stars, even when they are inhospitable and deadly. Yet the deeper we look down, into our own world, the more we recoil.

In fiscal year 2024, the National Oceanic and Atmospheric Administration received a total of $6.72 billion in funding. This seems respectable at first glance. Yet its Office of Ocean Exploration, the only federally funded program dedicated solely to ocean exploration, was allotted just $46 million, or roughly 0.7 percent of NOAA's budget.[18] Compared to the lavish sums poured into space exploration, that figure is almost laughably small. Still, there is a faint but growing glimmer of hope. Private initiatives to probe the deep ocean are on the rise. Even so, the money involved remains a mere drop in the bucket compared to what private companies funnel into space.

So, now that we've dissected the dollars, what explains this funding gap? The simplest answer is that most people don't even see it as an imbalance. America's fixation with space exploration is a legacy of the Cold War, a period when the United States and the Soviet Union waged what historians describe as a meta-proxy war. This was not fought on battlefields or oceans, but on the frontiers of technology, ideology, and propaganda.

Citizens of the U.S. have long enjoyed a wealth of films, books, and television shows celebrating the early years of spaceflight. Yet even the most balanced accounts tend to tip the scales toward glorifying the American program while giving short shrift to Soviet achievements. Many Americans recall John F. Kennedy's famous 1962 "We choose to go to the moon" speech as the dawn of the space age.[19] However, it was actually a bold response to the Soviets' earlier triumphs. By that point, our Russian counterparts had already...

- Launched the first artificial satellite, *Sputnik* (1957)
- Sent Laika the dog into orbit (1957)
- Landed the first human-made object on the Moon (1959)
- Sent Yuri Gagarin into space aboard *Vostok 1* (1961)[20]

These were stunning achievements worthy of celebration by all humankind. Yet in that era, the U.S. usually downplayed or buried such news, while tightly controlled Soviet media emphasized collective glory over individual heroism.

Both nations made phenomenal leaps in technology and human ingenuity during this generation. For the purposes of this book, it's worth looking at the motivations and turning points for each side. Both have shaped our enduring enchantment with the stars above while leaving the depths below largely unexplored. But this fixation on the sky didn't begin with *Sputnik* or *Apollo*. Long before superpowers raced for orbital supremacy, human beings had already crowned the heavens with their curiosity. The sky has always pulled at us. Not merely as a theater for scientific conquest, but as the earliest canvas for myth, meaning, and longing.

Many books and films depict early humans gazing skyward in wonder, tracing constellations, divining sacred origins, and worshiping heavenly deities. This isn't a romantic invention. Archaeological evidence abounds of ancient cultures with remarkably advanced astronomical knowledge.

One of the most striking examples dates to 4–5,000 BCE, on Egypt's Nabta Plateau. Here, a people constructed an archaeoastronomical stone circle that predates Stonehenge by more than two millennia. This prehistoric wonder aligned with the solstices and may depict the constellation Orion.[21] Even if one discounts Nabta entirely, a long line of early civilizations left well-documented records of sophisticated celestial understanding, including:

- Sumer
- Babylonia
- Egypt
- Olmec
- Maya
- Aztec
- Inca
- Hittite
- China

- India
- Minoan and Mycenaean
- Greece
- Persia
- Aboriginal Australia
- Polynesia
- Druid

From a modern perspective, the depth of their knowledge is astonishing and so advanced for its time that we still struggle to understand how they acquired it.

When JFK made his proclamation, it carried the weight of history. One can imagine those anonymous sky watchers, shamans, and star cartographers from forgotten eras collectively raising their fists in triumph. The innate human urge to look up in wonder and reverence is undeniable and woven into the very development of our species.

This book, however, asks a different question. Did any of these cultures devote similar attention or resources to exploring the hidden realm of the deep ocean? Not really. Many did become accomplished masters of sea travel in all manner of watercraft. But the first genuine effort to plumb the depths didn't come until 1872, with the launch of the *HMS Challenger* Expedition.

The *Challenger* was a British Royal Navy corvette retrofitted for a groundbreaking scientific mission. Its crew used weighted ropes to measure ocean depths. This method ultimately gave the Challenger Deep in the Mariana Trench its name. On this expedition, they mapped what was then the deepest known point in the ocean. They measured temperature at extreme depths, hauled up sediment and strange life from the abyss, and returned with voluminous new data on salinity, currents, and marine biology. Among the cataloged specimens, more than 4,700 were entirely new to science. This achievement is nothing short of extraordinary.[22]

Before we proceed, it's essential to acknowledge a notable contrast and a salient fact. The sky is ever-present above us, almost always visible in some form, while the deep ocean remains hidden. Entire landlocked cultures may have had no reason even to imagine its existence. Yet, a quick look at humanity's long relationship with the sea reveals a consistent attitude toward these invisible depths. It is one of caution, even dread. Across time and cultures, the briny deep has been cast as a vault of unbidden horrors, best avoided. Some of our oldest surviving accounts depict it as the inverse of the heavens, which were, as a rule, seen as the home of gods and the source of life. In the collective unconscious, the ocean became a place of danger and death. This was not without good reason.

The Sumerians (circa 4000 BCE) told of Tiamat, the goddess of saltwater chaos, a personification of the primordial sea itself. In the Babylonian creation myth, the god Marduk engages her in a titanic battle, ultimately defeating her and using parts of her body to shape the heavens and the earth. In this telling, the sea is cast as a cosmic adversary. It is all-powerful, hostile to humankind, and conquerable only by a god or a champion among men.[23] The stuff of legend, indeed. Does this sound familiar? Even without direct knowledge of Sumerian or Babylonian texts, most of us instinctively recognize the form of this struggle, or at least feel it stir something deep within.

This story recurs as a central motif in many archaic religions and myths. Scholars call it *Chaoskampf* or literally, "battle against chaos."

> *The Chaoskampf myth is a category of divine*
> *combat narratives with cosmogonic overtones ... in which*
> *the hero god vanquishes a power or powers opposed to*
> *him, often dwelling in, or identified with, the sea, and*
> *presented as chaotic, dissolutory forces.*
>
> "Distinguishing Wood and Trees in the
> Waters: Creation in Biblical Thought" (2022)[24]

This view of the ocean as malevolent, chaotic, and actively opposed to humankind recurs throughout our oldest myths, stories, and religious texts. Homer's *Odyssey* gives us Scylla and Charybdis. The *Tanakh* (Hebrew Bible/Old Testament) offers Leviathan and, of course, the Great Flood. In Norse mythology, there is Jörmungandr, the "Midgard Serpent." Chinese tradition recounts the stories of the Dragon Kings of the Four Seas. Hindu cosmology offers the Samudra Manthan, also known as the "Churning of the Ocean." Roman and medieval lore warn of the Kraken.

From primitive charts to medieval maps, the uncharted edges of the world were frequently embellished with sea monsters and mythic beings, serving as visual warnings of the peril and mystery that awaited beyond the known seas.[25] Across cultures and centuries, a clear consensus emerges. The ocean's vastness is not only beyond our capacity to explore but perhaps even impossible to comprehend. Time and again, we are cautioned about beings and realms that defy our most basic assumptions regarding the nature of reality itself.

When we compare these attitudes toward the sea with the views of the heavens held by the same cultures, the difference is glaring.

The myths and sacred texts regarding the sky are not uniformly positive, but they are far more varied and nuanced in their representations. The stars and heavens are in many cases tied to complex relationships with gods, spirits, and other visitors "from above." Across time and geography, the night sky is consistently linked to order, divinity, and eternity, whether as a source of wisdom or as the luminous destination of the afterlife. The prevailing sentiment might be summed up as *hopeful longing.*

By contrast, the thread that runs through these same cultures' vision of the abyss is one of *dreadful awe.* It is a recognition of something inscrutable, and ultimately indifferent to our fate.

These opposing lenses, one cast upward in hope, the other downward in dread, are not presented here to dismantle or discredit. They persist in the human imagination for good reason. Rather than challenge them, this investigation aims to probe the deeper truths that may lie beneath the myths, adding nuance to our ever-evolving understanding of the natural world. For as stories of old remind us, the "natural" realm we claim to master has always been, and still is, far stranger and more surreal than our most carefully reasoned visions could conjure.

By placing our brief human history against the backdrop of inestimable, vanishing epochs, and by reframing our inherited notions of what lies above and what lies below, we might ready ourselves for a new era of understanding that embraces our true origins. But how does one prepare for a shift in perception? Perhaps there is no preparation at all, only the resolve to face it and the quiet conviction of things hoped for... the certainty of things not yet seen.

That tension between seeing and believing is as old as humanity itself. Most of us know the phrase "Doubting Thomas," which is shorthand for someone who insists, *"I'll believe it when I see it."* The term originates from a biblical account in which one of Jesus' closest followers refuses to accept secondhand reports of the resurrection without direct proof.[26] Whether viewed as history, allegory, or myth, this moment captures something deeply human. Our reluctance to trust the unseen, especially when prior hopes have ended in disappointment. Across faiths and philosophies, we encounter this same struggle between skepticism and belief, between waiting for evidence and daring to trust without it.

The New Testament presents numerous portrayals of doubt. John the Baptist, Peter, and others remind us that uncertainty is a thread woven through the human story. In Judaism, figures such as Abraham, Moses, King David, and his son Solomon wrestled with ambiguity that tested their deepest commitments to honor, integrity, and faith. Siddhartha Gautama likewise faced this same inner turbulence. It was his encounter with doubt that fueled the transformation of thought, leading to the central tenets of Buddhism. Within Buddhist teaching, doubt is more than a stumbling block. It is also a catalyst, a springboard into mindfulness and self-inquiry.

Islam, too, offers a nuanced approach. Doubt is not only an expected outcome of reasoned human engagement with the world, but also a vital tool for navigating the tangled, contradictory tapestry of reality. Across traditions, this lesson recurs. What we believe to be true may be shaped by incomplete knowledge, misunderstandings, or blind spots so keenly ingrained that we fail to recognize them. We are all perpetually riding the swells between what *seems* true and what is, between what appears false and what truly deceives, whether by design or by accident.

Through civilizations past and present, the same wisdom emerges. Our convictions are never immune to distortion. We navigate a shifting frontier between perception and reality, belief and truth, forever vulnerable to what we have not yet learned or can't yet imagine.

From our vantage point within the walls of our own consciousness, where perception defines the limits of reality, objectivity glitters as a mirage. Like water shimmering on the horizon, it promises clarity if we can only reach it. But this, too, is an illusion. Human experience is inherently subjective. Objective truth, on the other hand, endures, unmoved and unbent, indifferent to our awareness of it. The gulf between the two is treacherous ground. This is the place where philosophical ankles are perpetually sprained.

This is the gap we must cross in the inquiries ahead. Not merely to seek what is hidden or distant, but to confront what we have never thought to look for. Together, we will consider that which hails from worlds unimagined and perhaps even unimaginable.

And so, having accepted that our reach will always outpace our grasp, we turn toward the places and things on this planet that exist far beyond that which is unseen. These are not simply hidden. For now, they remain beyond comprehension itself. Although our minds cannot yet contain them, their reality is undiminished. They are as much a part of Earth as the mountain ranges we cross or the fertile plains we cultivate. They have been so for far longer than we can begin to fathom.

Human history is a long chronicle of encounters with what was once unconsidered, unimagined, and unknowable. Yet we still consider. We imagine. We learn. In time, we extend our fingers of perception into the infinite depths. We stand firm upon ground that, in some earlier age, was as mysterious as what we now seek in hopeful longing or dreadful awe.

Whether the unknown greets us with a beckoning hand or a sudden shove, we move forward. Always forward, into the breach and beyond. This is our call now.

We must not falter at the threshold of what lies ahead. Navigating the known is brave. Exploring the known unknown is valiant. But when we arrive at the frontier of the "unknown unknowns," will we stop short? Can we afford to? Do even angels hesitate here?

No matter. Just as we once hurled our hearts and minds into the vast heavens above, so now we must grasp the handle of the door that opens into the yawning expanse below. Beyond it waits a darkness older than memory, a silence heavier than stone, yet threaded with the promise of revelation. Step through, and the world you knew will not follow.

# CHAPTER TWO

### Through a Human Lens Darkly: What the Mirror Misses

*"For now we see through a glass, darkly; but then
face to face: now I know in part; but then shall I know
even as also I am known."*

— *1 Corinthians 13:12, KJV*

These words of the apostle Paul resonate across centuries. They call us to the wonder and mystery of our place amid the ever-shifting tides of creation. They are an invitation, a clarification, an exhortation, and a warning. How do we truly understand ourselves and our purpose here? Can we? According to Paul, yes. But only through a process that involves not just placing our eye to the spyglass of existence, but recognizing that while we seek, we are also being sought.

We are not solely experiencing. We are being experienced. When we seek to *know*, we are in truth seeking to be *known*. This is among the most profound and enduring mysteries of human life and consciousness. In our quest to find our rightful place within the dance of creation, we may stumble, stepping on a partner's toe or losing the rhythm entirely. Yet just as two dancers merge into one, each surrendering selfish, solitary desires to serve the beauty of a shared choreography that no single person could create alone, so too does all of creation move as one. It does so in service of a far greater, far more beautiful order.

Paul's words remind us of this possibility. In their original context, they address matters of doctrine, morality, and insight. When Paul wrote this letter, Corinth was a wealthy, cosmopolitan port city. It was home to cultural diversity, a panoply of philosophical schools, and by some accounts, harbored a reputation for moral leniency and self-indulgence.[27] His letter, in part, served as an admonition. But that is not why his words appear here.

Instead, let us consider a less-obvious yet equally relevant reading of his message. We must cease to exclusively exercise our free will and agency as intelligent, inquisitive beings, and instead thoroughly and intrinsically grasp that the same free will and agency can be visited upon us. In that realization, our hearts and minds are freed, and we begin to move in rhythm with the universe. Only then can we truly know ourselves and our rightful place among the assembled multitudes of heaven and earth. As we will see later in this chapter, Paul's own life will offer a vivid example of what it means to meet that rhythm head-on.

From this vantage point, the author seeks to reattune our understanding of not only *who* we are and *where* we stand in the cosmos, but also of *how* we engage in the dance and *why* we sometimes waver or lose the rhythm. One need not adhere to a singular worldview or framework of faith to grasp this truth, whether through innate insight or deliberate discovery.

A more contemporary voice who forged his doctrine on the anvil of philosophy rather than faith is Thomas Nagel. His seminal essay *What Is It Like to Be a Bat?* offers as much to say on this subject as the Apostle to the Gentiles. In Nagel's view, the very act of perception confines us.

We are prisoners of our own point of view, unable to truly inhabit another mind.

*"Our own experience provides the basic material for our imagination, whose range is therefore limited. ... In so far as I can imagine this (which is not very far), it tells me only what it would be like for me to behave as a bat behaves. But that is not the question. I want to know what it is like for a bat to be a bat. Yet if I try to imagine this, I am restricted to the resources of my own mind, and those resources are inadequate to the task."*
- *What Is It Like to Be a Bat? (1974)*[28]

In Nagel's words, we find an essential echo of Paul's. Whether approached through the apostolic or the analytical lens, the takeaway is the same. There are truths about consciousness and existence that cannot be reached from a purely objective standpoint. Indeed, it could be argued that there is no such thing as a point of view that is wholly objective. Our human experience is cut from the fabric of subjectivity. Even if we could quantify and catalog every detail of a bat's neurology, physiology, behavior, and actions, we would still have no idea what it *means* to be a bat. "Bat-ness" or "human-ness" cannot be distilled into objective data points alone. There will always be an "other-ness" that slips beyond our grasp. Just as our own "other-ness" must remain elusive to any being who might study us as we, and Nagel, study the bat.

We cannot escape the confines of our human mental framework. Inevitably, this limits our imagination, leaving it biased and incomplete. Not only in truly knowing and understanding creatures of a different species, but even in fully comprehending other human beings. This limitation has no effect on the internal reality of the other creature or person. It simply means we have no direct access to it.

From this paucity of comprehension arises the idea that knowledge and experience, without relationship, are ultimately piecemeal and incomplete. When we deny our own vulnerability in perceiving the universe, we embrace an illusion. Bats do not live by blood alone. Paul and Nagel sound the same alarm within our psyche. In Paul's letter, the focus shifts from seeking knowledge of God to the deeper mystery of being known by God. Nagel echoes this elegantly, showing us that consciousness is necessarily a "first-person" experience. Any attempt to imagine it from outside ourselves will inevitably yield a result distorted by its deficiency.

Only when we set aside this outward-facing heuristic analysis and allow ourselves to be seen, known, measured, and regarded as part of a shared mutual recognition does the illusion shatter. In that moment, we begin to feel the rhythm and flow of the dance. It is only when we willingly admit that the lens of our spyglass is shadowed that our vision begins to clear. In this place, a new type of perception dawns in our hearts. We see not only what lies before us, but the truth that we are also being observed through this lens, darkly.

Carl Jung also grappled with this protean concept. It is an ontological calf that stubbornly resists the phenomenological rope. In his now-famous work *Modern Man in Search of a Soul*, he distills it to a single, piercing truth.

> *"In actual life, it requires the greatest discipline to*
> *be simple, and the acceptance of oneself is the essence of*
> *the moral problem and the epitome of a whole outlook*
> *upon life."*[29]

This idea is worth pausing on as we continue to follow the thread we have been pulling. Jung's insight rings true. Why does it resonate so deeply? What is it we fear in the act of accepting ourselves? It may be the parts we either refuse to see or cannot bear to face. If some aspect of you remains unimaginable to another part of yourself, how can you ever truly accept who you are?

Could this be the root of our belief that we *can* know the bat? For if we can project some part of ourselves onto the other, we avoid confronting our whole self in any meaningful way. We sometimes ascribe human traits to animals, or even map attributes of our own character onto other people. It is a comfortable delusion. But it is also one that may turn malignant in certain contexts.

One of the most common expressions of this phenomenon is known as anthropomorphism. Its etymology is straightforward, from the Greek *ánthrōpos* ("human") and *morphē* ("form").[30] We give "human form" to other entities constantly. The term is typically applied to animals, beings we believe reflect our own nature back to us.

The author encounters this impulse regularly. Picture a mother goose shepherding her goslings across a busy street as cars slow to let her pass. We might remark, *"What a good mom."* Indeed, the adult is clearly acting to protect each of her offspring. Yet if we return a week later and find one of the goslings lying dead along the shoulder, our opinion of Mother Goose may diminish. Perhaps not. Either way, the underlying pattern remains. We apply value judgments and form opinions not only from what we observe, but also from the internal framework we carry. We each have our own attitudes and understanding of the mother-child relationship.

We might ask ourselves, what constitutes a "good" mother? A "bad" one? What are my expectations of my own? Of those I encounter in daily life? How have I seen motherhood portrayed in the media? The answers are as varied as the individuals who give them. Some of us have never known our mothers, literally or otherwise. Yet all these ingrained elements of our psyche are active, naturally shaping our view of the world around us.

The author has no desire to refute or deny this impulse, which seems to be nearly universal among humans. It is real. We feel it genuinely. It is entirely fair to observe a parent in the animal world caring for its young and to feel whatever it stirs in you, whatever sentiments it gives birth to. Yet it is equally important to maintain a frame of reference that encompasses both *seeing/knowing* and *being seen/known*.

We cannot truly fathom what it is to be a goose, mother or not. The motivations and intricate inner workings of geese are theirs alone. When we incorporate this awareness into our worldview, it opens us to the understanding that the "other" remains, in every sense, *other*. We will never fully inhabit another's experience. Paradoxically, this is freeing. It is, in fact, essential to our experience as humans. From this stance, we can observe similarities and differences with a clearer, less subjective eye.

Have you ever encountered an animal in the wild that stopped to truly look at you? If so, you have felt this shift in consciousness. It is unmistakable.

One afternoon, while taking out the trash, the author opened a side door and encountered a juvenile black bear standing about fifteen feet away. The bear was briefly startled. It turned to gauge the situation, and in that instant, eye contact was made. Although young, the animal was still large enough to cause injury with a single swipe. Fortunately, it was ambling toward the trash cans.

Acting purely on instinct, a nickel from a pile of coins by the door was hurled to the ground in hopes that the jarring, metallic clatter would scare the bear away. A wiser choice would have been just to shut the door. But this moment matters. It was not abstract or self-indulgent. It was vividly, undeniably real. The brief meeting of eyes was exhilarating, bracing, beautiful, almost mystical. It was also laced with a terror deeper than physical danger. The fear was psychic. The sense that the bear saw this human in a way no other human ever could shook something loose inside. It's a feeling that remains impossible to articulate fully.

The nickel worked. The bear turned away and loped across the street toward a neighbor's house. As it passed their trash cans, it lifted a massive paw and knocked one over. This gesture looked unmistakably like aggravation. Or, perhaps, a message. *"I could do this to you, too, human. Stop throwing nickels at me."*

There is no way to know what the bear's actions meant to it internally. However, the author's humanity was expressed in a way that reinforced a basic survival instinct. *Bears will kill you and eat you. They will not take mercy or grant a second chance. A hungry bear sees any human as a potential meal.* In that moment, human characteristics were projected onto the bear. Yet contact was also made with its own *otherness.* Both the reflection and the reality were present, each essential to understanding the encounter. Had the bear been observed from a great distance, without it seeing or acknowledging the observer, the experience would have been entirely different. It was the direct mirroring of eyes, each gazing into the other, that created this chain of startled recognition and layered emotion.

Moments like this can take many forms, not just encounters with animals. Anthropomorphism, after all, is not reserved for the feathered, furred, or finned. We do it with clouds, mountains, the wind, even the sea. The dictionary gives us a standard definition. *"The attribution of human characteristics or behavior to a god, animal, or object."* We can overlay our own motives, thoughts, and dreams onto anything, weaving familiar shapes into the unfamiliar.

For those who reject the existence of gods, there can be comfort in believing that the only imaginations above our own are those we have hung there ourselves, crafted from mortal cloth and placed on high to be revered. It is hard to argue with this view. Especially when faith, religious or otherwise, is removed from the equation. The faith we refer to here is not only belief in a deity, but any conviction regarding the unseen and unknown.

Faith is intrinsic to the human condition. We have faith that the red light will turn green. That when we order lunch, it will arrive. That when we sleep, we will wake. Yet sometimes the light stays red. The meal does not appear. The morning never arrives. We know these ruptures are possible. Yet we live as if they will not occur. When that faith suddenly or unexpectedly fails us, it can alarm, frustrate, and even terrify.

The author had faith that opening the door to take out the trash would be a mundane act, unworthy of recording in memory. That faith was broken in an instant. This habitual conviction is an internal construct, a mental shortcut that allows us to move through life without constant anxiety. Trauma can tear this construct apart, leaving one unable or unwilling to trust the everyday cycles of predictability and order.

The bear encounter was not traumatic. But it did interrupt the regular programming with a special news bulletin in bold, screaming headlines. **BEAR! HOLY CRAP!** By upending the illusory but useful faith in the *next moment,* and coupling that disruption with the shock of being truly *seen* by the bear, the result was a jolt out of the normal flow of human expectation.

This book seeks to recreate that jolt. In the pages ahead, we will meet beings stranger than a bear, a goose, or a bat. To follow where this path leads, we must shed the instincts that drive us solely to observe, define, and expect. We must prepare to be observed, defined, and expected in turn.

Before we leave the realm of animal anthropomorphism, it's worth considering one more of our fellow Earth-dwellers. It is one that lives not on land, but in the ocean. For that is where we are heading. Every age has its obsessions. In our early 21st century, one of these is the octopus.

Humans have been fascinated by these masters of their environment since first encountering them. The fantastical beasts sketched on early maps may have been less invention than interpretation. Perhaps they were echoes of real encounters with these magical and formidable creatures.

Fossil records indicate that the octopus lineage dates back approximately 300 million years to the Carboniferous Period.[31] This era is notable for the emergence of amphibians and early reptiles, as well as life's first significant push onto dry land.[32]

The notion that an ocean-dwelling creature might, over prodigious stretches of time, evolve to leave the sea is remarkable. We'll return to that idea later. For now, it is enough to note that while amphibians and amniotes (ancestors of reptiles, birds, and mammals) were colonizing land, the octopus stayed put. Its evolutionary path had already proven strikingly effective for survival.

When scientists sequenced the octopus's genome in 2015, the findings were eye-opening. Octopuses have roughly 33,000 protein-coding genes. This is far more than humans, who have about 20,000. Among these are 168 protocadherin genes, compared to our 70. These are key players in brain development and neuron communication. Unlike vertebrates, octopuses do not produce myelin, the insulating sheath that enables long-range neural transmission. Instead, they appear to rely on this expanded set of protocadherins for highly efficient short-range signaling.[33] From our perspective, this might seem like a workaround. But if octopuses were the ones doing the studying, they might see it differently. They might point out our relative lack of these genes and suggest that humans have evolved myelin to compensate for that deficiency. The gaze cuts both ways.

Their neurological marvels don't stop there. Octopuses can edit their RNA extensively.[34] This ability enables them to adapt rapidly to changing environments without altering their genetic makeup permanently. It may reduce the incidence of harmful mutations while preserving extraordinary flexibility, especially in response to environmental shifts, such as temperature changes.

The implications are profound.

First, while the octopus genome's differences surprised the scientific world, perhaps they shouldn't have. Anyone who has seen an octopus, its shifting colors, fluid intelligence, and uncanny presence, instinctively senses its utter *other-ness*. These findings underscore that intelligence can arise through radically different evolutionary routes, in different eras, and under different conditions.

Second, almost all of our information about octopuses comes from species that inhabit relatively shallow, sunlit depths within human reach. Many are found between the surface and roughly 200 meters (660 feet) down.[35] If those we *can* observe are already so strange, what wonders await in the abyssal depths?

Consider *Grimpoteuthis*, the "Dumbo" octopus, which lives between 1,000 and 7,000 meters (3,300 and 23,000 feet) below the surface. With ear-like fins for propulsion and other adaptations for life in the deep, they hint at a hidden diversity of cephalopods we have yet to encounter.[36] Perhaps ones we cannot yet imagine.

Even in bright daylight, our perception can darken the lens through which we view the world. When that lens turns toward the lightless deep, what hope do we have of telling one thing from another, motion from stillness? To go beyond the realms we were born to, we must accept, as both Nagel and Paul remind us, that we are part of the natural order, not its overseers. We are not assembling the puzzle of life from above. We are pieces within it.

To place ourselves properly among those surrounding pieces requires courage, the kind shown by mythical figures such as Antigone and Prometheus[37], as well as those described in historical accounts. Joan of Arc, Galileo Galilei, Harriet Tubman, and Oskar Schindler were all giants of steely determination. They stood, not just *against* the forces of dogma, malignant orthodoxy, and darkness. They stepped boldly *into* that darkness. True bravery is opening the door to the unknown without any assurance of what waits on the other side.

Paul understood this kind of courage well. No biblical figure faced the sea more often than he did. His journeys were dangerous and exhausting. Still, his faith pushed him onward. Arrested in Jerusalem on false charges, he appealed directly to Caesar. This was a right granted to Roman citizens, ensuring that he would be sent to Rome for trial.

On the way, he warned the ship's crew that the season's weather made sailing dangerous. He was ignored, and soon the vessel was caught in a violent storm that lasted two weeks. Amid this chaos, Paul had a vision that steadied both him and his companions. After being shipwrecked, the survivors washed ashore on Malta. Paul was bitten by a venomous snake here, but miraculously survived. This event elevated him in the eyes of the locals. He stayed there for three months, healing the sick, before continuing to Rome.

Once in Rome, under house arrest but permitted visitors, Paul preached boldly. Although we don't have a biblical account of his trial, early Christian writers tell us he was released and continued his mission, traveling extensively before his eventual death.[38]

You do not need to share Paul's faith to feel the resonance of his story. Storm, snake, shackles, and darkness. These are not just trials but encounters. They are moments in which we are not only seeing the world but being seen by it. The waves, the poison, and the chains are metaphors for the doctrines, doubts, and fears that can fracture, infect,

and bind us. Yet we must board the ship, trusting that on the other side lies the most terrifying gift of all. To be known.

And so...

Downward now we go, dear reader, against all instinct, against all judgment. We will brave the unseen and the unknown together. Let's go deeper. Let's go all the way down.

# CHAPTER THREE

## Whispers From the Void: Evolution Without Witness

---

*"Too much sanity may be madness, and the maddest of all, to see life as it is, and not as it should be."*

*Miguel de Cervantes, Don Quixote*

---

From myth to molecule. From the gaze of a bear to the knowing eyes of God. We have been circling the question of what it means to be seen. Now, it's time to take that question somewhere few eyes, human or otherwise, have ever gone. Straight into the lightless reaches of the deep ocean.

The deepest part of Earth's oceans is so hidden and mysterious that even its very name reveals our uncertainty about it. This layer, characterized by four traits:

    1. Complete absence of sunlight
    2. Near-freezing temperatures
    3. Extreme nutrient limitation
    4. Crushing hydrostatic pressures

...has what could be called an identity crisis. Some sources call it the *Hadal Zone*, a nod to Hades himself. Others refer to it as the *Hadalpelagic Zone*. A few claim the two are interchangeable, but the truth is more nuanced. In many taxonomies, especially for regions between roughly 6 km (3.7 miles) and 11 km (6.8 miles) down, "pelagic" denotes open water segments, while steep trench walls are labeled "hadal".[39]

For clarity, and to avoid sending us both through linguistic decompression stops, we will use "hadal" throughout.

Regardless of which name is applied, the territory itself sits so far beyond our vision, experience, and understanding that even to contemplate it is to invite surprise, confusion, and sometimes, a flicker of dread. Less than 0.001% of Earth's deep seafloor, including the Hadal Zone, has been explored or studied.[40] Put another way, 99.999% of this mysterious realm remains not just unobserved but utterly hidden from human eyes. These depths are biological black boxes, sealed tight, waiting for someone or something to lift the lid.

At the risk of pounding the point like a depth charge, **99.999%**! It's a number that is jolting in its sheer, immovable magnitude. Why does so much remain out of reach? What keeps us from breaching these voids of science and understanding?

Part of the answer is physical, while the other part is psychological. We have already noted humankind's innate bias. We look *up* with an eye toward the divine, and look *down* with the expectation of monsters. The author is not here to argue against that instinct entirely, only to underscore this. Even if monsters do await us below, they dwell in a kingdom we've barely begun to map.

**99.999%.**

We have confronted this kind of remoteness before, places once thought forever beyond our reach. Until human ingenuity found a way to breach the barrier.

There was once another place, far from Earth, that shared the same air of mystery and inaccessibility as the Hadal Zone. That place was our former ninth planet, Pluto. For much of human history, Pluto, its very name borrowed from the god of the underworld, was seen as unknowable, unreachable, and forever veiled in cosmic distance.

But times change. Today, not only has nearly every part of Pluto's surface been mapped, but even its largest moon, Charon, has a Wikipedia page. It includes images so sharp you can zoom into individual craters from the comfort of your living room. Charon also carries a name steeped in legend, borrowed from the ferryman of the dead in ancient Greek lore. In truth, however, the name hides a warmer origin. It is an homage by the discoverer, James Christy, to his wife, Charlene. A private gesture wrapped in a public myth.

To date, humans have observed roughly 85% of Pluto's surface and a sizable portion of Charon's.[41] Even farther out, the same New Horizons spacecraft that surveyed Pluto later flew past and photographed Arrokoth ("cloud" in the Powhatan language). Arrokoth is a celestial body in the Kuiper Belt located some 6.6 billion kilometers (4.1 billion miles) from Earth. It also now boasts its own Wikipedia page, complete with an exceptionally clear image.[42]

Like many who are drawn to the spirit of exploration and discovery, the author regards such achievements with deep respect. These triumphs of imagination, persistence, and engineering are a testament to the brilliance and daring of those who made them possible. They stand as milestones in humanity's ongoing quest to explore the unknown.

And yet, for all our victories in charting the icy edges of the Solar System, there remain realms here on Earth that lie almost entirely beyond our gaze. These opaque and silent realms, unlike Pluto, are not light-years away. They are directly beneath our feet.

And so, the question returns. Why? Or perhaps more accurately, why *not*? Setting aside humanity's obvious reluctance to explore the uncharted, dizzying, and invisible depths of our own world, the most immediate reasons lie in three of the four traits mentioned earlier. The Hadal Zone's utter darkness, near-freezing temperatures, and inexorable pressures.

These conditions are as forbidding as they are formidable, making any attempt at exploration a staggering challenge. But are they truly more daunting than sending a spacecraft on a 6.6-billion-kilometer journey to photograph an object in the Kuiper Belt? Strangely, the answer is yes. While both environments share a particular hostility to human presence, they are, in many ways, complete opposites. Space offers a near-perfect vacuum, the Hadal depths a suffocating weight. Space is cold and airless, while the deep sea is cold and utterly saturated. Each is alien in its own way. Yet one of them is much closer and far less explored.

In space, the enemy is absence. Specifically, the total lack of pressure. Engineering a spacecraft begins with the problem of keeping that absence at bay. If you are a fan of Dutch filmmaker and unapologetic provocateur Paul Verhoeven, you might remember a particularly unforgettable scene from *Total Recall* (1990)[43]. In it, Douglas Quaid finds himself unceremoniously ejected onto the surface of Mars without a protective suit. What follows is a rousing blend of body horror and cartoonish absurdity. Distended veins, a grotesquely swollen face, and eyes bulging out like Tex Avery's Big Bad Wolf in *Red Hot Riding Hood*.

The accuracy of this sequence as a depiction of exposure to a vacuum is generous at best. However, the scene serves as a clear visual shorthand for what happens when an Earthborn biological specimen is suddenly thrust into space. In short, Arnold goes boom.

With this vibrant cinematic metaphor in mind, it's easier to appreciate the real-world demands that this perfect vacuum of space imposes on spacecraft. Every component must be extraordinarily light yet exceedingly strong. It must also remain stable over years and, in some cases, *decades* of exposure. These machines must function in a zero-pressure environment while also enduring both searing heat and bone-cracking cold, many times within the span of a single orbit.

Because space contains no air, heat cannot be transferred through convection or conduction. Instead, spacecraft depend on large radiators that release excess energy as infrared radiation into the vacuum.[44] Maintaining the right temperature is a constant balancing act, preventing electronics from freezing in the shadow or frying in direct sunlight.

Power presents a different challenge. Far from the Sun, solar energy fades to a trickle. Deep-space craft must therefore draw their strength from another source: the slow, steady warmth of radioactive decay.[45] This energy keeps instruments alive when sunlight can no longer reach them. Yet the same process that sustains them also endangers them. Ionizing radiation gradually corrodes materials at the atomic level, forcing engineers to design with specialized shielding, coatings, and alloys that can withstand that invisible assault.[46]

Navigation must be both redundant and autonomous. Communication with Earth can take anywhere from a few minutes to several hours, depending on the distance. And once a probe leaves our planet, the opportunity for repair is essentially zero.[47] For crewed missions, life-support systems must be closed-loop, endlessly recycling air, water, and other essentials inside a sealed habitat.[48]

These are the realities of surviving the void above. But if space is the realm of absence, of nothingness that must be kept at bay, the Hadal Zone is its inverse. It is a realm of immense pressure. Where spacecraft must resist the urge of the vacuum to pull them apart, submersibles must battle a relentless, all-encompassing force trying to squeeze them mercilessly. Both realms demand absolute precision, yet one wages war against emptiness while the other is saturated with force. The same ingenuity that carries probes through the vacuum must be inverted and reimagined to descend into Earth's ultimate gravity well.

When it comes to engineering deep-submergence vehicles (DSVs), the enemy is not the absence of pressure, but pressure itself, in a form so absolute it defies human intuition. Beneath the waves, the weight of the ocean stacks relentlessly, layer upon layer, until it exerts forces capable of crumpling steel like tinfoil.

Underwater pressure is measured in atmospheres, just as weight is measured in kilograms or distance in kilometers. Standing at sea level, you already bear one full atmosphere, the weight of the entire column of air stretching from your head to the edge of space. This is a force we're born into, so we rarely notice it. But change it, even slightly, and the body responds. Ears pop in airplanes as the eustachian tubes adjust. Joints ache during rapid climbs or descents.

The deeper you go, the more that invisible hand tightens its grip. At just 10 meters (33 feet) below the surface, the pressure doubles. By the time you descend into the Hadal Zone, the force is hundreds of times greater than what you experience on land. It is enough to obliterate an unprotected hull in milliseconds. Designing a craft to survive here means not just sealing against emptiness, as in space, but withstanding assault from every direction, every second, without a single point of failure.

As a DSV descends into the ocean's blackened corridors, pressure mounts with relentless precision. In the devastating depths of a hadal trench, it can reach

1,000 atmospheres. In other words, take the full weight of the atmosphere pressing on you right now, and multiply it by a thousand.[49]

Trying to imagine that kind of force almost feels cartoonish. Picture extending your thumb on a table and, very gently, lowering the Eiffel Tower onto it until all its weight rests on that single square inch. Ouch.

Need another example? Imagine 3,000 cars, each weighing about 3,000 pounds, stacked neatly on top of you. Or, if that still feels too abstract, picture three SUVs balanced on a single LEGO brick. Then multiply that by a thousand. Now you're starting to get the idea.

And that's just the *primary* obstacle. Like their spacecraft counterparts, DSVs must endure a gauntlet of environmental hazards, only in reverse. Where space requires the ability to survive in a vacuum, the deep demands the ability to withstand an all-encompassing, relentless pressure. Add to that the difficulty of navigation in near-total darkness, the need to generate power without solar input, and the requirement that materials remain watertight under constant strain. Taken together, the task becomes as daunting as any space mission, perhaps more so.

Next on the list are the deep ocean's *very* low temperatures. While not as extreme as the cryogenic chill of space, the cold here, typically between 1°C and 4°C (34°F and 39°F), is still a formidable adversary.[50] It won't flash-freeze a DSV in seconds. However, paired with the overwhelming pressures and other environmental hazards, it adds a persistent and exhausting strain on every material and system aboard. Fortunately, sudden or dramatic temperature swings are rare at these depths, offering engineers at least one small mercy.

But the deep has its own equivalent of space's ionizing radiation. It is the constant, corrosive force of saltwater. It infiltrates, erodes, and punishes the slightest imperfection in seals, joints, or hull plating.

Combine that relentless corrosion with the unimaginable pressure outside, and even the tiniest design flaw can spell catastrophe.

Finally, there are the twin challenges of communication and light. Or rather, the near-total absence of both. Just as spacecraft must operate far from real-time contact with Earth, a DSV descends into a realm where radio signals die almost instantly, and visual information is swallowed by absolute darkness. Both must be overcome through redundancy, autonomy, and the ability to function in isolation from the surface.

Communication between DSVs and their surface counterparts relies on an ingenious workaround, acoustic modems. Instead of radio waves, which seawater quickly absorbs, these devices encode data into sound waves. They then transmit them upward through the water column.[51] At the surface, the signal is decoded back into usable information. It's a system born of necessity, turning the ocean itself into a conduit.

And then there's the darkness. Roughly 1,000 meters below the surface, sunlight surrenders entirely, leaving the realm beyond in absolute blackness.[52] Here, artificial lighting becomes the craft's eyes. This illumination must be engineered to endure the trifecta of corrosion, incredible pressure, and bitter cold. DSVs and spacecraft share this kinship. Both must bring their own light into environments where it does not exist naturally.

These design requirements are formidable. Yet, if the builders have both the resources and the will, they are not insurmountable. It is that will which now demands our attention. While the engineering of the abyss is challenging, history shows that humanity's reluctance to descend into the ocean's invisible depths far outweighs its hesitation to soar into the heavens.

If humanity's technological limits are imposing, its psychological limits can be even more so. The mysteries of our impulses, why we leap toward some challenges and turn away from others, are sometimes best illuminated not by data, but by stories. For the purposes of this chapter, no literary figure offers a better guide than Miguel de Cervantes' ingenious, maddening, and timeless knight-errant, *Don Quixote.*

The quote that opens this chapter is not an ornament. It is a throughline, threading not only this section but the entire work you hold in your hands. Cervantes' full title, *El ingenioso hidalgo Don Quixote de la Mancha*, translates to *The Ingenious Gentleman Don Quixote of La Mancha*.

Even here, before a single page of narrative unfolds, we catch a glint of the book's marvelous alchemy. It is a mixture of earnest nobility and absurd folly, set against a world both mundane and strange.

Few works in Western literature have cast such a long and commanding shadow of influence. Its reach extends into art, philosophy, politics, and psychology, with interpretations as numerous as its readers. And yet, for all the layers that scholars have peeled away over centuries, one question remains the most debated and perhaps the most important for our journey. What truly motivates Don Quixote? The answer, or rather the many possible answers, shape the way we experience his story. By extension, they also shape how we understand our own quests into the unknown.

One of Cervantes' most noteworthy achievements lies in his steadfast refusal to define his creation with the neat clarity we expect from most works of fiction. From its very first pages, *Don Quixote* dismantles conventional narrative structures. The "author" we meet is not Cervantes himself, but an editor and translator of a found manuscript, supposedly penned by an earlier fictional chronicler. From the outset, fact and fiction are intentionally tangled. This forces the reader to navigate a shifting terrain where neither is fully stable.

This is not the work of an "unreliable narrator" in the traditional sense. Instead, it is the work of a narrator who continually calls attention to the artifice of storytelling itself. At several moments, Quixote and his faithful squire, Sancho Panza, encounter characters who recognize them because they have read about them, figures within the fiction who are *aware* of the fiction itself. This circular, self-referential loop dissolves the ordinary boundaries of story and reality, inviting us to question whether those boundaries exist at all.

Through this intricate lattice of narrative mirrors, Cervantes allows Quixote to speak with an authority that transcends his apparent madness. Time and again, the knight utters truths that pierce our entrenched assumptions, not just about what is "true" or "false," but about what is *real* or *illusory*. It is here, in the overlap of folly and insight, that Quixote becomes our unlikely guide. He is a figure driven not by the certainty of outcome, but by the irresistible pull of the quest itself.

In this way, his journey mirrors our own strange calculus when deciding where to aim our most daring explorations, toward the distant heavens or down into the crushing dark of the abyss.

Perhaps the most resonant distillation of Quixote's spirit comes not from Cervantes' own text, but from the 1962 musical adaptation *Man of La Mancha* by Dale Wasserman, in which the knight declares: *"Facts are the enemy of truth."*[53] The line has been interpreted in countless ways. Yet for the journey we are undertaking, it serves as a vital compass point. It reminds us that the "facts" we cling to are in many cases the very stones that determine our path. We trust them to bear our weight as we move forward. But those same stones can also anchor us in place, preventing us from seeing other possible routes, or even imagining that such routes could exist at all.

Only by loosening our grip on these certainties, by daring to test the ground beyond the lamplight, can we become aware of realms outside the narrow circle of illumination we have always known. Cervantes lights this way as clearly as any explorer's beacon. He is far from alone in doing so. One could fill an entire volume with voices urging the same truth. Our own journey cannot linger too long among aphorisms, but a few are worth carrying with us as we venture on.

*"Reality is that which, when you stop believing in it, doesn't go away."*

*Philip K. Dick, "How to Build a Universe..." (1978)*[54]

A reminder that truth is not something we can banish simply by averting our eyes.

*"One does not discover new land without consenting to lose sight, at first and for a long time, of every shore."*

*André Gide, The Counterfeiters (1925)*[55]

Exploration demands both distance and disorientation. The familiar must fade before the unfamiliar can emerge.

*"You must unlearn what you have learned."*

*Yoda, The Empire Strikes Back (1980)*[56]

Wisdom sometimes begins not with accumulation, but with subtraction, the dismantling of the frameworks that once defined our reality.

These thoughts strike at the very heart of complacency and obstinacy. For in sameness lies the illusion of comfort. In familiarity, the illusion of safety. It is only when some uninvited fragment of reality looms up from the darkness, mocking our most cherished conventions, that we are jarred into embracing this crucial, occasionally unsettling aspect of our existence.

And what as-yet-unconceived realities await in the 99.999% of the Hadal Zone's black void? Even now, they skulk and quiver, shift and stir, exist and persist in a realm we have not only failed to explore, but often even to imagine. Imagine we must, dear reader. For as humans are wont to do, we have already begun to tug at the handle of this gateway. The ambassadors awaiting us on the other side are merely the overture to a much richer, more redefining opera of reconciliation. It is one that will test our epistemological mettle in ways even Quixote might recoil from.

Let's meet a few of them now, in all their incongruous splendor.

As depth and pressure mount in the trenches and other abyssal domains, biological morphology begins to diverge drastically. Frequently in ways that defy even our most flexible expectations. We already accept that ocean-born life operates on a different set of rules than its terrestrial counterparts. Gills instead of lungs, fins instead of limbs, body plans sculpted by the currents rather than the wind. We anticipate a certain strangeness beneath the waves.

But the further we descend into the hadal reaches, the more that strangeness blooms into outright otherness. In this place, the sea creatures once sketched at the edges of old maps feel less like archaic superstition and more like field notes. The unforgiving atmospheric compression, perpetual darkness, and meager rations of this realm have acted on evolution with a sculptor's patience and a trickster's hand. It shapes life into forms that strain the boundaries of our imagination.

Yet these creatures are not solely the product of environmental extremes. They are also heirs to deep time itself. Epochs stack upon epochs. This lineage stretches back into a past so remote that even the human mind, with all its stories and symbols, can barely keep its footing.

Earliest estimates place the emergence of hominids around 6–7 million years ago.[57] This is the blink of an eye compared to the ocean's protracted evolutionary chronicle. Like the pressure gulf between sea level and the Hadal Zone, this chasm of time stretches our comprehension almost beyond reach. While the best evidence suggests life first appeared on Earth roughly 3.5 billion years ago, complex multicellular organisms likely emerged closer to 560 million years ago.[58] When we also account for the relatively recent discovery that life *thrives* in environments once deemed utterly "incompatible" with biology, we begin to grasp just how far beyond human norms we are venturing.

Even the debris here can astound. In 2021, a deep-sea expedition off the California coast uncovered the tusk of a Columbian mammoth resting 3,000 meters (10,000 feet) below the surface.[59] These massive Pleistocene wanderers vanished roughly 12,000 years ago, a mere heartbeat in geologic time. We see here plainly that even a relic from humanity's deep past is a relative newcomer. The find serves as a reminder. In these hidden realms, wonder doesn't just live in the present. It can erupt unexpectedly from antiquity, like a ghost surfacing from a locked vault of ancient marvels.

And this is only the beginning. Choosing a single example from the cascade of discoveries made in Earth's deepest reaches is almost impossible. Each new revelation seems to outpace the last in improbability. Down here, form and function are rewritten by forces few of us will ever experience. These include not only the unrelenting pressure, total darkness, and scarcity of nutrients, but also something more intangible. The absence of human eyes. Creatures shaped in such secrecy can seem less like products of evolution and more like emissaries from a reality just beyond our own.

One such emissary is the Humpback Anglerfish (*Melanocetus johnsonii*). It is a member of the shadowy fraternity known as the *Black Seadevils*. This peculiar specimen of adaptation has achieved an unlikely celebrity status among deep-ocean denizens. It first captured public imagination in 1995, grinning its toothy, lantern-lit grin from the cover of *Time* magazine's "Mysteries of the Deep" issue. Then, in early 2025, it stunned researchers and the internet when a team studying sharks spotted one swimming near the ocean's surface.[60] This behavior had never been documented before, making the encounter a bona fide viral moment. Despite its appearance, described as gruesome, otherworldly, or outright nightmare fuel, the Humpback Anglerfish once again proved that beauty and fascination are, at times, a matter of perspective.

Despite the flash of viral fame, the Humpback Anglerfish is far more than a curiosity. It is a masterclass in deep-ocean adaptation. Nearly every part of its anatomy tells the story of survival in a place where light never reaches, food is scarce, and pressure could obliterate most lifeforms into pulp.

The most immediately striking features are its gaping mouth, bristling with translucent, needle-like teeth, and the illicium and esca. This is an elongated "fishing rod" that rises from its head, tipped with a bioluminescent lure. Suspended in perpetual darkness, the anglerfish wiggles this glowing bait to draw unsuspecting prey close enough for a sudden, decisive strike. Once captured, escape is virtually impossible. A distensible stomach and an astonishingly flexible jaw allow it to swallow meals nearly its own size. This trick is reminiscent of certain snake species on land.

Its skeleton is lightly ossified, reducing body density and countering the tremendous weight of thousands of meters of seawater. The rest of its body is soft, gelatinous, and jet-black, perfect camouflage in an environment where shadows are the only backdrop. Lacking a swim bladder, it avoids the problems gas-filled organs face at extreme pressures. Its sexual dimorphism is radical. The female is about five times the size of the male, whose life's purpose is almost entirely reproductive.[61]

Perhaps most impressive are the sensory canals tracing the sides of its head and body, exquisitely tuned to detect even the faintest water movement. In a world without light, touch and vibration become stand-ins for sight. They enable anglerfish to map their surroundings and locate prey with remarkable efficiency.

These traits, the cavernous jaws, outsized teeth, and sensory refinements, are not unique to the Humpback Anglerfish. They are part of a shared toolkit for survival in the abyss, repeated with persistent regularity across unrelated species that have converged on similar solutions to the same brutal challenges.

At these depths, teeth are more than weapons. They are a strategy. The greatly elongated, needle-like fangs common among deep-sea predators don't just intimidate. They ensure that once prey blunders into range, escape becomes nearly impossible. Coupled with enormous, hinged jaws, they allow a predator to ambush, secure, and swallow meals in a nutrient-scarce world where you may not get a second chance to feed. Some go even further, lining the back of their throats with *pharyngeal teeth*. This extra set of inward-pointing spikes traps victims and ushers them toward digestion.

Teeth define not only survival here, but also identity. Many of these creatures proudly display their dental dominance in their names. Fangtooth. Viperfish. Dragonfish. The Fangtooth Fish (*Anoplogaster cornuta*) boasts the largest teeth relative to body size of any known fish in the ocean.[62] They are so long and formidable that they slot into special pockets in the roof of the mouth when closed. The Viperfish's fangs are even more outrageous, extending past its eyes when its jaws are shut, like sabers holstered for the next strike.[63]

From the grim geometry of teeth, we turn to another deep-sea adaptation. One that is less about brute force and more about bending light to one's will. In a world without sunlight, illumination becomes both lure and language, weapon and warning. Here, life has learned not just to endure the dark, but to *engineer* it, painting flickering constellations across the black canvas of the abyss.

Bioluminescence is hardly exclusive to the ocean. On land, we know it best in the lazy pulse of summer fireflies or the slow drift of "will-o'-the-wisps" through wetlands. At sea, we have grown increasingly familiar with it in the form of glowing plankton blooms. "Sea sparkle" or "milky seas" can transform whole stretches of coastline into shifting fields of blue fire. These blooms have become increasingly prevalent in the early 21st century. They are fueled by a complex mix of human influence, including fertilizer runoff, changing climate patterns, overfishing that removes plankton grazers, and urban sprawl along shorelines. Even the reach of global shipping can ferry these bioluminescent plankton across distant waters.[64]

In the hadal depths, bioluminescence is no idle decoration. It is a finely honed survival tool. The molecular machinery behind it is as poetic as it is functional. Luciferin is the light-emitting compound, and luciferase is the enzyme that unlocks it. Both names trace back to the Latin *lucifer*, "light-bearer."[65] It is a title that has traversed mythology and theology, accumulating shades of underworld association along the way. Here in the abyss, the bearers of this light do not dispel the darkness so much as weaponize it.

Deep-ocean species wield their glow in four primary ways, the first of which is defense.

Among the deep sea's most theatrical practitioners of defensive bioluminescence is a creature whose very name sounds like it should be whispered around a campfire. In the seemingly never-ending pageant of oddities we encounter here, few can compete with the strange and wonderful Vampire Squid. Like many of its abyssal neighbors, the ominous sobriquet overshadows its taxonomy. It is, in fact, not a squid at all, but the sole occupant of its own order, *Vampyromorphida*.[66] The name itself foreshadows the dark glamour of its form and function. Drill down to the species designation, *Vampyroteuthis infernalis* (literally, "vampire squid from hell"), and you have a fair preview of what's to come.

Although measuring only about 30 cm (1 foot) in length, its velvet-black to pale red skin and glaring eyes, which shift between red and blue depending on the light, would look perfectly at home in a Hieronymus Bosch triptych.

In keeping with this eerie stage presence, the Vampire Squid's body plan is as unconventional as its name. Like the Dumbo Squid, adults possess a pair of small fins that extend laterally from the dorsal surface. These serve as their primary means of propulsion, allowing them to glide rather than dart. This conserves precious energy in the nutrient-scarce abyss. Unlike true squid, whose arms drift freely in the water, *Vampyroteuthis* wears a trailing "skirt" of skin that webs all eight arms together. The inner surface of this skirt is an inky black. It is a void within the void, lined with fleshy, sensory-rich spines called cirri. These cirri function like tactile antennae, enabling the squid to locate and sort potential food, directing edible particles toward its mouth with precision.

Yet for all its anatomical eccentricities, the Vampire Squid's bioluminescent defense array is its true showstopper. Like most deep-sea inhabitants, it lacks a swim bladder. More notably, it also foregoes the ink sac that shallow-water squid use to vanish in a puff of darkness. In its place, *Vampyroteuthis* wields something far more theatrical. A glowing, weaponized slime bomb. When threatened, it can draw its webbed "skirt" up over its body, turning itself inside-out so that the cirri bristle outward. It's a sudden transformation from graceful glider to thorny nightmare. In the same instant, it can eject a cloud of shimmering blue mucus that clings to whatever disturbed it.[67] This "burglar alarm effect" not only blinds would-be predators but can also mark them like a thief caught under a spotlight, broadcasting their position to larger hunters nearby.

And just in case that isn't wild enough, the Vampire Squid has one more sleight-of-hand. It can cluster its glowing arm tips above its head, luring an attacker toward the decoy lights and away from its vital core.

With such flair for misdirection, *Vampyroteuthis infernalis* more than earns its infernal title, though "Squiddy from the block" might better capture the mix of attitude and survival savvy on display here.

But defense is only one facet of this luminous arsenal. Many species also use bioluminescence to attract prey. They dangle glowing lures, ghostly bait in the darkness, as the Humpback Anglerfish does with its signature esca. Others deploy it to find mates, flashing specific patterns or frequencies of light to signal their presence across the black expanse. And some, perhaps most surprisingly, use bioluminescence for camouflage through a technique called *counter-illumination*. They produce light on their undersides to match the faint glow from above, effectively erasing their silhouette from predators lurking below.[68] In the abyss, light isn't just illumination. It's deception, seduction, disguise, and communication, all rolled into one.

From theatrical tricksters to colossal curiosities, the Hadal Zone is nothing if not a showcase of extremes. Before we venture into its more surreal departures from terrestrial anatomy, we should pause to consider a phenomenon that is both counterintuitive and captivating. Deep-sea gigantism and its opposite, miniaturization, reveal that size itself can become an adaptation, sculpted by pressures both literal and ecological.

Scientists are still untangling the exact forces at play, but several theories surface repeatedly.[69]

1. **Frigid temperatures + slower metabolism** – Colder water can slow biological processes, which extend lifespan and allow for greater growth over time.

2. **Reduced predation** – In a realm with few hunters, large size can further reduce the odds of becoming someone else's meal.

3. **Extreme pressures** – Crushing depths may influence cell structure and physiology in ways that favor bulkier builds.

4. **Resource scarcity** – A bigger body can store more energy reserves and travel farther in search of food.

One notable example comes from even deeper than many of our other abyssal acquaintances, a "tiny giant" named *Dulcibella camanchaca*. Its genus honors Dulcinea, muse to Cervantes' Quixote, while its species name comes from the Aymara word for "darkness." Fittingly so, for this predator thrives nearly 8,000 meters (26,000 feet) below sea level. This is one of the most extreme habitats ever documented.

Most amphipods are tiny, measuring 1–10 millimeters, and subsist as scavengers. *Dulcibella* shatters that mold, quadrupling the typical size to a substantial 4 centimeters, and doing so as an active predator. Its discovery by the Woods Hole Oceanographic Institution required an audacious journey into one of Earth's most inaccessible realms.[70] *Dulcibella* stalks the darkness here as a rare giant among miniatures.

At these extreme depths, size variation becomes a recurring motif, on occasion in ways that defy our expectations. Alongside the "tiny giant" *Dulcibella*, the deep plays host to titans that put their shallower relatives to shame. There's the colossal squid, stretching a fantastic 14 meters (46 feet), with eyes the size of dinner plates.[71] The giant isopod, an armored pill bug, scaled up to 50 centimeters (20 inches).[72] And there are even larger amphipods than *Dulcibella camanchaca*, such as the mighty *Alicella*, which reaches 34 centimeters (13 inches) of armored exoskeleton.[73] The Japanese spider crab spans nearly 4 meters (12 feet) from leg tip to leg tip[74], while giant polychaete worms writhe through the abyss with unsettling grace. Their segmented bodies have adapted to a world without sunlight.

These beings feel both other and eerily familiar. They are an evolutionary echo of creatures we recognize, magnified to austere grandeur. Yet even they are only the foothills before stranger peaks. Press on, we must, for as Cervantes reminds us:

> *"The truth may be stretched thin, but it never breaks, and it always surfaces above lies, as oil floats on water."*[75]

When approaching the boundaries that now loom in our immediate path, we must surrender notions of either madness or sanity. We must resolve, no matter what lurches forth from the watery tomb of darkness to behold this angel or devil as it is. Not as we would have it be. More importantly, we must always be aware that we have also been beheld.

The summons draws us deeper still. It beckons beyond the seafloor, past the last reach of daylight and science. We are invited into realms where our familiar logic does not merely falter but begins to fold back upon itself, an Ouroboros devouring its own tail. Here, the end of one world is always the doorstep of the next. To cross it is to consent to be changed.

# CHAPTER FOUR

## Breathing Stone and Soft Machines: At the Collapse of Pattern

*"Stories and secrets fight, stories win, shed new
secrets, which new stories fight, and on."*

*China Miéville, Embassytown*

The Sapir-Whorf hypothesis, also known as linguistic relativity, proposes that the structure of a language shapes the speaker's perception of reality.[76] That the very fabric of spoken, written, and gestured expression does more than describe the world. It *defines* it.

Using language itself to probe the ontological scaffolding *built* by language may seem like an act of futility to some. However, utilizing the powerful and ancient symbol of the Ouroboros as our criterion[77], let us persevere and even extend this hypothesis further.

There is yet another facet of language, one often missing from the discussion. The *interior tongue*. This is the silent, ceaseless flow of thought that precedes expression. It is from this private ocean of cognition that speech, script, and sign are born. They rise, not unlike a tube worm unfurling from the seabed beside a hydrothermal vent.[78] This incorporeal broth, this etheric stew, births our audible utterances and visible pantomimes. And if the outer forms define perception, then surely the inner ones shape the self.

When humans confront something never before imagined, it is this internal medley, the simmering stew of private thought, to which we first turn. But what happens when the broth lacks the right ingredients to absorb this new and unfamiliar flavor?

We face a choice.

Do we reject the unimagined outright? Do we forage for new ingredients to make the strange more palatable? Relegate it to the back shelf, out of sight and out of mind, among other jars too strange or bitter to taste? Or do we light a new fire beneath a different vessel altogether?

A skilled potager can tend to a consommé and a chowder simultaneously. Should our standards for constructing reality be any less versatile?

That question, like all truly important ones, belongs to the individual. But for the next leg of this journey, the author urges you, dear reader, to fling open the pantry. Prepare new stockpots. Strange ingredients are inbound. And though their taste may be unfamiliar, sample them we must.

The sea keeps its own pantry, stocked with ingredients few of us could imagine.

Take the polychaete worm, a member of the segmented annelid family whose name derives from the Latin *anellus*, or "little ring."[79] You have likely seen its terrestrial cousin wriggling onto sidewalks after a summer storm. But in the crushing depths of the ocean, these segmented worms evolve into something far more extraordinary.

Among the most varied and visually startling invertebrates on Earth, deep-sea polychaetes defy our conventional mental stew of what life *should* look like. Their physiology doesn't just depart from the familiar. It seems to parody it. Feathery appendages, iridescent bristles, pulsating sacs, and jaws that can evert from the mouth like a nightmare glove puppet.[80] These are forms that invite both dismissal and enchantment, aversion and allure.

Children encountering an earthworm recoil with a shriek of "Yuck!" before stepping in closer, transfixed. Deep-sea polychaetes elicit a similar paradox. Their binomial names, like their forms, reflect our meager attempts to categorize what barely fits within comprehension. And yet we try. Because something in us still insists, if we can name it, maybe we can know it.

Consider, for example, *Osedax mucofloris*.[81] It is known colloquially (and with alarming accuracy) as the "bone-eating snot-flower." Yes, it is more or less *exactly* what it sounds like.

This remarkable worm features a feathery tendril at its distal end, delicate enough to resemble a bloom, and a bulbous, mucus-like base that allows it to burrow deep into the bones of dead whales.[82] From this unlikely perch, it extracts fats and proteins locked inside the skeletal remains and digests them as its primary source of food.

That alone would qualify it as a radically foreign ingredient in the broth of our assumptions. But we're just getting started. Its abilities only grow stranger from here.

Like many of its Hadal Zone neighbors, the "snot-flower" relies on a strategy that recurs frequently in these depths. Symbiosis. At its simplest, symbiosis is a sustained relationship between two distinct organisms, sometimes mutually beneficial, sometimes less so. In the case of *Osedax*, we're dealing with a specific variant called endosymbiosis, in which one organism lives *inside* the other.[83]

*Osedax mucofloris* hosts specialized bacteria within its tissues that help break down and absorb nutrients from bone.[84] These bacterial partners, in turn, receive structural proteins, such as collagen, from the worm.[85] This is a mutual exchange that sustains them both. And this isn't a one-act play. As the whale bone continues to decompose, other microbial allies take the stage. *Sulfurimonas* bacteria, for example, step in during later stages of decay to shield the worm from harmful by-products that emerge as the process unfolds.[86]

It's a multi-phase, biochemical ballet that is still being studied.[87] However, it clearly showcases the intricate choreography of survival at great depths.

Due to their unconventional method of nutrient extraction, *Osedax* worms lack what we would consider a traditional stomach, mouth, or anus. Instead, their bodies consist of three core parts. An ovisac (housing their reproductive organs), a soft trunk, and root-like structures that burrow into bone and host their bacterial partners.[88]

But digestion isn't the only category where *Osedax* breaks the mold. Like the Black Seadevil and other denizens of the deep, this genus exhibits extreme sexual dimorphism. In a reproductive arrangement that seems more fable than fact, hundreds of microscopic males, sometimes referred to as a "harem," live inside the gelatinous tubes that surround a single female.[89] Their only job is to produce sperm on demand. She uses this to fertilize her eggs, which sometimes number in the hundreds.[90]

This unorthodox setup of near-invisible males clinging to a single, much larger female isn't just a biological footnote. It serves particular and highly strategic purposes.

First, this arrangement eliminates any competition for food between males and females. A vital advantage in a world where food is scarce and never guaranteed. Second, it removes the need for females to expend precious energy searching for a mate. In the isolated, sunless void of the deep sea, proximity is everything. Sexual dimorphism allows *Osedax* to enjoy the evolutionary luxury of a built-in, ever-ready cache of reproductive partners, none of whom will ever ask what's for dinner.[91]

Taken together, the attributes of *Osedax*, including the eccentric *mucofloris*, are not just bizarre by human standards. They challenge the very framework of how we define life. These distant, bone-burrowing cousins of the humble earthworm are strange indeed, but still not the most surreal. Not by a long shot. The curtain is just beginning to rise on what lies beyond the liminal edge.

As we peruse and putter inside this underwater midnight pantry of the newly discovered, poorly understood, and fluidly defined, consider the gorgeous, deadly, and difficult to categorize world of siphonophores. Like the polychaete worms, this group has a well-known emissary familiar to most land-dwellers. *Physalia physalis*, better known as the Portuguese man o' war. With its iridescent sail, delicate tendrils, and infamously wicked sting, it has long inspired fascination and apprehension in equal measure.[92] And that's just a surface-dweller. If this species, which drifts along the ocean's uppermost layers, can provoke such fear and beguilement, what strange wonders, what exquisite and otherworldly designs await us among its abyssal kin?

Siphonophores, like many other deep-sea life forms, utilize symbiosis, but on a scale and structure that is emergent in its complexity. They are not a single creature, but rather a colony. A collective of specialized, interdependent individuals that live and function as one.[93] Chief among these are the zooids, which divide labor between feeding, movement, and sensory input. Nectophores provide propulsion. Gonophores are responsible for reproduction, while structures such as bracts and pneumatophores help regulate buoyancy and orientation.[94] These definitions are, admittedly, the barest of summaries. An entire book could be devoted to the complex and engrossing interdependency of colonies that make up siphonophores. Still, even a brief glimpse reveals how these ghostly, bioluminescent drifters showcase a more recondite truth. Nature, when given the freedom to bloom in utter darkness, invents structures far stranger than we're taught to imagine.

*Praya dubia*, the Giant Siphonophore, drifts between 700 and 1,000 meters (2,300 and 3,300 feet) below the surface, and earns its name by growing up to 50 meters (160 feet) in length.[95] That's nearly the size of a blue whale. But unlike whales, *Praya dubia*, as with most siphonophores, possesses a body so delicate and gossamer that if brought to the surface, it would suffer the same explosive fate as Douglas Quaid on the surface of Mars.

This fragility is tied to one of its most curious adaptations, a hydrostatic skeleton. Unlike our internal bone design, this structure is made of pressurized seawater and relies on the coordinated action of circular and longitudinal muscles.[96] This allows the colony's translucent form to ripple, undulate, and drift through the bathypelagic zone in elegant, slow motion.

As with many deep-sea organisms, *Praya dubia*'s name reflects both its mystery and our uneasy regard. It is loosely rendered as "dubious prayer."[97] Is this shimmering thread of life a holy vision or a haunted omen? A blessing or a warning? Fortunately for us, the answer remains benign. It poses no threat to humans. But what must we seem to this drifting cathedral of light? Do they even take notice of our colony of soft-shelled bipeds, flailing through the void? Which of us is truly more fragile? We remain unmoored from even the outer limits of its fathomless world, peering in from a distance, detached and astonished.

And still, stranger marvels await.

One such wonder, *Syringammina fragilissima*, Latin for "very fragile sandpipe," is undeniably biological, yet so anomalous in form and behavior that it continues to defy complete classification.[98] First discovered in 1883, it remains a taxonomic enigma. Its strange architecture and behaviors resist our tidy categories and logical constructs.

While it exhibits some recognizable features of its subphylum and clade, this organism pushes classification to the brink. It belongs to the clade Xenophyophorea, a group already considered iconoclastic by biological standards. These organisms are unicellular, yet multinucleate.[99] A single cell with many nuclei. That alone makes them profoundly strange in the context of terrestrial life.

Over the decades, xenophyophores have been miscategorized as sponges or protozoans, and even proposed as belonging to their own distinct phylum. Today, they are tentatively housed within the Foraminifera.[100] But certainty still eludes us. These ambiguous lifeforms extend delicate tendrils called pseudopodia to construct a fragile external casing, known as a test. The pseudopodia of xenophyophores closely resemble those of amoebas and white blood cells.[101]

And fragile they are. Like so many deep-sea inhabitants, xenophyophores shatter at the touch, making them exceedingly difficult to collect, preserve, or study in any traditional sense.[102]

In other words, we know just enough about *Syringammina* to know how much we *don't* know.

Where *Syringammina fragilissima* truly stands out is in its size. It is the largest known xenophyophore, with specimens reaching up to 20 centimeters (8 inches) across. We're talking about a *single-celled* organism, encased in a hardened outer shell, roughly the size of an *adult human hand*. That's right. One cell, the size of a palm.[103]

Visually, it resembles a roughly shaped ball of sediment or sand, laced with hundreds of branching, hollow tunnels. The organism itself forms these channels as it retreats ever deeper within the structure during growth. Over time, this creates a labyrinth of tiny passageways, some of which serve as shelter for other small deep-sea creatures.[104]

As for its diet, we're still speculating. It may graze on bacteria found within sediment. Or perhaps it's a filter feeder, drawing nutrients in through its branching tunnels. These are educated guesses. Because, like much about *Syringammina*, its full life cycle remains an unsolved riddle.[105]

Another standout in the realm of radical adaptation is *Chrysomallon squamiferum*, better known as the Scaly-Foot Snail. This gastropod has been observed only near deep-sea hydrothermal vents in the Indian Ocean, and it could rightfully be described as *hardcore*.

For starters, it incorporates iron sulfide not only into its shell but also into the armor-like scales on its foot, essentially making it a mollusk clad in metal. But the surprises don't stop there. In its esophageal glands, lives a population of symbiotic bacteria that enables the snail to nourish itself through chemosynthesis. Incredibly, the Scaly-Foot Snail doesn't need to eat in any traditional sense.

There is still no clear consensus on which side of this relationship benefits more, the snail or the bacteria. Perhaps they are true co-conspirators in survival. Adding to its list of oddities, *Chrysomallon* is a simultaneous hermaphrodite, possessing both sets of reproductive organs throughout adulthood. And as a final twist, it has an unusually large heart. It is proportionally about 4% of the snail's body mass, compared to a human's 1.3%. Again, researchers debate whether this oversized circulatory (and by extension, respiratory) system primarily serves the snail or its microbial passengers.[106]

Other examples of this borderline-Seussian biological wizardry abound in the deep. Entire volumes could be, and probably have been, written about nothing more than these expectation-exploding life forms.

Take *Bathypterois grallator*, the Tripod Fish. Yep, it actually has long, bony extensions trailing from both its tail and pelvic fin. These "tripods" extend beneath it like a clown on stilts while it waits patiently for prey to drift by. As a bonus, *Bathypterois* is also a hermaphrodite. It is happy to mate with others of its kind, but also fully capable of handling the job solo, thank you very much.[107]

Or consider the elusive Ipnops, cousin to the Tripod Fish and member of the *Ipnopidae* family, which has only recently been observed in the wild. Instead of traditional eyes, it boasts two large, flat membranes. These pale, translucent discs resemble glow-in-the-dark stickers. Their flattened, yellowish structure may help detect bioluminescent prey. Or they may glow themselves. We don't know yet. Either option is strangely beautiful.[108]

And then there is the record-breaking *Abyssobrotula galatheae*, a species of cusk eel discovered 8,370 meters (27,460 feet) deep in a Hadal trench near Puerto Rico. Little is known about this ghostly champion of depth, beyond its vestigial eyes and egg-laying habits. But it holds the crown, so far, as the deepest-living fish ever found.[109]

This array of noteworthy, and at times downright baffling, traits among life in Earth's deepest ecosystems is breathtaking in scope, variety, and its apparent disregard for human-derived expectations. We've explored a wide range of biological oddities. So now feels like the right moment to step back, take stock, and organize them into a more coherent framework. Especially with the speculative leaps we're about to take. Broadly speaking, these adaptations can be grouped into five main categories.

1. *Structural & Physiological Adaptations*

### Reduced or Gelatinous Bone Structure:

Rigid bones don't stand a chance in the crushing depths. Many deep-sea creatures minimize or eliminate skeletal material entirely, replacing it with cartilage or gelatinous frameworks better suited to extreme pressure.[110]

### Loss of Swim Bladders:

Traditional gas-filled buoyancy sacs would implode in the Hadal Zone. Instead, species tend to rely on waxy fats or gelatinous tissues to maintain vertical positioning within the water column.[111]

### Soft, Flaccid Bodies:

Floppy is functional. These pliable internal structures aren't just pressure-resistant. They reduce energy demands and distribute force evenly throughout the body in high-compression environments.[112]

### Amorphous Exteriors:

Blobby, blunted silhouettes aren't design flaws. They're survival tools. Flexible outer forms help these organisms navigate tight crevices and ride the chaotic microcurrents of the abyss without snapping like brittle shells.[113]

### 2. *Sensory Input & Output*

## Bioluminescence:

Nature's flashlight, beacon, and trap all in one. Deep-sea organisms utilize this biological glow for various purposes. To communicate,

lure in prey, signal potential mates, and even to disappear by blending into the faint residual light that still filters down near the upper reaches of the deep.[114]

## Highly Sensitized Mechanoreceptors:

In a world where sunlight soon vanishes entirely, vibration becomes vision. These finely tuned sensors can detect the tiniest shifts in pressure or movement, ripples in the dark that might signal danger, dinner, or both.[115]

## Reduced or Absent Eyes:

When light is truly extinct, eyes become obsolete. Many deep-sea species forgo them entirely, evolving alternate pathways for perception, strengthening other senses to compensate for the void.[116]

## Oversized or Hyper-Sensitive Eyes:

On the flip side, some organisms go all in on the ocular route, sporting enormous, finely tuned eyes capable of detecting even the faintest flickers of bioluminescent shimmer across the black canvas of the abyss.[117]

### 3. *Feeding Adaptations*

**Expandable Stomachs & Jaws:**

Why settle for a snack when you can swallow a feast? Many deep-sea predators are equipped with elastic, distensible jaws and stomachs that allow them to consume prey equal to, or larger than, their own body size. When meals are few and far between, go big or go hungry.[118]

**Slow Metabolism:**

Time slows in the deep. With so little energy input available, these organisms downshift their metabolic rate to a crawl, conserving resources and stretching meals across prolonged, foodless stretches of time.[119]

**Symbiotic Chemosynthesis:**

Who needs sunlight? Some deep-sea dwellers have outsourced digestion entirely, forming partnerships with bacteria that convert chemicals such as hydrogen sulfide and methane into usable nutrients. A quiet miracle of energy alchemy, thriving where photosynthesis dares not tread.[120]

**Sediment & Filter Feeding:**

For the more passive diners of the deep, sifting through mud or filtering water for plankton, bacteria, or organic debris is the name of the game. Why hunt when you can harvest the soup around you?[121]

### 4. *Reproductive Strategies*

## Hermaphroditism:

No partner? No problem. In a world where neighbors are infrequent and contact is scarce, many creatures carry both sets of reproductive tools, ensuring the species lives on, even in solitude. Self-sufficiency, Abyss-style.[122]

## Extreme Sexual Dimorphism:

Why bother with dating drama when you can reduce your mate to a microscopic sperm-pod? Some species exhibit such radical size and role disparities that the male becomes a permanent, parasitic accessory, an efficient reproductive USB-stick fused to the female.[123]

## Brooding & Large Yolked Eggs:

When you can't afford to lose a single offspring, go all in on each one. Investing heavily in fewer, well-nourished eggs gives hatchlings a fighting chance in a perilous world, allowing them to emerge strong and well developed.[124]

## Larval Drift Strategy:

The great elevator of survival. Some species send their larvae toward the light, into shallower, food-rich waters. Here they grow and feast before diving into their adult lives in the abyss below. A strategic detour through the buffet line of life.[125]

## 5. *Ecological & Behavioral Adaptations*

### Sit-and-Wait Predation:

When food is scarce, patience is a weapon. Many deep-sea hunters conserve precious energy by going full ambush. They hunker down in the darkness, blending into the gloom, and using glowing or chemical lures to draw in the unlucky.[126]

### Symbiosis with Microbes:

Many creatures outsource critical life functions to microbial partners, especially near hydrothermal vents and cold seeps. These tiny roommates can help digest toxic chemicals, extract energy from minerals, or manufacture nutrients from nothing but the chemical broth of the abyss.[127]

### Colonial Organization:

Think modular. Siphonophores and similar organisms break the mold by operating as entire colonies of specialized units, each doing its part to function like a single, elegant organism. It's biology by committee, and it works.[128]

### Incorporation of Metallic/Mineral Structures:

Some species don't just wear armor. They build it into their flesh. Iron sulfide scales, mineral-infused shells, or even potential oxygen-generating compounds become part of the body itself. Evolution went full biotech.[129]

An impressive array of evolutionary tendrils, to be sure. And it takes only a modest leap of imagination to trace how some of them might extend even further, into stranger, more speculative forms. But one category remains conspicuously absent from our list. It's one whose implications stretch beyond biology and into the very heart of consciousness itself.

### Communication.

We've already established that humans carry implicit bias like a favorite pair of sunglasses, tinted lenses through which we filter the world. That's not always a flaw. Sometimes it protects us from being overwhelmed by the full spectrum of perception. But it's a double-sided filter. Too much self-recrimination over our biases is as limiting as blind adherence to them. Like anthropomorphism, this impulse is just part of the toolkit. Rather than place it on an altar or burn it in effigy, we might do better to observe it dispassionately, like an earthworm wriggling on the sidewalk after rain. Better still, we can imagine ourselves *as the worm*, considered in turn by a lifeform so distant from us in cognitive structure that we become the inscrutable object of its gaze.

It is into this speculative reversal that we now step.

We're not the first to crash this philosophical party. In fact, it's a rather lively soiree, long attended by inquisitive minds and conceptual acrobats. Chief among them is the visionary author China Miéville, whose words opened this chapter. His award-winning novel *Embassytown* doesn't just explore what language *is*. It pushes at the edges of what language, and by extension, communication, *could become* when disentangled from the limits of human structure and expectation.[130]

Miéville's *Embassytown*, like much of his work, ruptures and reassembles our understanding of things we thought familiar. It's not only a meditation on language. It's also an excavation of the foundations beneath it. His book probes the deep architecture of linguistics, how language is weighed, measured, and judged, how meaning is both built and broken. That framework is deeply relevant to our present discussion.

We won't attempt to map the entire conceptual territory of *Embassytown* here. Instead, we'll draw selectively from its most striking ideas, ones that speak directly to the speculative direction of this work. The goal is to use Miéville's insights not as a summary, but as a springboard. A kind of narrative phosphorescence to help light the trench we're swimming through.

Note: If you haven't read this phenomenal novel, be warned, mild spoilers may surface ahead.

In the story, humans are tasked with establishing communication with the non-human Ariekei while residing on their home world, Arieka. This interaction presents not only a logistical challenge but also becomes the central source of tension in the narrative. The core issue lies in the almost insurmountable differences between human and Ariekei modes of verbal communication.

To bridge the gap, genetically engineered human pairs must be custom-designed solely for this purpose. These biologically manufactured human "twins" must be perfectly calibrated in every aspect, both physically and mentally, as well as linguistically. If their synchronization wavers by even the slightest degree, communication fails entirely.

Miéville uses this conceit to illuminate a more bottomless chasm, the gap between implication and inference, communication and understanding. It's a clever, unsettling allegory for something we all carry within us. The presumption that *our* way of perceiving and conveying meaning is the default. The Ariekei don't just *speak* differently. They *experience language differently*. And in that gap, Miéville shines a light on the shadowy terrain of implicit human bias.

Our shared, spoken method of using mouths to communicate and ears to understand is foundational and seamlessly embedded in our experience of the world. So much so, in fact, that we rarely pause to imagine alternatives or consider what might lie entirely outside that framework. Of course, we're familiar with the more surface-level limitations of verbal communication. Different languages, dialects, and accents can all disrupt the smooth transmission of meaning. But even these are still operating within the same paradigm.

The global community of Deaf and hard-of-hearing individuals, often underestimated or misunderstood by so-called 'Hearies,' offers a living example of a rich, nuanced linguistic experience that flourishes outside the bounds of vocalized sound. Their use of spatial and visual language systems reveals that speech is not the sole path to conveying meaning.[131]

Then there are the fringe cases. Stories of "feral" children (whether anecdotal or not) who grow and adapt without exposure to traditional linguistic development.[132] Or the cultural and religious echoes of the Biblical Babel myth, in which divine intervention shatters a once-unified human language and scatters the population across the Earth. An allegory, perhaps, for the fear of losing shared understanding.[133]

Yet in the context of this book, it is Miéville's human-Ariekei interplay that offers the most illuminating parable. The true conflict lies not merely in differing communication styles, but in the unexamined, possibly *unknowable*, assumptions each side brings to the encounter. It is this shared blindness, this mirrored failure to see the limits of one's own linguistic architecture, that ultimately leads to anguish and ruin on both sides.

So, keeping Miéville's premise in mind and returning to our catalog of unexpected outgrowths born of evolutionary necessity in the deep, what's missing?

We've seen creatures that glow, devour bone, reproduce with colonies of microscopic males, and bypass eating altogether. But do they *speak*?

Perhaps not in words that human ears perceive. But does that mean they don't communicate? And if they do, who among us is listening properly?

Do the physically blind miss more than the perceptually blind? Surely not. Do wonders that slip past the slumbering heart make themselves any more available to the sentinel on the tower? Who is better positioned to receive the message, the dreamer or the guard?

Life in the deep does not lack communication. On the contrary, it appears to teem with messages. If only we knew how to read them. Among the most obvious examples are whales, whose haunting, sonorous calls traverse thousands of kilometers.[134] Could the Arieki and whales converse? If humans fail to decode such language, is it because the transmissions are too complex or because our interpretive bandwidth is too narrow?

Many fish, as well as some mammals, communicate using infrasound, which is a frequency vibration far below the threshold of human hearing.[135] The ocean hums with grunts, pulses, pops, and drumming rhythms. Each is likely saturated with meaning that dissolves the moment it meets the shoreline of our understanding.[136]

Sound is only one language of the deep. We also witness flashes of bioluminescence, blinking in rhythmic or staccato bursts.[137] But are these signals purely functional? Camouflage? Lures? Such conclusions arise not from evidence, but from interpretation. And interpretation always reveals more about the interpreter than the interpreted. If a message was never meant for us, do we lose nothing by missing it? Or do we lose a mirror in which we might have seen something *essential*?

Chemical cues, too, are everywhere below. They are ignored by us, but omnipresent to others. Entire bacterial communities engage in quorum sensing, a signaling mechanism that utilizes population density as a means of communication through molecular language.[138] These exchanges are not just nonverbal. They are *environmental*, woven into the very atmosphere of the deep.

Pressure differentials become messages of movement and presence in the abyssal dark.[139] Perhaps most tantalizing of all is the concept of electrical signaling. Subtle currents generated by biological fields are utilized by certain creatures not only for detection and orientation, but also for active communication.[140]

The lenses of our bias-colored glasses are not just dimmed. They are cut along lines that favor the linear and discrete. These boundaries comfort us, shape our expectations, and define the known terrain. And comfort, while not the enemy, is a seducer. It draws us back toward familiar ground, toward "known knowns," and away from windmills we've yet to recognize as such. But in the deeper zones of the ocean, even the simple deceit of refracted light, like a paintbrush bent in a vase of water, fades into absence. Here, communication does not travel in straight lines. It blooms along ambient paths, diffuses through unseen channels, and echoes in symbol sets we haven't yet learned to decipher. Or perhaps ones we were never meant to.

Our exploration of biological variance in the deepest ocean layers has taken us not only through radical forms of life but across immemorial strata of evolutionary time. Of the two, the latter may be even more unapproachable to the human mind. It is within these yawning epistemological canyons, those gulfs between what we know and what we cannot yet conceive, that we orient towards the unimaginable. Yes, there are deeper layers than "beyond the bottom." We'll go there in Chapter Six. But first, the sky.

Already guessed that, haven't you? Good. We've lingered in the deep long enough. Let's lift everything we've gathered here, every adaptation, every assumption, and aim it skyward. For there, too, in the air above, strange forms stir. Are the parallels we'll find a coincidence?

Is anything?

# CHAPTER FIVE

## Contact Without Context: Foundering on the Shore of Recognition

*"There are more things in heaven and earth,
Horatio, Than are dreamt of in your philosophy."*

*Hamlet, Act I, Scene V*

F lawed assumption. Is there anything more dangerous? To reason? To progress? To sanity? Perhaps. But charging boldly forth on a mistaken belief has been the ruin of many beautiful and terrible dreams alike.

History and fable are thick with warnings. The soldiers of Troy learned this lesson well.[141] When an unanticipated tribute arrives at your doorstep, especially one crafted with exquisite care to suit your sensibilities, *beware*. Your undoing may be nestled in its womb. All too often, humankind plants its heraldic flag in the shifting sands of time and circumstance, only to cry out in shock when that proud imperial standard tips, falters, and washes out to sea on a tide of indifference. RMS *Titanic*[142], Daedalus and Icarus, Oedipus, Dorian Gray, Faust, Captain Ahab, and even Dr. Henry Wu from Michael Crichton's *Jurassic Park*.[143] All choked on the bitter tang of hubris.

But perhaps the all-time champion in this department is Shakespeare's paramount punch to the gut, *Hamlet*.[144] We need not plunge fully into the labyrinth of plot and character, but a brief overview of the Prince's downfall will serve us well before we proceed. What assumption does he make, and how is it flawed? (Apologies to scholars of the Bard. If any liberties have been taken, blame not the stars but the author.)

The interwoven themes of *Hamlet* are a testament to Shakespeare's craftsmanship.[145] Somewhere, right now, a young actor or student is marveling for the first time at the revelations of heart and soul that Shakespeare wrested from the universe's eternally unyielding grasp. His works endure for the same reason all timeless creations do. They speak profound truths to the deepest and most persistent aspects of humanity, those qualities that outlast even death and echo through generations.

When we think of *Hamlet*'s central themes, the usual suspects arrive first. Revenge, corruption, death, and madness. Let's set those aside for now and instead turn our gaze to perhaps the play's most formative motif. The tension between appearance and reality, truth and deception, thought and action.

Before the curtain rises, the Prince's father, King Hamlet, lies murdered by his brother Claudius, who wastes no time in marrying the queen, Gertrude. This union enables Claudius to claim both the throne and the power that accompanies it.

The opening scene unfolds just before dawn. A ghost appears to three men stationed at the gates of King Hamlet's home, Elsinore Castle. Two of them are soldiers on watch. The third is the Prince's friend, Horatio. Here, we encounter a crucial and fateful thread in the intricate mosaic of the story that follows.[146]

The soldiers, spooked and keen to keep their distance, ask the erudite scholar Horatio to address the apparition on their behalf. They have their reasons. First, they trust his literacy and learning to better equip him for a conversation with whatever this is. Especially since it appears in the form of the late king. Second, they believe his critical mind will help guard against being swayed by any supernatural influence. And finally, Horatio is Hamlet's friend. Translation? *That thing out there, ghosting it up? It looks just like your buddy's dad, Horatio. We're not touching this one.*

When Horatio reluctantly complies and addresses the figure, the ghost retreats without a word. The soldiers note that this is not the first time. *"Thus twice before, and jump at this dead hour, with martial stalk hath he gone by our watch."* They speculate about what its repeated appearances might mean, correctly suspecting it heralds misfortune. (Good instincts, fellas.)

As they debate, the ghost appears again. Once more, they try to communicate, and once more, it withdraws. This time, at the crow of a rooster, signaling the coming sunrise.

It's important to note that none of the men ever *confirms* the ghost as King Hamlet himself.[147] They comment only on its resemblance. This detail is key, both to the plot of *Hamlet* and to the framework of the book you hold in your hands. Before we follow that thread further, let's step back and examine a small detail that seems to be overlooked, even by the Horatios of our time.

Very early in the scene, Bernardo remarks: *"Last night of all, when yond same star that's westward from the pole had made his course t'illume that part of heaven..."*[148] In modern English: *"Last night, when that star to the west of the North Star had traveled across the night sky to the point where it's shining now..."*

When we search the stars west of the North Star, Polaris, we find the constellation Cassiopeia.[149] This constellation was well known and richly mythologized long before Shakespeare penned *Hamlet* at the turn of the 17th century[150] (circa 1599–1601).

But here's the curious part. Shakespeare doesn't have Bernardo refer to a constellation. He points to a single star, west of Polaris, that *"had made his course t'illume that part of heaven,"* directly tying this to *"yond same star... westward from the pole."* In Bernardo's account, a specific star has been moving into position just west of the North Star, night after night.

This celestial aberration is subtly yet unmistakably linked to the ghost's appearance.[151] The heavens shift, and something unearthly walks the ramparts.

These events are, at least in the minds of the watchguards, connected. Ordinarily, stars and other celestial bodies do not move into position and then halt. Especially not repeatedly over the course of several nights. Planets *can* appear to stop and change direction. However, this is an optical illusion created by the intricate dance of planetary orbits. When a planet enters such a state, known as apparent retrograde motion, the shift is gradual.[152] It is never abrupt from our vantage point on Earth.

What exactly is Bernardo describing? This strange celestial behavior is offered almost casually. Yet it sits in the text like a quiet, shimmering omen. Is Shakespeare conjuring a heavenly specter to mirror the earthly one that Horatio implores?

We cannot know. And perhaps that is why Shakespeare endures. Glittering, winking details resist final answers. They whisper to us across the centuries, urging us to turn them over and over in our minds, like rare stones in the hand.

The heavens have always played host to strange visitors, some charted, some not.

In the United States, where this is being written, UFO/UAP lore is usually said to begin in July of 1947 with the now-infamous "Roswell incident."[153] We won't rehash that story here. It has been reported, argued, and mythologized more than almost any other. But what of the sightings and reports that came before? They are numerous, provocative, and widely recorded.[154]

For our purposes, whether these events were genuine UAP or some other phenomenon not easily explained by known physical processes matters less than the fact that they happened. Such reports were well known even in Shakespeare's time.[155] Could the Bard have been channeling something stranger into his narrative? Perhaps. As with so many things in great literature, and in the anomalous affairs of the heavens, it is up to each of us to decide.

Like Hamlet, we are free to render our own verdict, unassisted. Or we may adopt the judgments of others, whether through ignorance, confusion, insecurity, grief, or blind faith.

Descending into the labyrinthine inner workings of Hamlet's heart is beyond this author. However, his own words provide a clear enough portrait. Eventually, the Prince arrives at the castle gate, having been told of the strange events unfolding there. Just before the moment the ghost reveals itself to him, Hamlet is mid-soliloquy.[156] He is bemoaning his uncle Claudius's drunken revelry and indulgence of base appetites. The Prince laments the stain this behavior leaves on Denmark's reputation, tacitly agreeing with foreign critics who see it as a mark of national decay.

It is in the middle of this acidic meditation that the ghost reappears. Hamlet sees the figure for the first time, and the words spill out of him:

*Angels and ministers of grace defend us!*

*Be thou a spirit of health or goblin damned,*

*Bring with thee airs from heaven or blasts from hell,*

*Be thy intents wicked or charitable,*

*Thou comest in such a questionable shape*

*That I will speak to thee. I'll call thee "Hamlet,"*

*"King," "Father," "royal Dane." O, answer me!*

*Let me not burst in ignorance, but tell*

*Why thy canonized bones, hearsed in death,*

*Have burst their cerements*[157]

Overwhelmed by the apparition, whose form so perfectly mirrors that of his father, Hamlet does what the guards and Horatio never dared. He *names* it. *Hamlet. King. Father.* He assumes identity before evidence. Meaning before proof. And from that moment of premature certainty, the rest of his tragedy will unfurl in ruin and despair. [158]

As we set out to explore and untangle the phenomenon now labeled "UAP," we would do well to keep Hamlet's tragic misstep firmly in view.[159] His failure to pause, to interrogate, and to interrupt his own chain of thought before declaring meaning. Should we rush too swiftly past our own uncertainty, naming before knowing, we risk the same fate. This admonition is born from personal experience. The author has already stumbled into that trap, falling prey to faulty assumptions long before any errant star or spectral visitor had the chance to "harrow up mine soul. "

This lesson is not abstract. Many of us have been drawn to accounts of mysterious, unidentified aerial phenomena (née "flying objects"). We are captivated by their fantastical details and arresting narratives. Among the most compelling are two separate incidents reported in 1967 at Malmstrom Air Force Base in Montana.[160] Until researching for this book, the author, like many readers, had incorrectly believed them to be a single event. Even if that had been true, it would still make a worthy starting point for investigation. But discovering that the attending set of circumstances was, in fact, far more nuanced and complex only deepens the intrigue. This is especially true, given the specific details and the credibility of those who reported them.

If assumptions can so easily mislead, then context becomes our strongest safeguard against error. Malmstrom Air Force Base offers plenty.

Before it bore its current name, Malmstrom was known as Great Falls AAB (Army Air Base). From the moment it opened during World War II, it was an installation of strategic importance to both Montana and the United States military. Located roughly 10 kilometers (6 miles) east of the town of Great Falls, the base officially began airfield operations when the first B-17 bomber touched down in November 1942.[161]

Great Falls quickly became a critical training ground for bombardment groups.[162] These squadrons would go on to play decisive roles in raids over Germany, clearing the way for Allied precision bombing campaigns. Expansion came swiftly, and with it, a pivotal role in the United States' far-reaching 'Lend-Lease' program. This program was a crucial wartime lifeline. It supplied badly needed food, oil, and military equipment to Allied nations throughout the war.

The program's scope defies a concise summary, but its significance is beyond dispute. As historian Richard Rhodes records, Lend-Lease was not limited to crates of matériel. It included shipments of strategic and scientific materials, including uranium compounds, graphite, and other elements later recognized as essential to atomic research. In addition, top-secret intelligence reports and schematics for cutting-edge aviation and electronics were transferred. It also served as a transit corridor for a surprising number of Soviet agents. Each passed through Great Falls on their way to or from missions whose details, in many cases, remain obscure.[163]

Because this same ground later became the site of some of the most debated UAP reports in military history, it's worth remembering this was a crossroads for nuclear secrets, cutting-edge technology, and high-stakes espionage. In other words, the stage had been set long before the curtain ever rose on the events of 1967.

This storied, classified legacy naturally set the stage for Malmstrom's next transformation, into a crucial linchpin of Cold War operations. With the birth of the United States Air Force, the base was renamed Great Falls AFB, its mission shifting in step with the geopolitical tensions of the era.

It was during this period that another, perhaps even more consequential chapter unfolded. For a brief span in 1953 and early 1954, the 582nd Air Resupply and Communications Wing operated from the base. Their mission roster included activities officially labeled "PSYWAR," or psychological warfare. Alongside other classified operations, specifics remain obscured.

The military's own definition, however, is almost unnervingly concise. *'Activities intended to convey selected information and indicators to audiences to influence their emotions and reasoning.'*[164] Clinical words for a thoroughly human art. It is one that blends strategy, perception, and control.

Keep this concept in mind. We will return to it later, and its relevance will become clearer when we examine the events that made Malmstrom a recurring waypoint in UAP history.

Not long after the PSYWAR chapter closed, the base endured the tragedy that would bestow its present name. Colonel Einar Axel Malmstrom lost his life here during a routine flight in a T-33 Shooting Star. The Colonel was a decorated war hero, respected leader, and by all accounts, a man of uncommon decency.

A veteran of World War II, Malmstrom was as admired by his fellow airmen as he was by the local civilian community. Early reports after the crash portrayed his death as an act of selfless courage, suggesting he chose not to eject, steering the failing aircraft away from a nearby town. Later Air Force records, however, indicate that a sudden medical emergency may have rendered him unable to escape.

Whatever the cause, Malmstrom's sacrifice and character left a lasting mark. Great Falls AFB became Malmstrom AFB, a tribute not only to military valor, but to the kind of personal courage that resists calculation.[165]

The name has remained ever since, a reminder that history here is shaped as much by human choice as by geopolitical strategy. The forces at work on this ground have long been a mix of the visible and the unseen.

In the years that followed, the base's strategic profile continued to deepen. Aircraft Control and Warning Squadrons were stationed here, along with some of the most advanced radar arrays of their time. These systems formed part of the United States' early warning shield. Designed to detect and track aircraft, they served as a first line of defense against a potential attack.[166]

As technology evolved, so did the role of Malmstrom AFB. It became a node in NORAD's continental defense network, and later a key component of Strategic Air Command. Its mission now extended far beyond Montana's borders.[167]

That legacy, combined with Malmstrom's specific geographic position, made it an ideal choice for deployment of the U.S. military's state-of-the-art, three-stage solid-fuel missile, the Minuteman. This intercontinental ballistic missile (ICBM) was stored in vertical, cylindrical silos constructed in clusters underground. Each silo was sealed by a massive blast door and linked to a centrally located launch control center.[168]

These were no monuments to history. They were kept in a constant state of readiness and poised to launch at a moment's notice, should the order ever come. It was within this high-stakes environment, with the machinery of immediate nuclear response humming silently beneath the Montana plains, that events would soon unfold that defied both explanation and expectation.

The next pivotal moment in Malmstrom's history came in late 1962, when the world stood at the brink during the Cuban Missile Crisis. This epochal standoff accelerated the nuclear arms race. For Malmstrom, it meant heightened readiness and rapid missile upgrades. Throughout those tense weeks, and in the years that followed, its silo arrays remained on constant alert, each one a silent guarantor of deterrence.[169]

It is these very silos, primed and fortified in that crucible of Cold War brinkmanship, that would later stand at the center of some of the most-debated UAP encounters in military history: the "Echo Flight" and "Oscar Flight" incidents.[160] What happened next would challenge not only the reliability of cutting-edge defense systems but also the assumptions we make about what moves through our skies.

The first officially recorded incident at Malmstrom AFB came on the morning of March 16th, 1967, at precisely 8:45 a.m. In the Echo Flight Minuteman I array, all ten missiles, each housed in its own underground silo, dropped into "No-Go" status within seconds of one another.[170] In missile terminology, "No-Go" means precisely what it sounds like. The weapon is either unavailable for launch or completely non-operational.

Each missile was positioned in its own Launch Facility (LF), and each LF was equipped with a critical diagnostic system called the Voice Reporting Signal Assembly (VRSA). This device allowed crews to monitor key aspects of their silo's operation, run diagnostic queries, and manage malfunctions detected by the Launch Control Center (LCC).

When queried during the Echo Flight event, the VRSA returned fault codes for all affected silos. Each code pointed to the same issue. The guidance and logic systems both experienced failures, specifically at channels 9 and 12.[171] The shutdown had been sudden, simultaneous, and, by all known operational standards, impossible.

All ten missiles were eventually brought back online, without replacing a single piece of hardware or performing any physical repairs.

As one might expect, an exhaustive investigation followed. It uncovered a few incidental anomalies. A diesel generator malfunction occurred at one Launch Facility, and a faulty actuator was found at another. But these minor issues could not account for the simultaneous failure of all ten missiles.

Investigators from Strategic Air Command, Boeing, Autonetics, and OOAMA (the Office, Ogden Air Materiel Area, a logistics, support, and maintenance contractor) pored over every scrap of data. Their findings, drawn from lightly redacted official documents, read as follows:[172]

- **No evidence** of sabotage, personnel error, or component replacement.

- **No unusual weather** conditions. No electromagnetic interference, lightning strikes, or atmospheric disturbances.

- **Confirmed**: the shutdowns did not result from any standard or known "No-Go" logic or command.

- **Testing**: a 30-microsecond pulse through the Logic Coupler *could* cause channels 9 and 12 to return the same fault codes, but only under controlled conditions.

- **Most likely cause:** *"Some type of external noise pulse, possibly electrical or electromagnetic."* However, the probability of such a pulse disabling *all ten* logic couplers simultaneously was deemed vanishingly remote.

The official report goes out of its way to dismiss any rumors of UFO or UAP involvement, ruling out the possibility outright.[173] Mobile strike teams reported nothing unusual on radar during the incident. No contacts. No returns. Nothing that could be pointed to as an intruder in the sky.

Remember that absence. It will matter later. The lack of radar confirmation during the Echo Flight shutdown will recur in another context, one just as unsettling.

So, what can we conclude from the report? Only that it was an extremely anomalous and thoroughly troubling event, investigated exhaustively, yet never fully explained. Strategic Air Command, defense contractors, and technical specialists all weighed in. The best they could offer was a narrowing of possibilities to an *external* pulse, electrical or electromagnetic in nature. Even then, no definite source for such a pulse was identified.

Once this troubling vulnerability was addressed, no further official incidents of this kind were recorded. Officially, the matter was closed. Unofficially, however, Malmstrom's strangest days were still ahead.

One week after the well-documented Echo Flight shutdown, reports surfaced of a strikingly similar incident in the separate Oscar Flight array. But here, the trail grows thin. No comparable official record exists for this second event. No declassified report or internal memo exists in the archives.[174]

What we have instead is testimony. Captain Robert Salas, Colonel Fred Meiwald, and Robert Jamison each recount what they say occurred.[175] A cursory review of Salas and Jamison reveals a spectrum of opinion about their credibility, from staunch belief to outright skepticism.

Because there is no surviving contemporary documentation, this account cannot be approached in the same way as Echo Flight. Still, certain elements can be confirmed. Those details, sparse though they may be, are telling.

Captain Robert Salas's service at Malmstrom AFB is not in question and is well documented. His training took place at Chanute and Vandenberg Air Force Bases. This was followed by assignment to Malmstrom as a missile launch officer responsible for ten Minuteman I ICBMs.

After leaving the military, Salas continued working in aerospace engineering, holding positions with Martin Marietta and Rockwell International. Both firms were involved in missile contracts.[176] On paper, he is precisely the kind of person whose testimony on missile site security would carry weight.

And yet, when it comes to the events of March 24th, 1967, the trail goes cold. In Salas's account, he was on duty at the Oscar Flight LCC when reports of UAP began coming in from security personnel in the immediate area. He claims these sightings coincided with all ten missiles under his command abruptly becoming nonoperational.[177]

No official documentation supports this version of events. Salas maintains that no record exists in the public domain because the incident was suppressed or "scrubbed." He suggests this was done by his chain of command, leaving no trace in FOIA-released materials or other channels.[178]

Both the Echo and Oscar Flight accounts tempt us toward the same mistake that undid the Prince of Denmark. We may be drawn to name the ghost before we understand it, to declare its nature while it still stands in shadow. Let us resist that urge. It is better to hold our judgment and gather more light before deciding whether we are confronted by an angel, a devil, or merely the reflection of our own expectations.

And while the "Oscar Flight" incident leaves us with no official record, such absence is not the end of the trail. Do other military and government reports from the same era, though avoiding the term UAP, reveal that sightings extended well beyond on-site security personnel? Indeed, they do. In fact, there are many such reports.

A verified, declassified NORAD Command Director's Log, paired with Strategic Air Command reports from October to November 1975, details repeated sightings and direct military responses to UAP over multiple U.S. Air Force nuclear facilities. Malmstrom is among them.

Hold on to your Hamlets. The stage is about to grow considerably more crowded.

The 1975 reporting chain extends beyond Malmstrom. Loring AFB in Maine, Wurtsmith AFB in Michigan, and Minot AFB in North Dakota all appear in the log. Each is a nuclear site, either a missile field or a storage facility.[179]

Between November 7 and 10, 1975, Malmstrom generated multiple sighting reports through official channels alone. The declassified log reads like a series of stage directions.

**Spotlight:** Bright orange-gold objects, some dotted with smaller lights, hovering or maneuvering near launch facilities.

**Cue effect:** An airborne object emits tubular black shapes. Smaller craft, perhaps, or pieces are jettisoned into the air.

**Scene change:** Objects repeatedly rise into the sky, vanishing at sunrise.

**Enter security:** Teams dispatched to investigate each report.

**Stage blackout:** No radar or surveillance confirmation of airborne objects, only anomalous returns, flickering at the edge of the instruments.

Specifically, on November 8 and 9, multiple fighter intercepts are launched, including F-106s scrambling from their hangars.

**Stage cue:** Unidentified objects cut their lights the instant fighters arrive, appearing as dark silhouettes against the sky.

**Re-light:** Once the interceptors retreat, the lights return as if nothing had happened.

**Movement:** Some objects race across the radar scopes at extreme speeds. Others hang motionless in the air. Altitudes vary from great heights to near ground level.

**November 9th–10th:** The cast expands. FAA and civilian pilots report seeing the same objects, described as "arc welder's blue" or "starlike," but clearly maneuvering under control.

Reports from Loring, Wurtsmith, and Minot match the performance. Objects were reported to be hovering, advancing, and withdrawing at calculated intervals. They reacted directly to interception attempts and always, *always*, left no conventional radar confirmation. Their silhouettes varied. Discs, tubes, or shapes that do not fit into the known airframe catalog.[180]

One final note in the declassified log. Reference to an additional document, withheld on the grounds that it was "properly and currently classified and is exempt from disclosure." The signature at the bottom is that of Colonel Terrence C. James, Director of Administration, U.S. Air Force.[181] Curtain down, for now.

Less than a decade after Captain Salas' reported Oscar Flight incident, which mirrors the officially documented anomalous shutdown at Echo Flight, we find something fascinating. Verified records describe the same aerial behavior reported by Salas in 1967. Unidentified objects are witnessed maneuvering near missile sites, reacting to approach, and exhibiting characteristics suggestive of electromagnetic interference. Yes, the events are separated by roughly eight years, but the sheer number of reports and their consistency across both military and civilian witnesses is conspicuous.

If these were isolated accounts, they might be dismissed as curiosities. But alongside the Echo Flight shutdown, which concluded with a suspected *external* "electrical or electromagnetic" pulse disabling 10 ICBM silos within seconds, they begin to form the outline of something larger. Look closely, and a pattern starts to emerge.

We should pause here to be transparent about sources. Every detail cited so far comes from military or government channels, which are officially documented and verified, whether declassified or unclassified.[183] And while "unclassified" may sound innocuous, these were still *internal communications*, never intended for public release. In practice, they were as inaccessible as the "classified" files until now.

It's worth noting that the very law that enabled us to see any of these documents, the Freedom of Information Act (FOIA), did not exist at this time. Though signed into law on July 4, 1966, FOIA did not officially take effect until July 4, 1967.[184]

Its chief architect was U.S. Representative John Moss (D-CA), who began developing the framework shortly after being elected to the House of Representatives in 1953. Moss chaired subcommittees on Foreign Operations and Government Operations. He also served on other committees that dealt with National Security, atomic energy, water, power, and oceans.[185] Moss's tenure spanned the dawn of both the atomic age and modern deep-ocean exploration, two realms whose boundaries and secrets often ran in parallel.

That parallel is instructive here. As a rule, nuclear power plants are built near large bodies of water, drawing on them for cooling their reactors.[186] Nuclear weapons production, meanwhile, carries a long and well-documented history of contaminating those same waters.[187]

These twin realities of water and nuclear infrastructure are as tightly bound as any engineered coupling. They are also bound to verified internal reports of UAP phenomena near nuclear facilities worldwide, not just in the United States.[188]

As covered earlier in this chapter, the Great Plains of the 1960s and '70s were no stranger to official reports of anomalous airborne phenomena.[179] Accounts of objects behaving in ways that defied the expectations of both military and civilian pilots, as well as ground-based witnesses, were common.[180] However, these accounts were not limited to that decade or that region. The pattern stretches far wider.

Some notable cases:

- **1965 – Edwards AFB (Kern County, California):** Multiple luminous objects observed over the base and surrounding airspace for several hours. Radar tracked the activity, and an F-106 interceptor was scrambled from nearby George AFB.[189]

- **1965 – Warren AFB (Cheyenne, Wyoming):** Security personnel and missile crews observe unknown objects with light displays near missile sites. Some encounters include anomalous instrument readouts.[190]

- **1966 – Ellsworth AFB (Rapid City, South Dakota):** A security alert team watches as a glowing object approaches, hovers, then departs at high speed.[191]

- **1975 – Loring AFB (Caribou, Maine):** Unidentified objects repeatedly observed over the nuclear weapons storage area, prompting security alerts and pursuit by Air Force aircraft. The sightings continued over several nights and initiated a series of similar incursions that autumn at Wurtsmith, Minot, and Malmstrom AFBs.[192]

From base to base, state to state, the stage directions are eerily consistent. Approach, linger, retreat, in defiance of known flight capabilities, and always near nuclear assets.

These accounts are only a sampling, a small slice of official, documented cases spanning different locations, eras, and operational contexts. Yet taken together, they reveal a clear, discernible, and well-documented pattern. We see UAP sightings and encounters in direct proximity to nuclear weapons sites, nuclear power plants, and missile facilities.[188] Not only in the United States, but worldwide.

This pattern persists across decades and national borders. It extends from Manhattan Project-era incursions at Hanford, Oak Ridge, and Los Alamos[193], to the incidents already discussed at strategic missile bases, nuclear storage facilities, and bomber airfields. Also included are activities clustered around nuclear power plants, such as Indian Point (NY) and Fukushima (Japan), along with more recent reports involving French, Belgian, and U.S. plants in our present era.[194]

The historical record also reveals "clusters" of activity during periods of heightened nuclear tension, most notably around the Cuban Missile Crisis.[195] This was a period when the stakes for any interference, human or otherwise, could not have been higher.

The consistency of specific details across these reports is obvious. The inhibition or outright shutdown of launch systems, as well as the triggering of intrusion alarms within secure facilities, all recur at the same locations over spans of decades.[196] These patterns are not just the domain of fringe researchers. They are acknowledged, at least in broad terms, by military officials and even present-day governments.[197]

So, why does this matter? Like Hamlet, confronted with the ghost, we might be tempted to react with immediate certainty. We can easily be drawn to name a thing before we understand it, letting the thrill of recognition override the discipline of inquiry. Our aim, however, is to follow Horatio's example. We must resist the lure of premature conclusions, whether they flatter our preconceptions or intrude suddenly and uninvited, demanding form before definition.

If we set aside appearances and even apparent functions, what remains? What insights might emerge from this sustained, almost obsessive focus on the machinery of nuclear power and weaponry? That is the question before us now. For if we claim to be creatures of reason, we are duty-bound to turn and face the unassailable reality of the pattern itself. It is conspicuous, enduring, and impossible to ignore.

It is not the author's aim to suggest that all reports, official or otherwise, originate from a nonhuman or supra-sapien source. Rather, it is the sheer weight and persistence of the recurrences in this pattern that compel us to ask, what possible driving impetus might such expectation-defying objects possess?

Repeated, identical behaviors over long stretches of time suggest, though they do not prove, intentionality. From a human vantage point, the apparent goal could be interpreted as a kind of pushback or warning, aimed at preventing the potential consequences of tampering with the atomic structure of fissile materials. In much of the research and writing on this subject, that pushback is cast as largely benevolent.[198]

Instead of being aggressive, the interactions have the character of a "light touch." A simultaneous shutdown of the entire launch complex resulted in no physical damage to hardware or harm to personnel.[172] What it did leave was confusion, unease, and the unshakable knowledge that someone, or something, had demonstrated control. It is as if these acts broadcast both superior capability and deliberate restraint.

Such behavior is far more sphinxlike than an outright attack. It hints, without proving, at qualities we associate with empathy and even wisdom. And in that ambiguity lies perhaps the most unsettling question of all. What kind of intelligence chooses to show its power in this way?

From motive, we turn to origin. Here lies perhaps the ultimate question. It is not only for those who witness these events, but for any who seeks to understand them.

It is the author's contention that we, as a species, have erred in our judgment. Not through malice, but reflex. We have been hailed by apparitions and rushed forward, crying *"Alien!" "Extraterrestrial!" "Visitor from the heavens!"* Perhaps one of these is true. But perhaps not.

Which is more palatable to us, more accessible to our imaginations? To suffer the slings and arrows of outrageous fortune, or to take arms against a sea of troubles? The author's goal is not to redefine, but to set aside rigid definitions in favor of an as-yet-unfamiliar possibility, one perhaps less outrageous than the misguided missiles with which we currently surround ourselves.

Let us arm ourselves, then, not for battle but in anticipation of an ocean still unfathomed. For it is to the ocean we now return. The abyss that is ever-present yet ever-hidden. 99.999% of it remains beyond our sight.[40] Is it possible, or even probable, that these repeated visitations originate much closer to home than we have dared to imagine?

It is tempting, even comforting, to imagine this knock at our door coming from across the stars. Distance makes it easier to romanticize, to mythologize, to tuck the unknown into constellations forever beyond our reach. But what if the visitors do not arrive in silver ships from the heavens? What if they rise silently from below instead?

What care would a traveler from the stars have for human affairs? Would they make note of our loftiest ambitions, or our ugliest transgressions? Are the patterns of behavior described here not more suited to co-inhabitants of our earthly abode?

Who knocks on the door at 1 a.m. when the celebration or altercation spills past polite decorum?

The downstairs neighbor. That's who.

# CHAPTER SIX

## Legion of Circumstance: At the Threshold of Inception

*"If you do not expect the unexpected, you will not find it; it is trackless and unapproachable."*

*Heraclitus, Fragment 18 (Diels–Kranz 22B18), as preserved by Clement of Alexandria, Stromata 2.17.96.4.*

W e return now to the deepest ocean environments, regions yet unexplored and in many cases, unimagined by human experience.[199] As promised, we will descend still further. We will travel beyond not only the ocean floor but also what we, as a species, have typically allowed ourselves to believe possible.

Our journey to this waypoint has been steady and deliberate. However, beyond this, we must loosen our grip on the inherited worldview that shapes not only how we perceive the world around us, but also how we perceive ourselves. In its place, we are called to lay aside the shadowed lens, both outward and inward-facing, through which our ideologies filter reality.

As previously discussed, there is no contempt here for the natural limits of human comprehension, any more than a seafarer resents the vessel that carries them safely through shifting currents of water and air. These limits shield us from forces we sense instinctively yet cannot, or choose not to, fully invite into our experience. The reasons for such guardedness are many. Foremost among them are fear, self-preservation, and a general or specific reluctance to engage in that most delicate of acts. The willing suspension of disbelief.

Yet such barriers are not fixed. They can be overcome, perhaps not by every species on Earth, but surely by humanity. We have confronted heights once thought intractable, even insurmountable, and have found ways to ascend. Each peak we conquer becomes not a terminus but a vantage point, a temporary refuge from which we push onward into the unknown.[200] These summits, distant and dazzling, eventually fall to humanity's collective capacity to see boundaries not as walls, but as waystations.

There are, as always, many opinions about this particular trait of our species. No matter. Much like the oft-repeated fable of the chicken crossing the road, humanity tends to make it to the other side. Perhaps this is because, somewhere deep within us, in a place without a name, there resides a quiet conviction. *It is better to seek out your fate with bold determination than to wait for it to kick down the door of a stagnant or hibernating mind.*

The path ahead is speculative. But speculation need not float untethered. The author aims to anchor it, if not in universally accepted truths, then at least in plausible reasoning. We search here for patterns of behavior and circumstantial evidence that refuse to be ignored. Many of the elements have already been laid out. What follows is a cohesive arrangement of how they interlock and what shapes they may reveal. In its briefest form, the argument is this:

Owing to the following separate but wholly interconnected factors:

1. The as-yet-unexplored and unknown ecological state of 99.999% of deep ocean environments (depths of 6,000-11,000 meters).[40]

2. The relatively continuous environmental stability of these zones across deep time (~500 million years).[201]

3. The continued and consistent discovery of thriving life in places long dismissed by humans as uninhabitable or irrelevant.[202]

4. The growing body of evidence revealing far greater cognitive complexity in nonhuman organisms than previously imagined.[203]

5. The ever-expanding understanding of how life on Earth adapts and evolves in ways we have consistently failed to anticipate.[204]

We may now hypothesize, with reasonable confidence, that life forms recognizable to us as *sapient*, or perhaps even *hyper-sapient*, may currently exist within deep ocean environments. They may even have done so for a span of time that eclipses the timeline of our own species.

This is not a fanciful or outlandish supposition. Especially when we consider what we have learned about our planet, our fellow earthbound organisms, and ourselves in recent decades. Let's now take each point in turn to see how they converge toward this possibility.

(1.) We have already established that humans have explored an infinitesimal fraction of the deep ocean. As of this book's publication date, the statistic still holds. Roughly 99.999% of all deep ocean environments remain completely unseen and unexplored. While this author will spare you from complicated mathematical proofs, for all practical purposes, 99.999% is functionally 100%.

Consider the following scenario. You and a friend both order the banana pudding at a roadside barbecue shack. You step away briefly to wash your hands. When you return, would you notice if they had eaten just 0.001% of your dessert? With banana pudding... maybe. But when we are talking about the sheer extent of deep ocean habitats still untouched by human eyes, the reality is unavoidable. For now, we remain essentially ignorant.

(2.) This is perhaps the least overtly evidentiary pillar of the argument. The author freely acknowledges the difficulty of quantifying the complex interplay of variables involved. Still, we can infer a great deal from the data we *do* have.

Let's begin with the geological and oceanographic perspective. The key conditions that define life in the deep ocean have existed in some form since at least the early Paleozoic era, roughly 500 million years ago. These include perpetual darkness, near-freezing temperatures, and pulverizing hydrostatic pressure.[201] While specific ocean basins are geologically younger due to the dynamic churn of plate tectonics, deep marine basins themselves have been a continuous feature of Earth's biosphere since the mid-Paleozoic. The destruction and rebirth of individual basins do not negate the broader ecological continuity of abyssal zones over deep time.[205]

Fossil records support this continuity. Organisms such as echinoderms, foraminifera, and various worm-like species exhibit lineages that extend across hundreds of millions of years. Interestingly, they have survived multiple mass extinction events. This resilience suggests an enduring, stable ecological niche in the hadal depths.

Biological evidence strengthens the case. Living deep-sea organisms, such as coelacanths, xenophyophores, and certain species of deep-sea corals, have been genetically traced back to divergence points in the distant geological past. Their origins also date back hundreds of millions of years.[206] These organisms are not evolutionary newcomers. They are survivors across vast temporal landscapes.

Finally, we turn to the chemistry and physics of the deep ocean itself. The very qualities previously mentioned that make it inhospitable to surface life also render it comparatively immune to the planetary upheavals that have ravaged shallower ecosystems. Events like asteroid impacts or climate-induced thermal shifts are far less consequential at such depths.[207] Moreover, the deep ocean's global circulation system has likely been in operation since at least the Paleozoic. This density-driven circulation, known as thermohaline circulation, has continually linked depths across the globe in an enduring planetary-scale loop of fluid memory.[208]

(3.) Of all the pillars supporting this hypothesis, this one may be the most quietly profound. Time after time, we find that the world is not so much hiding its secrets from us as it is waiting, sometimes impatiently, for us to notice.

Humans have a curious habit. Typically, we fail to extrapolate from our own data. Presented with conditions that should strongly suggest the possibility of life, we either dismiss them outright or greet life with stunned surprise when it reveals itself.

Consider early 2025, when an iceberg the size of Chicago broke from an Antarctic ice shelf. Upon exploring the ocean floor beneath, one scientist remarked, "We didn't expect to find such a beautiful, thriving ecosystem."[209] Why not? In nearly every locale previously declared too hostile, barren, or isolated for life to exist, we have found it flourishing in forms and patterns that defy our expectations.

This recurring astonishment betrays something more deep-rooted than simple anthropocentrism. Humans don't just place ourselves at the center of the living world. We mistake our own familiar environment for the only valid room in a sprawling, many-chambered mansion of existence.[210] And having made ourselves comfortable, we rarely bother to check what lies behind the next door.

Subglacial Antarctic lakes, buried beneath more than a kilometer of ice and once thought sterile, teem with life. This includes bacteria, archaea, crustacean remnants, and the indestructible tardigrade. These lakes exist without sunlight and with minimal nutrients, yet they host entire hidden biospheres.[211]

Even more unbelievable, in 2019, slow-moving but metabolically active bacteria were discovered inside rock samples drilled 2.5 kilometers (1.6 miles) beneath the seafloor in the South Pacific Gyre. This is arguably the most remote marine environment on Earth. With no landmass for thousands of miles and nutrient-starved, they persist. Ancient, alive, and utterly unbothered by our absence.[212]

On the surface of the Atacama Desert in Chile, one of the driest and most irradiated places on Earth, we find another thriving microbial community. This environment is so extreme that it is compared to that of Venus. These extremophiles survive by scavenging atmospheric gases in an environment that should render any hope of biological complexity impossible.[213] Meanwhile, just offshore, in the Atacama Trench, life continues to astound in the hadal depths.[214]

Even the skies are alive. Bacteria float through the lower and middle atmosphere, some actively dividing as they drift through frigid clouds bombarded by UV radiation.[215] Life endures, far above our heads.

And if all of that weren't enough, the deepest boreholes and mines have yielded thriving microbial ecosystems 5 kilometers (3.1 miles) beneath Earth's surface.[216] Even in sites of extreme radiation, such as nuclear waste storage pools, we find specialized microbes that not only endure the toxicity but actually help clean it up.[217]

So why do we continue to treat life as the exception when all evidence suggests it is the rule?

(4.) So many once-sacred benchmarks of human cognition, those presumed to distinguish our species as uniquely intelligent, have now been observed in a startling array of nonhuman animals.[218] It raises a question. *If we wonder about them, might they also wonder about us?*

No longer cloistered behind the imagined fortress of our own exclusivity, humanity finds itself face-to-face with a far more thrilling and perhaps unsettling truth. Our species is not alone in our capacity for emotion, logic, or discernment. We are entering a revolutionary era of thought about thought *itself*.

Even our ingrained tendency to anthropomorphize animals may hint at a deeper acknowledgment of this continuity. We see ourselves mirrored not just in great apes and whales, but in birds, cephalopods, insects, and beyond. The growing body of scientific evidence now confirms what our intuition has long whispered. Other creatures think, feel, strategize, grieve, remember, and adapt in ways far closer to our own than we were once willing to admit.

This paradigm shift began in earnest with one groundbreaking moment. Jane Goodall's quiet, patient observation of wild chimpanzees in Tanzania. In 1960, on her first night at Gombe Stream National Park, Goodall set in motion a transformation in scientific thought. At that time, it was still widely held that the use of tools was a defining trait of *Homo sapiens*, a supposed dividing line between "us" and "them." Goodall soon documented chimpanzees not only using tools but also *modifying* them for specific tasks. They selected twigs, stripped the leaves, and fashioned them into termite-fishing instruments.[219] The chimps demonstrated foresight, intention, and craftsmanship.

The response from famed paleoanthropologist Louis Leakey captured the gravity of this revelation:

> *"We must now redefine man, redefine tool, or*
> *accept chimpanzees as human."*[220]

To their credit, many in the scientific community eventually did redefine "man," though not without resistance.[221] But this redefinition did not diminish humanity. On the contrary, it expanded the possibilities of what the world could be, and what our shared place within it might look like.

When we loosen our grip on once-exclusive traits, we do not surrender our identity. We enrich it. Like discovering a wild plant growing freely in the open that we once thought was ours alone to cultivate, the revelation that intelligence blooms across the tree of life does not reduce our worth. It elevates our understanding.

Each connection we trace across species is not a subtraction from the human soul, but an aeration of it. These insights till the soil of thought, fertilize empathy, and clear new space for growth in the fecund landscape of sentience.

Since Dame Goodall's foundational work, examples of nonhuman animals creating and using tools have become so commonplace that they now appear in documentaries and even children's books.[222] Capuchin monkeys and sea otters use stones to smash open hard-shelled food. Crows craft hooks from sticks to extract insects from tight crevices.[223] Humpback whales blow intricate "bubble nets" to trap krill with surgical precision.[224] Even insects like the assassin bug employ sticky resin to turn their bodies into biological flypaper.[225] The diversity in type, form, and function is staggering.

And still, just as we tend to assume certain environments must be devoid of life, we continue to underestimate the intellectual landscape of lifeforms that are *not us*. A typical rebuttal to the possibility of sapient life evolving in an alternative biosphere, such as the ocean, is to point to the human hand with its opposable thumb. The usual claim is that this form factor is a prerequisite for intelligence. But is our so-called "meat hook" design truly superior to the octopus's tentacle? Or is it instead, one possible answer to an evolutionary equation?[226] Might there be other solutions we haven't even imagined?

Consider:

$$x + y = 5$$

Humans plug in 2 and 3, then dust off their primate palms in satisfaction. But the solutions here are infinite. Acting as if we've solved for all variables is not just shortsighted. It's dangerous. One false assumption at the outset can cascade into an entire architecture of faulty reasoning. Given enough distance, a single degree off course will land you on a different continent.

And this fallacy, the presumption of cognitive superiority, has given rise to more of the same.

In addition to tool use, animals across a broad phylogenetic range demonstrate abilities we once thought were uniquely human. A clear example of this is "mental time travel" or future planning. We now know that corvids such as ravens and crows don't just cache food. They plan *when* and *how* to use tools to retrieve it. Studies show they anticipate future needs based on past experiences. This is a concept once thought exclusive to humans. More striking still, these birds change their behavior when observed by potential thieves. They will delay or alter their food storage based on their perception of other birds' intent.[227]

Chimpanzees, too, have demonstrated this social awareness. They can distinguish between a human who is unwilling to offer a treat and one who is simply unable to do so.[228] That distinction alone requires an awareness of other minds that many still deny to nonhuman animals.

Another illusion of human uniqueness that's steadily dissolving is that of self-recognition. Once the province of poets and philosophers, the "mirror test" is now passed not only by great apes and dolphins but also by elephants, certain birds, some species of fish, and even ants.[229]

Language-like communication using symbols? Documented.[230]

Cultural transmission across generations? Check.[231]

Empathy, altruism, metacognition, and the awareness of one's own knowledge gaps? Present and accounted for.[232] (Even Donald Rumsfeld would be impressed.)

Perhaps most jaw-dropping of all is the growing evidence for numeracy among nonhuman animals. This includes not just counting but also mathematical abstraction. The humble honeybee, when trained to select the lower of two numbers in exchange for sugar water, consistently chose zero over one and performed above chance in repeated trials.[233]

So wherever one might draw the line between "us" and "them," the gap is far narrower than we imagined. In fact, in many areas, that line may not exist at all.

**(5.)** The final tentpole in this pentangle of supposition overlaps meaningfully with earlier points, particularly #3, but warrants its own spotlight. The ways in which other life forms have adapted to their specific environments, and that of the planet, are seemingly as various as the forms of life themselves. They are wondrous, not only in their variety but also in their variance and filigree. We find them curling and spiraling along fault lines in ways that humans can at times fail to appreciate, accept, or even comprehend. As Dr. Ian Malcolm so eloquently put it: "Life, uh... finds a way."[234]

Let us begin with the clearest, most heretical truth in this discussion.

**The basic requirements for** *human* **life are not universal requirements for** *life*.[235]

Not on Earth. Not anywhere.

Sunlight? Optional.

Oxygen? Extraneous.

Organic carbon? One recipe among countless alternatives. Many microbes don't rely on organic carbon at all but instead metabolize inorganic sources that were once thought to be toxic or inert.[236]

Even water, the so-called 'life-giver,' may be necessary only in trace amounts or at sporadic intervals for certain Earth organisms.[237]

Consider some of the most extreme examples. Certain life forms are so divergent in constitution that, if discovered on another planet, they would spark worldwide headlines, even interstellar press conferences.

As noted in earlier chapters, deep-sea hydrothermal vents are among the most remarkable discoveries of the modern scientific era. These submerged volcanic chimneys host thriving ecosystems entirely independent of sunlight. Here, life endures radical pressure, near-boiling temperatures, and an utter absence of solar energy. But this is only the beginning. The miracle lies less in life's existence here than in its persistence and metabolic ingenuity.

Instead of photosynthesis, these organisms rely on chemosynthesis, the process of converting chemical compounds, such as hydrogen sulfide, into usable energy. Chemosynthetic bacteria serve as both the foundation of the food chain and symbiotic partners to larger life forms.[238] For example, giant tube worms lack mouths and digestive systems. They survive entirely by hosting chemosynthetic bacteria within specialized internal organs.[239] This ability effectively delegates digestion to an internal farm of chemical alchemists.

Other residents of this world, such as deep-sea clams and mussels, have developed cavernous gill structures to house these bacteria. Some mussels boast gills up to *20 times* the size of their edible, shallow-water cousins. To put it in perspective, a single deep-sea mussel can harbor up to *one trillion* symbiotic bacteria. This exceeds the bacterial load in 264 gallons of seawater and is over *100 times the current human population of Earth.*[240]

The deeper we drill into this biochemical wonderland, the more marvelous its revelations become. But remember, this chapter promised to go deeper than the sea floor.

Let's begin by flipping a stone.

Who among us hasn't, as a child, overturned a rock in the dirt, driven by boredom or curiosity, only to startle an entire microcosm of life beneath? The worms and insects inhabiting this hidden realm squirm and undulate, recoiling from the light. Their protest, in that moment, mirrors the way we might object to a light being flipped on in the middle of a deep slumber. But it is in those under-rock spaces, in the crevices beneath our perception, that Earth's strangest and most resilient life so frequently waits.

In late 2024, a team of researchers exploring an undersea volcano off the coast of Central America sent a remotely operated vehicle to investigate the ocean floor. While maneuvering through terrain, the ROV disturbed a section of seafloor, flipping over a large chunk and exposing what lay beneath. What did it reveal? Life. *Flourishing* life. Again.

Beneath the already lightless layers of the abyss, tucked inside a previously unknown cave system, the team discovered thriving colonies of giant tube worms, snails, and other resilient creatures. This type of revelation shouldn't surprise us. Yet it always does. In the deep, where we expect emptiness, we find abundance. This subterranean discovery beneath the seafloor itself challenges us to ask, just how extensive *are* these cave systems?[241] And more intriguingly, *how expansive are the systems of life they support*?

If these caverns have been functioning as part of Earth's biosphere for millions of years, beyond light and beyond reach, what other hidden chambers and ecological vaults still elude human imagination? Hamlet may have had the right of it. *"There are more things in heaven and earth..."*[242]

Let's go even further down.

Across the extremes of environmental hostility, life continues to make its case. In scorching heat, freezing permafrost, intense radiation, toxic acidity, and oxygen-starved voids, it endures.

Microbial life has been found not just beneath glaciers but also deep inside *solid rock*. It thrives in microscopic pores by feeding on hydrogen and sulfur, sometimes for *millions of years*. Regions once believed entirely "abiotic" are now proven to be microhabitats teeming with extremophiles.[243]

You would think radiation 1,000 times higher than levels capable of obliterating human DNA might put an end to the party.[244] Think again.

Acidic conditions that should dissolve cellular structures on contact? That's the playground of *acidophiles*, some of which are now employed by humans in mining operations.[245]

High alkalinity? *Alkaliphiles* shrug it off, possibly using acidic polymers in their cell walls to create molecular armor.[246]

No oxygen? Not a problem. Anoxic zones of the ocean, once referred to as "dead zones," are anything but.[247]

Yet all of this is still just the appetizer. The living world at every scale is filled with revelations. Some adaptations go beyond clever and into the realm of the truly uncanny.

Take *Elysia chlorotica*, a sea slug gracing the shallows of the U.S. eastern seaboard. It doesn't just *eat* algae. It *steals* their chloroplasts, integrating them into its own digestive system and using them to perform *photosynthesis*.

This elegant little shoplifter becomes both animal and plant. It turns sunlight into energy and, in the process, transforms its body into a vibrant green. This serves as an evolutionary camouflage and survival hack rolled into one. Scientists refer to this process as kleptoplasty, and it's as unusual as it sounds.[248]

But Elysia, odd as it is, may still pale in comparison to a smaller and even more bewildering creature, the *Bdelloid rotifer*.

Rotifers are known as cosmopolitan, meaning they are found almost everywhere. One class of rotifers, known as Bdelloidea, primarily inhabits freshwater environments. What makes them exceptional is their short-term adaptability and their ability to persist across epochs.

To begin with, Bdelloids have the stunning ability to enter a desiccation-induced dormancy at any stage in their life cycle. If the pond they live in evaporates, they do not perish. Instead, a Bdelloid dries out, shrivels into a durable stasis, and *waits*. Even more amazing, it can be carried by the wind like a spore, rehydrate in a fresh environment, and restart its life anew.

This is a strategy we recognize in dandelion seeds or fungal spores. But the Bdelloid isn't a plant. It's a full-bodied organism with a *head*, *trunk*, *tail*, *esophagus*, *teeth*, and even species-specific *lip shapes*. All in a body that is invisible to the naked eye. Moreover, these tiny machines aren't just playing the short game.

Bdelloid rotifers have been successfully revived after more than *24,000 years* frozen in Siberian permafrost. Let that sink in. They didn't just survive. They came *back to life* and resumed their business. Few lifeforms on Earth boast that kind of biological resume.

But even that might not be their most mind-bending feature.

*Bdelloidea* are entirely female. No male has *ever* been observed by humans. They reproduce via parthenogenesis. This is a form of asexual reproduction in which embryos develop without fertilization. (You may recall this method appearing among some deep-sea creatures mentioned earlier.)

This mode of reproduction, combined with desiccation-induced dormancy, gives them an edge in evading parasites. A parasite cannot survive the dry spell. Airborne travel may not destroy the threat, but it at least ensures geographic escape for the rotifer.

But the Bdelloid's crown jewel of adaptation is a trick that borders on science fiction, horizontal gene transfer.

During desiccation, their DNA literally shatters. Upon rehydration, the DNA reassembles itself. While reforming, it can incorporate foreign DNA fragments into its own genome. Bacteria, fungi, and even plant genetics can be absorbed. These exogenous snippets become permanent upgrades, like swapping parts on a modular spacecraft.

The result? Bdelloids have a unique resilience to radiation, the ability to repair extensive DNA damage, and possibly a workaround to the genetic "recombination noise" that can arise during reproduction in other organisms. They adapt not just over generations, but *within* a single lifecycle.[249]

In essence, these rotifers function less like individual creatures and more like biological USB drives, plugging into the environment and downloading software updates on the fly.

All these examples offer rich food for thought as we approach the more speculative dimensions of this inquiry. However, we now arrive at what may be the most tantalizing and paradigm-shifting idea of all. The existence of sensory modalities that lie *entirely outside* the human perceptual frame.

If our senses shape our world, and they most certainly do, then it follows that other creatures, equipped with radically different ways of perceiving, inhabit entirely different realities.[250] Most humans rely on the familiar five. Sight, hearing, touch, taste, and smell. Even within that range, our capabilities vary. (The author, for instance, has a relatively weak sense of smell but hypersensitive hearing.)

It has long been understood that many animals possess those same senses, but with far greater acuity. Dogs can smell 10,000 to 100,000 times more acutely than we can.[251] Birds, insects, fish, and snakes can detect wavelengths of light that we cannot see, ranging into the ultraviolet and infrared regions.[252] But beyond those amplified versions of human-like senses lies something else entirely. There are modalities of perception that we do not experience, cannot simulate, and in many cases, have only recently been discovered, *even to exist*.

Take the electric eel, for example. This formidable creature can unleash jolts of up to 600 volts, easily enough to incapacitate or even kill a human. But its more perplexing ability is subtler. It uses *weaker* electric pulses for navigation and communication. These eels "ping" their surroundings to build a mental map, detect prey, and coordinate with other eels.[253] This functions like a biological blend of radar and a taser. It's logical, yes, but also *unimaginable* in experiential terms. We can understand it intellectually, but we can't *inhabit* it. There is no direct analog in our sensorium.

The surprises don't stop there.

Some bacteria possess internal chains of magnetic iron crystals, allowing them to align with and move along Earth's magnetic field. This magnetic sensitivity is known as *magnetotaxis*. It acts as an internal GPS, guiding bacteria to favorable environments in water or sediment. This sensational ability, which we had no inkling of until the 1970s, reveals the degree to which even Earth's background forces are alive with informational possibility.[254]

Certain birds take this to an extraordinary level. They may, in fact, *see* magnetic fields. Specialized receptors in their eyes are sensitive to the Earth's magnetic field, allowing them to perceive the Earth's magnetic lines visually. We might compare this ability to seeing gravity itself. This could help explain the dramatic navigational feats of migratory birds, which are performed with precision across great distances and over generations.[255] We are only beginning to understand what else they might perceive.

And then, of course, we must acknowledge one of nature's most delightful anomalies, the platypus.

Despite its meme-worthy appearance, the platypus is an evolutionary marvel. It wasn't until the 1980s that scientists fully grasped its ability to sense electric fields. It employs this ability so effectively that it closes its eyes, ears, and nostrils while diving, relying solely on electroreception. It can detect the minute electric currents emitted by the muscular contractions of prey and use them to determine the direction and distance of the source.

Among mammals, only the platypus and one species of dolphin possess this ability.[256] Their example proves something critical. Mammals, ourselves included, wield the amazing capacity to potentially evolve new sensory tools.

And with that, we arrive at a profound truth. Humans do not just underestimate what life is. We underestimate what life can *perceive*. Evolution has shown no compulsion to color inside the lines of our expectations. The boundaries we imagine between ourselves and other life forms may be more porous than we think. Not only in physiology or behavior, but in consciousness itself.

Does any of this prove that a sapient species has evolved in parallel with us, hidden in the depths of our own planet's oceans? Perhaps not. However, when we consider the radical divergences that evolutionary biology has already taken across morphology, perception, and communication, and then connect these to the pillars of this broader argument, one might ask, how could such a thing *not* be possible? Indeed, for any mind open to discovery and tempered by scientific humility, the truly compelling question becomes, how could such a lifeform *not* exist in the yawning 99.999% of ocean we have yet to explore? Life has greeted us in every place we have dared to look, often where we were certain it would not be. The author freely acknowledges that these unplumbed depths may hold no sapient life form. But can the reader, in good faith, completely dismiss the possibility that one *might*?

Even beyond the prospect of sapient life lies something more provocative still. Any being forged in a realm so inhospitable to us as the deep ocean would be as inscrutable as the ecosystem that shaped it, perhaps even more so. Our known world has already delivered startling evolutionary paths. However, when we factor in immense spans of uninterrupted time, shielded from surface cataclysms, paired with the ceaseless innovation of biology, unconcerned with human expectations, we must entertain a haunting possibility. That the door we now face is neither locked nor closed. It has already been opened *from the other side.*

It is not the absence of visitors that keeps us blind, but the lack of imagination that leads us to believe we are the only ones doing the seeking. We are beheld by nature itself. The mystery now is not whether we are alone, but who, and what, is watching. For if our emotional and psychological reality is shaped by dreams, fears, and shared myths, what might *another* sapient species yearn for, dread, or instinctively flee? We fancy ourselves the sovereigns of this sphere, knocking upon doors with curiosity or command. But what if we are not the only ones who knock?

Will we answer? And what neighbors will we meet, standing on our own welcome mat?

Let's find out.

As always, together.

# CHAPTER SEVEN

## Inexorable Vicissitude: Metamorphosis of Perspicacity

*"Those who dwell among the beauties and mysteries of the earth are never alone or weary of life."*

*Rachel Carson, The Sense of Wonder*

On the evening of April 25, 2013, at approximately 9:20 p.m. local time, a FedEx cargo plane was preparing for takeoff at Rafael Hernández Airport in Aguadilla, Puerto Rico. That takeoff was abruptly paused. An unidentified object had entered restricted airspace, flying low across the runway without clearance or a transponder signal.

At that very moment, a *U.S. Customs and Border Protection* (CBP) DHC-8 turboprop aircraft was conducting a routine patrol nearby. On board were four crew members and one of the most advanced surveillance tools available, the *Wescam MX-15D* thermal imaging system.[257] This high-performance, multi-sensor platform is utilized for search and rescue, border security, and reconnaissance operations across various branches of the government. These include CBP, the Coast Guard, the Navy, and drones such as the *MQ-9 Reaper*.

It is not a toy.

The MX-15D is a state-of-the-art system designed for precision tracking under the harshest conditions. Understanding what was witnessed that night requires understanding the full scope of what this system is capable of.

Equally important is acknowledging the caliber of personnel operating it. These are not idle observers. They are highly trained professionals entrusted with the surveillance and protection of both military and civilian life. The men and women who serve in this capacity deserve deep respect. These jobs demand skill, responsibility, and constant vigilance, typically without recognition. The people operating the MX-15D that night were every bit as badass as the technology they commanded.

The MX-15D is part of a powerful family of stabilized multi-sensor turrets manufactured by L3Harris Technologies under their Wescam division. These systems are designed for military-grade reconnaissance, surveillance, and targeting operations. The capabilities of the MX-15D are impressive and include the following:

### 1. High-Resolution FLIR (Forward-Looking Infrared):
FLIR technology detects the infrared radiation, or heat signature, emitted by objects. The hotter an object is, the brighter it appears in FLIR imagery. The MX-15D can deliver thermal imaging at a resolution of up to 1280x1024. This allows operators to track targets in total darkness, dense fog, or over open water with superb clarity.

### 2. Military-Grade Laser Rangefinders and Illuminators:
These onboard systems measure precise distances between the sensor and a target object. Depending on atmospheric conditions, they are effective at ranges of 20-25 kilometers (12-15 miles) with an accuracy of 1 to 3 meters (roughly 3 to 9 feet). That's an extremely tight margin from a moving aircraft.

### 3. Gyro Stabilization:

High-speed tracking from a moving aircraft requires a rock-steady image. Gyro stabilization ensures the turret remains smooth and locked. This is crucial for capturing consistent and reliable footage during rapid maneuvers.

### 4. GPS/INS (Global Positioning System/Inertial Navigation System):

By combining satellite-based GPS with internal inertial motion tracking, the MX-15D maintains real-time positioning data that's faster and more accurate than GPS alone. This allows for precise location mapping and target geolocation during flight.

### 5. Zoom and Auto-Tracking Systems:

The turret's advanced optics and onboard software allow operators to lock onto targets and automatically track them across long distances, whether they're crawling across terrain or flying through the sky.

### 6. Color and Monochrome CCD Cameras: [258]

Camera nerds, this one is for you. CCD stands for Charge-Coupled Device. These sensors use an array of light-sensitive pixels embedded in a semiconductor. When light strikes the array, each pixel generates an electric charge proportional to the light's brightness. That charge is then transferred, or coupled, into a readable digital signal. This forms the basis of the final image.

By 2007, CCD technology had begun to lose ground to a newer sensor type, the CMOS (Complementary Metal-Oxide Semiconductor).[259] The author will refrain from delving into the intricacies of CMOS architecture. But suffice it to say that CMOS gained dominance due to two key advantages: lower power consumption and faster data processing. That said, in 2013, the military still widely preferred CCD for several critical reasons.

### 1. Superior Low-Light and High-Contrast Performance:

At the time, CCD sensors continued to outperform CMOS in long-range, low-light scenarios. This was especially true in high-contrast environments, such as shorelines, cloud cover, or open water. CCDs are also integrated more seamlessly with infrared systems, making them ideal for thermal imaging. This was an essential capability in military-grade surveillance and reconnaissance, particularly when tracking small or low-signature targets.

### 2. Global Shutter Advantage:

Unlike CMOS sensors of the era, which used a "rolling shutter" (capturing the image line by line), CCDs offered a true global shutter. This means the entire image is captured simultaneously. When tracking fast-moving objects from a moving aircraft, rolling shutters can introduce distortions that affect the accurate interpretation of speed, orientation, or maneuverability. In 2013, CCD was still the go-to for capturing clean, undistorted frames of high-velocity targets. (Modern CMOS sensors now include global shutter variants, but did not back then.)

### 3. Established Thermal Imaging Integration:

The military's advanced thermal imaging platforms were already meticulously integrated with CCD arrays. Swapping them out for CMOS would have required major system overhauls. This is not a trivial task when mission-critical reliability is at stake.[250]

To summarize, the surveillance setup in place was ideally suited for capturing exactly this kind of anomalous object. And that's exactly what it did. The target was recorded using one of the most advanced thermal tracking systems available at the time. It was an aircraft-mounted, mission-specific thermal system with long-range precision, high power output, and battlefield-grade image processing. Every detail outlined here will become highly relevant in just a moment. Stick with me.

In addition to the four-person CBP flight crew, personnel in the Aguadilla airport control tower were also observing. Initially, this target appeared as a "reddish/pink light" approaching from over the ocean. Just before entering the airfield's controlled airspace, the object extinguished its visible light. At this point, it was only detectable via its heat signature on the MX-15D.

What followed is where things get... strange.

The target flew without lights and maneuvered at treetop level altitude through residential areas at estimated speeds up to 120 mph. It then submerged into the ocean with no splash, no visible deceleration, and no disturbance on the water's surface. As if that weren't enough, thermal footage appears to show it splitting into two while reemerging.

Additionally, the MX-15D's thermal video timestamps and GPS coordinates were directly cross-referenced with radar tracks from a long-range array. The installation was stationed 145 kilometers (90 miles) east-southeast of Aguadilla, at an elevation of 1,042 meters (3,417 feet). This radar can detect objects at altitudes as low as 122 meters (400 feet).

The radar in question was operating with a transponder code reserved by the FAA for military and special operations, confirming its authority and precision.

Crucially, the radar data not only matched the flight path of the CBP DHC-8 aircraft as seen in the thermal video, but it also logged 50 primary radar strikes (i.e., no transponder) northwest of the airport before the DHC-8 even took off. Those strikes showed a mean velocity of approximately 270 km/h (168 mph).

Then, just before the object appeared in the thermal footage, it vanished from radar entirely.

The most likely explanation? It dropped below the radar's minimum detection threshold. As noted, this is 122 meters (400 feet).[261]

Another compelling aspect that reinforces the credibility of this event is the nature of the reporting and analysis that followed.

On October 21, 2013, the thermal video was obtained by the Scientific Coalition for UAP Studies (SCU). The source was someone who was not physically present during the incident but was in direct contact with a primary witness. This individual was thoroughly vetted, verified as credible, and remains anonymous for personal reasons. All involved parties signed non-disclosure agreements to protect the integrity of the investigation.

In 2015, the SCU released a 161-page technical report on the incident.[262] Its authors included credentialed scientists, engineers, and former defense analysts. It was a multidisciplinary team with wide-ranging technical and analytical backgrounds.

Their key findings included:

- The object exhibited no visible wings, rotors, exhaust, or any identifiable means of propulsion.

- It moved independently of prevailing wind conditions, performing stationary hovers and sudden accelerations.

- It entered the ocean at speed, with no observable deceleration and no splash. This behavior is inconsistent with any known natural or artificial object.

- As a result, simple explanations such as balloon, bird, drone, or insect were systematically ruled out.[263]

Despite the abundance and quality of available evidence, the Aguadilla incident has remained largely unknown to the public. The SCU submitted a formal request for detailed technical data to the All-domain Anomaly Resolution Office (AARO). This is the very U.S. government body tasked with evaluating and explaining incidents of this nature.

AARO's official response?

They stated that they were "not yet prepared to fully respond" but may do so in the future.

When pressed for a conclusion, AARO determined that the object captured in the Aguadilla video was a Chinese sky lantern, allegedly released during a nearby wedding.[264]

To be fair, we do not currently have access to the complete data set that AARO may be using for this interpretation. What we do have, however, is a detailed record of the observable phenomena and a growing list of counterpoints to the "sky lantern" theory.

So, let's focus on the data that *is* available, along with the pushback it has received from experts and investigators outside official channels.

The object observed in the Aguadilla footage does share superficial similarities with a traditional sky lantern. These flame-powered, airborne paper balloons are customarily released during celebrations. They are crafted from thin, oiled, or coated paper, held aloft by a lightweight frame with a large opening at the base. A persistent flame, either a small candle or a fuel cell, heats the air inside. This creates a density difference with the surrounding atmosphere. That difference allows the lantern to rise, much like a miniature hot air balloon.

However, unlike helium balloons, these lanterns are inherently hazardous, particularly as fire risks.

And here's where things start to break down.

Sky lanterns have been explicitly illegal in Puerto Rico since at least 2012, and likely earlier. Public safety campaigns dating back to 2010 featured fire officials and local media warning against their use. Especially around holidays and weddings, the very occasions most often linked to lantern releases. Beyond the risk of brush fires, their use near populated areas or airports carries serious aviation and safety violations, with potential fines or even criminal charges.[265]

Given these facts, the idea that a sky lantern could be launched so close to an international airport like Rafael Hernández is exceedingly difficult to believe. Particularly without anyone noticing, reporting, or facing consequences.

Could someone have released a sky lantern near Aguadilla? In theory, yes.

But they would have had to launch it extremely close to restricted airspace around the airport. In many jurisdictions, including Puerto Rico, launching one so close to restricted airspace would constitute a major safety violation and carry substantial legal penalties. It's possible that someone disregarded the law and proceeded anyway. A particularly brazen wedding planner, perhaps.

So, let's entertain the possibility. Could a sky lantern mimic the behavior observed in the Aguadilla footage?

This is where we need to tread carefully. The claims being made about the object's behavior are, without question, superlative. Claims of this type demand rigorous scrutiny. It is essential that we welcome critical analysis, not avoid it. If the goal is a truer understanding of encounters like this one, then it is only through serious, good-faith examination of *both* the evidence and the counterarguments that we will get there.

Let's now consider if a paper lantern, glowing softly on a humid night, could really account for what the MX-15D captured.

What would a paper lantern look like when viewed through a high-grade, aircraft-mounted FLIR system?

There are some superficial similarities. The object witnessed over Aguadilla and a traditional sky lantern share a few basic traits. Both glow and are silent in their movement. Neither exhibits any obvious means of propulsion. They also share a roughly rounded or oval shape and could easily be mistaken for one another at first glance by an untrained observer.

But once we move beyond surface features, the differences become too stark to ignore. In fact, when you start digging deeper, the comparison collapses. Upon examining behavior, dynamics, and interaction with the environment, what we're left with isn't a benign paper vessel carried by a warm updraft. It's something else entirely.

The object observed over Aguadilla wasn't drifting. It was moving. At times, it traveled at speeds beyond 100 mph, executing sharp turns, path reversals, and even moving against prevailing winds. The SCU included meteorological data in its report. This data confirmed that local wind conditions could not account for the object's motion.

Second, consider how a paper lantern behaves thermally. Lanterns operate with a central heat source, while the surrounding material remains appreciably cooler. But the object captured by the FLIR system shows a uniform heat signature across its *entire body*. No central hot spot, no flickering, and no erratic thermal movement.

It appears consistently warm, but not nearly hot enough to suggest an open flame.

Third, and most compellingly, the object is seen to enter the water, submerge, and then re-emerge along the same trajectory. There is no visible splash, no deceleration, and no disruption to its movement. The FLIR system was operating in "black hot" mode, meaning that warmer objects appear darker on the screen. The object remains consistently dark, smooth in motion, and thermally visible at all times.

There are no anomalies in the surrounding frame. No infrared distortions, no video artifacts, and no lost frames or "ghosting." Its point of reappearance lines up precisely with where it should be if it had continued its path beneath the surface. This is not consistent with a lantern, balloon, or any object known to behave erratically in water. Let alone one that survives immersion, still burning.

There is no evidence of frame tearing, noise bursts, or sensor bloom. The video remains intact. Only the object disappears. The rest of the frame remains stable. This strongly argues against equipment failure.

In theory, a vapor plume or surface reflection might cause a temporary loss of infrared contrast. However, in this case, there is no visible environmental interference to explain why only the object vanishes while the rest of the scene remains stable. If the object were a lantern with an internal flame, submersion in water would almost certainly extinguish that heat source. Yet the object reappears with the same consistent thermal signature.

It's also important to reiterate that paper lanterns are illegal in Puerto Rico and are strongly discouraged in many countries due to the fire hazard they pose. Given the proximity to a commercial airport and strict local ordinances, it is unlikely that a lantern would be launched here unnoticed or unpunished.

It is through analysis of this kind that we can effectively rule out airborne lanterns, balloons, or other wind-borne objects. The observed behavior doesn't match.

The SCU report also effectively rules out other ordinary explanations, such as birds, drones, or any known type of aircraft. Ball lightning or plasma phenomena fall short for similar reasons. These phenomena would likely be extinguished upon submersion. They aren't known for persisting over sustained timeframes, and there's no evidence of ionizing discharge in the footage.[266]

But perhaps the final nail in the coffin for all conventional interpretations is the object's clear and unmistakable splitting behavior. It appears to divide into two distinct halves, both of which retain motion and structure.

Next, we turn to specific criticisms of the SCU's methodology. In particular, the accuracy of their evidentiary analysis. One recurring claim is that the SCU overstates the precision of radar returns. However, the SCU report itself explicitly acknowledges this limitation. The estimated accuracy of 1/8 mile comes directly from official FAA documentation. The SCU is clear that this radar data is not used to establish precise 3D positioning of the object.

Instead, the radar is used solely to confirm two basic facts: the DHC-8's flight path and the presence of an unidentified object. While the radar system certainly has inherent limitations, its use here is supportive rather than foundational. It corroborates the existing data captured by the aircraft and reported by the control tower. Nothing more, nothing less.[267]

Another critique leveled at the SCU following publication of their report was their reliance on a YouTube-uploaded copy of the video, rather than the original raw file. This is a valid concern, and one that the SCU acknowledges directly. They were not granted access to the raw file and were transparent about this limitation.[268]

Importantly, SCU does not use this copy to estimate absolute temperatures, only relative contrast and movement. The object's behavior is consistent across multiple zoom levels and contrast modes. This strongly suggests that it is not a camera artifact. Until SCU or another group gains access to the raw footage, the movement patterns, flight characteristics, and submersion behavior observed in the available recording remain firmly in the realm of the anomalous.

Another generally raised critique is that the report fails to properly account for parallax or camera limitations. Infrared cameras can indeed introduce parallax. Especially those mounted on moving aircraft or operating at oblique angles. However, the SCU analysis explicitly factors in the movement of the DHC-8, including simulations of relative motion. The report notes that the object's lateral movement and speed are not consistent with wind drift, even when parallax is considered. That said, the authors are clear that range and speed estimates do carry a margin of error.

Another area of criticism involves how the SCU interpreted and applied thermal gradient averages to compare against the object. But a close reading of the report reveals a careful, multivariable approach. The team compared the object's thermal signature with a wide range of environmental features. These included asphalt, roads, open pastures, and even cows. Crucially, no single source was used in isolation to extrapolate conclusions. This adds credibility to their methodology.

Some have dismissed the widely discussed "splitting" of the object as a possible camera distortion or limitation of the infrared technology itself. While sensor error is always a potential factor in any video analysis, that explanation seems inconsistent with what's visible here. The object clearly divides into two, with both halves maintaining independent motion across multiple frames and consistent thermal contrast. As previously noted, a typical IR anomaly manifests as image flicker, signal dropout, or uneven contrast. Not as a symmetrical split with identical motion vectors and stable IR signatures.[269]

The most effective way to confirm or rule out sensor-related anomalies would be through analysis of the raw video file. To date, this unfortunately remains unavailable to independent researchers. As mentioned earlier, SCU has formally requested that AARO release this data.[270] If the object truly is something as benign as a drifting lantern, then providing the raw footage would help settle the matter definitively. This is especially true regarding the splitting behavior, parallax effects, and potential image artifacts.

Some have argued that the Aguadilla case is "resolved" and requires no further investigation. However, AARO has not confirmed any definitive identification of the object.[271] The Scientific Coalition for UAP Studies has formally disputed the lantern hypothesis.[272] As the evidence outlined in this chapter demonstrates, this case remains anything but settled.

That leaves us in a diagnostic position familiar to medical professionals. A scenario in which the cause is unknown and must be diagnosed by exclusion is referred to as an "idiopathic" condition.[273] When all known explanations are ruled out, only the unknown remains. That is where we find ourselves now. We have established what this object is most likely not. And we've evaluated which theories collapse under scrutiny. So, the question becomes, what are we left with? In the author's view, three core possibilities remain.

First, we must consider the prospect that the Aguadilla object is a man-made craft. A classified airborne device unknown to the public, and perhaps even much of the military itself. It may be so far beyond current civilian understanding that we're unable to classify or recognize it. The evidence for this hypothesis is strong and deserves serious consideration.

Take, for example, the F-117A Nighthawk. Though it first flew on June 18, 1981, it had already been in classified development for years. It followed the Have Blue prototype and was the first operational aircraft to feature stealth technology, dramatically reducing its radar profile.[274] Even years after its debut, pilots encountering it for the first time, such as RAF test pilot Dave Southwood, described it as looking *"...impossible, something out of science fiction."*[275]

If that's what cutting-edge aerospace engineering looked like in the late 1970s, what might today's most advanced technology look like decades later? Materials science, aerodynamics, computing, and propulsion have all advanced exponentially since then.

Could the Aguadilla object be part of a secret government, military, or private aerospace entity project that is being tested? It's entirely possible. Might it someday show up in a declassified documentary, featured in a History Channel special, or parked outside a Presidential Library? Is it a weapon? An experiment? A proof-of-concept? We don't know. But we can be sure that many things exist beyond our current expectations, both in the U.S. and abroad.

One telling detail here is AARO's casual dismissal of the object as a mere paper lantern. That response might mean it is indeed innocuous. Or perhaps it's something classified. More unsettling still, it could mean they don't know either.

Two other possibilities remain, one extremely remote, the other far more feasible. The first is that the object in question is a manufactured craft of non-terrestrial origin. To fully consider this category, we must include all speculative sources. Exoplanetary, exodimensional, and exotemporal. Each has its advocates. Each inspires wonder and fascination. However, this author will leave those hypotheses to others who are better equipped and more inclined to pursue them.

Still, they cannot be entirely dismissed. While not impossible, such origins are, by every measure available to us today, extraordinarily improbable. Were one ever confirmed, the author would be among the first to be cautiously thrilled. Such a revelation would constitute a seismic shift in human consciousness, redefining our understanding of reality itself.

A far more likely explanation, as outlined in the previous chapter, is one of biological origin, terrestrial, yet not man-made. This would be a lifeform that evolved beyond the reach of human perception and over time scales longer than we are accustomed to imagining. Shaped by pressures and parameters we barely grasp. In the author's view, this is the most compelling and plausible explanation for what was seen over Aguadilla, and for many other encounters like it.

Lacking a proper name and wishing to avoid the clumsy overuse of placeholders like "object" or vague pronouns like "it," we will formally christen this terrestrial neighbor as *Pelagomorph noetica*.

The name, derived from Greek, is carefully chosen.

- **Pelago** – from *pelagos* ( π ε λ α γ ο ς ), meaning "deep sea."
- **Morph** – from *morphē* ( μ ο ρ φ ῆ ), meaning "form" or "shape."

Together: "a form born of the deep sea."

To this we add **Noetica**, from *noēsis* ( ν ό η σ ι ς), meaning "mind" or "intellect." The full name, *Pelagomorph noetica,* thus translates to "a deep-sea form possessed of intellect."[276]

We will explore the more esoteric implications of this moniker in subsequent chapters. However, for now, let us focus on their observable characteristics and behaviors.

At first glance, the notion that this object might be a terrestrial, biological, and sapient life form may seem even more far-fetched than the more familiar theories of extraterrestrial or exo-temporal origin. But the thought that keeps circling back in this author's mind is *99.999%.*

The more closely one explores the possibility, the more conceivable, and even probable, it becomes.

Perhaps the primary conceptual hurdle lies in imagining the biological structure and morphology of such an entity. Fortunately, the reader now carries a deeper appreciation for the granular details laid out in earlier chapters. We have explored an evolving portrait of the complexity, adaptability, and astonishing unpredictability of evolution within the hadal zones and other deep-ocean environments.

The next chapter will bring these strange and abundant biological iterations fully into focus.

For now, we look to the Aguadilla *Pelagomorph* encounter not as a final answer, but as a starting point. It is a guide that can help us begin classifying the possible behavioral patterns of these enigmatic ambassadors from the deep.

As with most aspects of unclassified or anomalous phenomena, consensus is elusive. But one trend stands out with distinct consistency. A large percentage, perhaps even a majority, of all encounters occur in direct proximity to bodies of water.[277] Whether it's an ocean, lake, or river, the pattern holds. Repeatedly, we find accounts describing objects or entities that emerge from, disappear into, or otherwise interact with aqueous environments in ways that defy conventional explanation.

Equally common are reports of craft or bodies that cross the air-water boundary effortlessly. They exhibit movement through both media with no observable disruption or change in behavior. This unimpeded traversal appears to be unaffected by the physical properties of either medium and is a hallmark of many such sightings.[278]

This pattern does not offer definitive proof that *Pelagomorph* is of oceanic origin. An entity from an "exo" source, be it extraterrestrial or otherwise, might well possess the capability to navigate any environment with equal ease. However, the sheer frequency of water-associated encounters suggests something else to this author. It strongly implies an origin within Earth's oceans, rather than beyond them. This would better account for the overwhelming aquatic pattern observed across decades of global reporting.

This theory, along with further supporting data, will be explored in more depth in the next chapter.

Not only do an enormously high number of sightings occur near bodies of water, but certain aquatic "hot spots" appear to generate these encounters with unusual frequency over time. One of the most well-documented and consistently active regions is the area surrounding Santa Catalina Island, off the coast of Southern California.

As with other locations steeped in long-running histories of anomalous activity, the number of specific incidents tied to this region is far too extensive to catalog here. Reports span decades, stretching back to the 1940s and possibly earlier. These come from both military and civilian witnesses in equal abundance.[279]

Even the two publicly acknowledged encounters reported by former U.S. President Ronald Reagan both occurred in this vicinity. Each involved airborne objects that seemingly defied the laws of physics.[280] Over the years, consistent reports, credible testimony, and sheer volume have made the Santa Catalina region one of the most compelling and enduring zones of interest in modern UAP investigation.

On July 15, 2019, the USS *Omaha* (LCS-12), an Independence-class littoral combat ship of the U.S. Navy, detected up to 14 unidentified radar contacts over the course of several hours. Built for speed and agility in both coastal and open waters, the *Omaha* is equipped with some of the most advanced sensors, communications, and command-and-control systems on Earth.

This encounter was not just observed by trained officers and sensor operators, but also recorded by multiple onboard systems. These included FLIR (Forward-Looking Infrared) video, sophisticated radar arrays, and ship-to-ship radio communications. The objects, which were visually and instrumentally confirmed, exhibited exceptional speed and maneuverability. Recorded velocities reached up to 138 knots (150 mph) and demonstrated abrupt directional and altitude changes that defied conventional flight profiles.

Importantly, crew members monitoring the systems confirmed these contacts were unlike any known aircraft, either domestic or foreign.

At least one of the *Omaha* contacts was visually recorded using the ship's FLIR system. The thermal footage shows what appears to be a spherical object flying above the ocean before descending and vanishing into the water, without producing any visible splash. The target, consistent across all reports, displayed no wings, exhaust, or conventional flight surfaces. Once it entered the ocean, it was never recovered or observed again.

After the video was leaked to the public, the Pentagon officially confirmed its authenticity, while declining to identify the object or provide any explanation for its nature or origin. [281]

Reporters of these encounters span a wide range, from active-duty military personnel to everyday civilians and even prominent public figures. Chapter Five's account of Malmstrom AFB offers a compelling thematic link. We repeatedly see proximity to nuclear sites, whether involving weapons, reactors, or storage facilities. That pattern holds true here.

The coastal corridor stretching from Long Beach to San Diego is dense with nuclear infrastructure. Point Loma, for example, is home to Submarine Squadron 11, which includes Los Angeles-class fast-attack submarines powered by nuclear reactors.[282] Just north lies Naval Air Station North Island, which reportedly stores aerial nuclear ordnance.[283]

During environmental cleanup at the decommissioned Long Beach Naval Station, radioactive materials, including Radium-226 and Strontium-90, were discovered.[284] (Notably, as explored in Chapter Three, Strontium-90 is used in long-duration, remote-energy applications, including deep-space probes.)

Further south lies the San Onofre Nuclear Generating Station (SONGS) in San Clemente. Though now inactive, SONGS still stores nuclear fuel in large canisters, just steps from the Pacific Ocean. Viewed on a map, SONGS is a mere 44 miles from Santa Catalina Island.[285]

As was also noted in Chapter Three, presence around nuclear-related facilities and sites is so prevalent that it would be difficult to explain in any way except as an ongoing and sustained concurrence. This phenomenon has yet to be adequately characterized. The behavior of *Pelagomorph* around these sites is especially intriguing, as it appears to follow consistent patterns that may offer glimpses of intention or preference.

Owing to our earlier discussion around anthropomorphism and epistemological inertia, the author remains cautious about projecting motive onto the unknown. But persistence of behavior across geography and decades is suggestive, to say the least. In a moment of eerie reinforcement, it was only after drafting the analysis of the Aguadilla encounter that the author discovered the existence of the BONUS (Boiling Nuclear Superheater) reactor facility. This decommissioned nuclear site is located just 17 miles south of Aguadilla, near Rincón, Puerto Rico.[286] More consequential to the framing and context for this chapter, though, is the fact that Rafael Hernández Airport is the site of decommissioned Ramey Air Force Base.[287] Now that we have a deeper understanding of what happened on this site in 2013, it's essential to recognize not only the history of this location but also how it ties directly to the bigger picture.

In Chapter Five, we explored the storied legacy and provenance of Malmstrom Air Force Base in Montana. Among its roll call of storied historical bullet points, we will note again that it was a critical node in the U.S. Strategic Air Command. In Aguadilla, we again find a location now fatefully tied to a debated but yet-to-be-fully-defined aerial phenomenon directly over another node of the SAC. Indeed, Ramey Air Force Base has a history that not only closely ties to Cold War-era activity but also mirrors that of Malmstrom AFB in its deployment of highly sensitive detection and surveillance arrays. At Malstrom AFB, these were advanced ground-based radar arrays in conjunction with air squadrons. Ramey AFB was home to equally advanced underwater

SONAR systems. The first generation was called SOSUS (Sound Surveillance System) and later became IUSS (Integrated Undersea Surveillance System).[288]

Another portentous link between the two bases is the presence of nuclear material. We observed in Chapter Five how Malmstrom AFB was a critically important location for the installation of ICBM launch facilities. The history of atomic weapons at Ramey AFB is a bit more challenging to uncover. But recent archival findings and declassified case files now make clear that Ramey Air Force Base was not just a remote Cold War outpost, but an active node in the nuclear infrastructure itself. Official U.S. Air Force memoranda, cited in a Department of Veterans Affairs radiation-exposure case, confirm that Ramey was a "nuclear-capable unit." Technicians stationed there in the 1950s and 1960s routinely serviced early, unsealed-pit nuclear systems used by Strategic Air Command's B-47 and B-52 squadrons. Their work involved direct maintenance of fissile material, generating measurable ionizing-radiation doses that the Air Force later reconstructed in detail for compensation claims.[289]

Veterans' testimonies align with those records. Former Air Police and munitions specialists recall guarding a "special weapons maintenance facility" on base and escorting Mk-28 thermonuclear bombs to transport aircraft during an emergency in 1964. This was a tense episode, some later dubbed "the day Puerto Rico almost disappeared." Maintenance officers from that period described periodic cleaning of plutonium pits and the disposal of radioactive residue into general waste areas such as the so-called Caliche Pit. These accounts, corroborated by multiple witnesses, match the documented presence of the 21st Munitions Maintenance Squadron and 72d Bombardment Wing, both standard custodians of nuclear ordnance within SAC.[290]

Taken together, these official and eyewitness records establish that Ramey Air Force Base functioned as a fully operational nuclear-weapons handling and storage site during the height of the Cold War. It supported not only airborne deterrent forces but also experimental programs such as Operation Argus (1958) and the previously mentioned Navy's SOSUS undersea-surveillance array. These formed a triad of air, land, and sea systems that tethered Puerto Rico directly to the global nuclear architecture. What now serves as Rafael Hernández Airport was, for two decades, one of the most heavily fortified and strategically sensitive pieces of real estate in the Western Hemisphere.[291]

While this chapter is not specifically about the Hadal Zone and deepest environments of the ocean, we do find a connection here as well. Just off the northern Coast of Puerto Rico, where Rafael Hernández Airport is located, we find the Puerto Rico Trench, the deepest point in the entire Atlantic Ocean. At its lowest point, the trench descends nearly 8.5 kilometers (5.5 miles), forming the boundary between the North American and Caribbean tectonic plates.[292] It is a place of tremendous geological tension and energy, where subduction and faulting have generated earthquakes exceeding magnitude 8.0 and tsunamis that have reshaped the island's coastline.[293]

NASA's geophysical surveys have also detected a localized gravitational anomaly beneath the trench. This is an invisible mass concentration so dense that it measurably distorts Earth's gravity field and even affects orbital readings.[294] This convergence of tectonic instability and gravitational peculiarity marks the region as one of the planet's strangest natural frontiers.

Adding yet another layer of intrigue, this same area sits at the southern vertex of the so-called Bermuda Triangle, the stretch of ocean long associated, rightly or wrongly, with navigational anomalies and unexplained disappearances.[295] Whether coincidence or correlation, it is striking that the Aguadilla UAP encounter occurred precisely at this intersection of deep-ocean abyss, Cold War infrastructure, and lingering mystery. It is a point on the map where the known and the unknown seem to share the same coordinates.

Patterns of coincidence are easy to romanticize, but they also invite closer scrutiny, particularly when those same patterns have a long history of anomalous interactions with our own technology. Therefore, it would be wise to factor in another aspect we see repeated in many UAP encounters. Report after report mentions the manipulation of electronics by unknown objects. Is the video analyzed by AARO, SCU, and numerous other researchers an accurate accounting of what occurred in 2013 over Aguadilla? We have no apparent reason to doubt its accuracy. However, we do have reason to exercise caution in the reading of any interpretation. While AARO and many other skeptics have written Aguadilla off as either a paper lantern, a balloon, a bird, or some other mundane airborne object, the author would encourage anyone curious about this incident to watch the video for themselves. Read about the MX-15D, parallax effects as they relate to airborne video capture, CCD technology circa 2013, and any other related topics. Also, however, keep in mind reports over decades and consistent firsthand accounts of unidentified objects directly interfering with recording equipment, sensors, and other related devices.[296]

Across numerous separate instances and over extended periods, witnesses have described the disruption or even outright disabling of complex technological systems. These include nuclear missile launch platforms, aircraft avionics, radar and radio systems, vehicle engines, electrical grids, perimeter security systems, alarms, fences, and even photographic equipment.[297]

The author would humbly suggest that such consistency and persistence over decades point to more than just ambient electromagnetic interference or incidental anomalies. A more coherent explanation is that it is intentional manipulation. But to what end?

Communication seems the most plausible answer. These interferences appear targeted and deliberate. But notably, they lack aggression. At no point are humans harmed or directly threatened. If a sapient entity possessed the ability to disable advanced weapons systems and precision-engineered machines, it would stand to reason that it could just as easily disrupt biological systems. Yet they do not.

The persistent focus on water and nuclear sites could suggest reconnaissance or data collection. However, it may also serve as a form of symbolic signaling. When events repeat in the same locations, with similar signatures, over time, that pattern *is* the message.

What would a parent, or any caring adult, do upon seeing a toddler reaching for the spinning blades of a fan, unaware of the danger? Would they shout? Smash the fan? Or would they reach over and switch it off before the child could be hurt?

If this happened repeatedly, the child might begin to understand a wordless, two-part message: *"This shouldn't be touched,"* and more subtly, *"I care enough to protect you. Not just in this moment, but in all the moments to come. I want you to learn the fan is dangerous when spinning, not to fear it blindly."*

If the child were especially inquiring or willful, and the caretaker unusually patient and kind, this might happen over and over, each time with hope that understanding would eventually dawn.

Another analogy might be how we feel when witnessing dolphins or whales beach themselves. We don't know why they do this. We fail to grasp any deeper motive or drive. But we instinctively run to push them back into the water. We say to ourselves, *"Why are you doing this? You're a whale. You belong in the ocean."* We intervene not because we comprehend, but because we care. We see a magnificent creature struggling to breathe and respond with compassion.

Is that how *Pelagomorph* views *us*?

We don't know. Perhaps we *can't* know.

This author firmly believes that we simply don't know *yet*.

Are there other ways communication may have been attempted?

It would certainly seem so.

Across numerous reports, we see repeated use of consistent shapes and light formations.[298] Could these features be intentional, selected specifically for their communicative potential? These appearances exhibit patterns in both their timing and location, as well as in the behaviors they display.

Some encounters include actions that human observers have reported as playful, even *theatrical*. Picture a human and a gorilla at a zoo. Without a shared language, each resorts to gestures, postures, or mimicry. Curiosity made visible through pantomime. Could *Pelagomorph* be engaging in a form of interspecies signaling? Might this behavior be a silent, kinetic message? *'Look what I can do. Can you do this too?'*

Another regularly reported phenomenon is the sudden disappearance of the entity, either from visual sight or from sensor readouts. This often occurs immediately after they have been clearly perceived.[299] Is this vanishing act itself a kind of signal? A punctuation mark to the encounter? An act of misdirection, or misinterpretation?

We don't yet know. But these patterns are not random.

The questions they raise deserve to be asked.

To the author's mind, it is far more likely that a terrestrial neighbor is causing these events. It is one that we have yet to fully recognize as a fellow 'Earthling,' rather than a visitor from the stars.

The instinctive urge to assign a cosmic origin to the unknown stems, at least in part, from our self-designation as the dominant life form on this planet. We imagine ourselves as the penultimate expression of biological possibility. But what if that's not true? What if it's not even close?

Our self-imposed uniqueness may ultimately stem from the inherent limitations of our perception and cognition. These constrain how we interpret the world around us. We've already examined how many of Earth's creatures experience entirely different slices of reality, informed by senses and processing mechanisms we lack. This is neither good nor bad. It merely is.

Another sapient being, born of this planet, wouldn't need to be radically more intelligent than us. They would only need to *perceive differently*. Should they employ a sufficiently foreign configuration of sensory and cognitive tools, this would render them functionally invisible to us.

Still, give the human mind its due. Once it latches onto a framework of understanding, it tends to double down. Pressed against a presence of such enormity, you may not recognize it as a sovereign entity. But once it stirs, or you step back, it dominates your entire field of vision.

*Pelagomorph*, perhaps, is trying to help us see.

Maybe they are just as mystified by our behavior as we are by the bubble nets of whales or the stone gardens of octopuses.

Then again, perhaps they understand more than we think.

This book, and its author's perspective, are not meant to convince the reader of any singular truth about nature or reality. The purpose is simpler. It is to remind us that we *can* change our minds.

We do it all the time, individually, collectively, even as a species.

Certainty offers comfort. It gives us a place to rest as we hurl ourselves through an unpredictable, unknowable cosmos. But comfort is not clarity. And clarity, when it arrives, frequently demands discomfort first.

Each of us walks a tightrope between opposing forces. Truth and illusion, complacency and fear, perception and blindness, innocence and understanding, empathy and indifference.

Sometimes, progress is nothing more than leaning ever so slightly toward the side that lets in more light.

Other times, though, change isn't just desirable. It's survival. Or it arrives unbidden, like a hard shove through a doorway we didn't even know existed. In such moments, belief in the door becomes irrelevant. Once thrust upon us, the change itself confirms its own presence.

Many times, it's only once we've fully crossed the threshold that we begin to comprehend what was waiting for us there all along.

What if we don't find a hostile alien or some unfathomable "other" on the other side, but instead a fellow traveler? Not an invader from beyond the stars, but a cousin. A relative long estranged by time and circumstance, whose presence, once recognized, feels strangely familiar.

And what if that long-lost kin were waiting to greet us, not with menace, but with joy?

*"Welcome! We have been beckoning for so long. We're so glad you have arrived. Let us share our lot with you now. The path ahead may feel strange, but we have walked it before. Let us be your guides. Take comfort. Your battle has been joined. Your journey has truly just begun."*

# CHAPTER EIGHT

## Estranged Foundation: Fathomless Aeons Unfettered

*"The only true voyage of discovery, the only*
*fountain of Eternal Youth, would be not to visit strange*
*lands but to possess other eyes, to behold the universe*
*through the eyes of another, of a hundred others, to*
*behold the hundred universes that each of them beholds,*
*that each of them is."*

*Marcel Proust, In Search of Lost Time, vol. 5: The*
*Captive, trans. C. K. Scott Moncrieff & Terence*
*Kilmartin, rev. D. J. Enright (New York: Modern*
*Library, 1999)*

**W**e have arrived. Together, no less.

It cannot be overstated that the first necessary step to seeing from a new perspective or a heretofore unexplored angle is intent. When one determines to occupy a space devoid of presence, there is an inherent risk. There is also a potential reward.

When the seeker deems the underlying peril to be outweighed by the theoretical benefit, intent may bloom into action. Yet inevitably, the seeker is warned. Those who lack intent, or who abhor the risk, will fatefully exhort the claimant of unfulfilled destinies to beware the most ruinous of hidden outcomes.

Still, the seeker seeks.

So, we press on, dear reader. We venture into a vacuum which, once approached, draws us out and away. Not with violence, but with gentle, ineluctable pressure. A positive knowledge differential. Like the proverbial moth drawn to a porch light, so too are humans drawn to knowledge still clenched in the indifferent fist of the cosmos.

We *must* know.

The first step to knowing is *looking*.
The second, *seeing*.
The third, *beholding*.
The fourth, *comprehending*.
The fifth, *accepting*.
The sixth, *incorporating*.

It is between the third and fourth steps that we most often stumble.

Let us both endeavor here to proceed carefully, with purpose and *intent*.

For many years, the popular term applied to *Pelagomorph* encounters was "UFO", Unidentified Flying Object. Most are familiar with this label. The official shift to "UAP" (Unidentified Aerial Phenomena) occurred in 2000. It appeared in a now-declassified report from the UK Ministry of Defence titled *Unidentified Aerial Phenomena in the UK Air Defence Region.*[300]

The term had previously been used interchangeably with "UFO," but this appears to be the first documented use in a classified setting. Earlier uses may remain buried in still-classified material. In any case, the timing is less relevant than what the reframing of nomenclature represents.

The shift in language, from "flying" and "object" to "aerial" and "phenomenon," suggests more than a cosmetic change. It hints at a more profound transformation in how these events were being interpreted and classified internally.

Much of the discussion surrounding this change, both online and in print, focuses on the apparent aim of destigmatizing sightings to encourage broader reporting.[301] However, given the decades-long and seemingly coordinated effort to portray those who reported such encounters as unsophisticated, delusional, or mentally unstable, this rationale seems ironic.[302]

Alphabet-soup rebranding aside, it is the behavior observed that matters most. So why the rebrand?

It is far more plausible to suggest that this de facto reclassification was not driven by PR strategy, but by verifiable, accumulating data.[303] If we imagine a quiet migration away from belief and toward knowledge, even of the limited and fragmentary kind, then this deliberate rechristening begins to make real sense.

A *manufactured* craft, propelled by stable fuel, is a flying object. A *biological* entity, lifted by an evolutionary adaptation unknown to humans, is an aerial phenomenon.

It could, of course, be something else entirely. As noted in the previous chapter, many "exo" origins remain possible. But as we move forward, the author hopes to bring one origin into sharper focus. One that is not only plausible, but increasingly probable.

If the terminology changed because the phenomenon defies conventional categorization, not as a craft but as a creature, then a more difficult question arises.

How?

How could a life form evolved in the lightless depths of Earth's deep oceans traverse not only the surface, but the sky above? And perhaps even the void beyond?

There are many hurdles to overcome in imagining such a possibility. The idea of trans-medium traversal, from sea to air to space, stretches biology to its speculative limits.[304] And yet, as outlined in previous chapters, we have already observed many extraordinary paths by which evolution has achieved the seemingly impossible.

How many more are still unknown? In what permutations and what preternatural combinations?

Here lies the key to the ascendancy of *Pelagomorph noetica*. They are formidable in form and miraculous in function.

Let us behold them, as they have beheld us. This is the only path to comprehension.

How does a life form, borne of the most extreme pressures on Earth, avoid suffering the same fate as Douglas Quaid on the surface of Mars? Would they not get the bends from rapid ascension? Or worse, explode? To glimpse what pressure can do, let's examine one unassuming ambassador from the deep.

The unfortunate title of "ugliest fish" is sometimes unfairly affixed to the blobfish. Its appearance at sea level is drastically different from its natural form at depths of 600-1,200 meters (2,000-4,000 feet) below the surface. The very adaptations that allow it to thrive in the abyssal depths are what betray it when brought topside. Without the pressure that shapes it, the blobfish's anatomy collapses. This results in the misshapen, gelatinous caricature we see memeified.[305]

This is not unique to the blobfish. Many deep-sea species undergo dramatic changes or disintegrate entirely when removed from their native pressure environments.[306]

The author proposes that the secret to *Pelagomorph*'s adaptability lies in a long, unbroken evolutionary timeline. One that has permitted the emergence of adaptations wholly unknown to us. Where other species deform, dissolve, or die, *Pelagomorph* persists.

One of the most celebrated examples of how a long evolutionary timeline can yield singular adaptations is the mighty tardigrade.

We, humans, stand in admiration of its capabilities, and not just because we struggle to explain them. In truth, we are still grappling with how even to *define* them within familiar biological categories.

The tardigrade's feats are numerous. One of the most unexpected is its ability to enter a dormant state known as the tun. When faced with an environment that lacks sufficient water or is too frigid, hypoxic, or saline, the tardigrade draws inward, curling into a microscopic sarcophagus. In this state, its internal proteins and sugars form a glass-like matrix that locks the organism into molecular stasis. Unbelievably, its metabolism appears to *cease*.

Because tardigrades can be revived from this state, even after extremely long periods, scientists believe metabolism doesn't halt entirely. Instead, it slows to a level that mimics death. In this way, the tardigrade becomes something like a living fossil, a creature paused in time.[307]

The author does not suggest that *Pelagomorph* mimics this exact capability.

Instead, the point is that the bounds of existence *we* conceive of are not the bounds that life is required to respect.

As Proust urges, we must "see with other eyes." We must reason beyond our own reckoning. Therein lie the solutions to any conundrum of thought.

So, if the tardigrade can all but stop time, what possibilities lie hidden in the occulted experiments of deep time?

Could a biological life form, adapted to the most extreme pressures known on Earth, evolve not only to withstand those pressures, but to become entirely *independent* of them?

Is it possible that *Pelagomorph* are externally agnostic? Perhaps instead of relying on ambient compression to maintain structure, they use internal regulation, modular containment, or even tunable tension systems.

The first possible adaptation we will conjecture is a kind of internal matrix, either hydrogel or colloidal in nature.

Some deep-sea species already utilize gel-like internal structures with minimal compressible gas. Hadal amphipods are a good example. These matrices do not rely on external pressure to maintain their shape. Instead, they maintain integrity through viscosity and internal cohesion.[308]

If we imagine that *Pelagomorph* possess bodies filled with non-compressible gel, then a sudden pressure drop, even into a near-vacuum, would not cause explosive decompression. No gas, no expansion. Or perhaps there is a slower, more controlled expansion. One that is buffered by additional adaptations yet to be named.

Another possible feature might be a flexible, non-rigid outer membrane.

If a hydrogel or colloidal internal matrix were paired with a highly elastic membrane, the result would be a formidable dual adaptation. A membrane that swells and contracts in rhythm with its surroundings could make survival itself a fluid art.

It is feasible to imagine such a membrane functioning as a buffer against external pressure variables. A viscoelastic skin could preserve internal integrity without structural failure. It might flex outward, absorbing shock, maintaining shape, and enabling passage between drastically different environments.

Viscoelastic materials behave in atypical ways. They can deform like a liquid, yet snap back like a rubber band. Familiar analogs include memory foam or even the cartilage in the human spinal column. They are firm yet forgiving, built for resilience.[309]

Added to this combination, a system of dynamic pressure modulation could allow for rapid internal adjustments. This would be a kind of redistribution network within the gel matrix. By utilizing nested compartments, *Pelagomorph* might be able to shift internal fluids between chambers in response to external pressure changes. These biological ballast tanks would maintain consistent internal tension, even under rapid depressurization.

This would provide yet another mechanism for absorbing the jolt of sudden environmental transitions.

And crucially, no internal gas pockets.

This absence would be key to the kind of movement and behavior so frequently reported in *Pelagomorph* encounters labeled as UAPs.

Without gas to expand or contract, there would be no risk of ruptures, bubbles, or buoyancy failures.

If any gases are present within *Pelagomorph* physiology at all, they may exist in the form of metal-rich oxygen carriers. These would be akin to hemocyanin, the copper-based respiratory molecule used by arthropods and mollusks.[310]

Echoing our earlier discussion of the tardigrade, another incredible trait we might hypothesize would be the ability to produce a self-sealing or phase-adaptive outer membrane.

Upon experiencing a substantial pressure drop, *Pelagomorph* may trigger a protective phase shift in their outer covering. This type of dynamic barrier, formed on demand, might resemble a mucosal sheath or even a temporarily hardened shell.

What if they could also selectively reduce metabolic activity to this outer shell, similar to the tardigrade entering its tun state? This kind of metabolic downregulation would minimize the risk of rupture, conserve energy, and stabilize form.

This adaptation wouldn't just shield against decompression. It could also provide protection against radiation during prolonged exposure, whether in the upper atmosphere or elsewhere.

Where we might expect a catastrophic blowout, *Pelagomorph* instead finds equilibrium.

The author hypothesizes that *Pelagomorph* is a biological entity evolved without lungs or air pockets. They are thus free from the vulnerabilities normally associated with life forms attempting such a transition.

If their tissues are filled with colloidal gel instead of gases, then their structure relies on fluid tension, not a skeleton. There would be no need for pressure equalization in the way human physiology demands.

Perhaps the immense pressure of the hadal zones has not anchored *Pelagomorph* to their environment. Instead, it has forged their freedom from it.

When exposed to a sudden vacuum or decompression, *Pelagomorph* does not burst like Douglas Quaid. They bloom into a different state of composition.

Adaptation to pressure is only part of the mystery. As we have seen time and again, surprises abound with life evolved in these deep zones. Some are subtle, others startling.

What other tricks might *Pelagomorph* have up its self-sealing, phase-adaptive membrane? Let's start with one of the most captivating.

How would a creature that evolved in Earth's hadal trenches be able to fly? The best answer is, they don't. At least not in any way we are familiar with, or primed to interpret as "flight."

*Pelagomorph* may instead pull off a much niftier trick.

But before we get there, let's set a baseline. How do airborne vehicles achieve flight?

This is accomplished through one of three methods. A propeller/reciprocating engine combination, a turbine or jet engine, or rocket propulsion.

The reciprocating engine is a piston-driven system, similar to that which powers an average motor vehicle. It drives propellers, allowing the aircraft to take off and stay aloft by generating thrust through continuous rotation.

Turbine engines, which come in several varieties, operate on a different principle. Air is sucked into the turbine, compressed, and then mixed with injected fuel, which is ignited. The resulting combustion creates high-speed airflow that is expelled out the back of the engine.

Newton's Third Law of Motion tells us that for every action, there is an equal and opposite reaction. This process of suck, squeeze, bang, and blow generates the lift and propulsion required for flight.

Rocket engines also rely on Newton's Third Law, but approach it differently. Instead of drawing in external air, they carry both fuel and an oxidizer in onboard tanks. These components are pumped into a combustion chamber where they ignite, producing an extremely hot, high-pressure state. The resulting gases are then expelled through a nozzle in a tightly controlled, directed burst. This generates thrust as long as the fuel-oxidizer reaction can be sustained.[311]

Of course, not all flight relies on mechanical engines or fuel combustion.

Nature has its own repertoire.

Biological airborne propulsion takes two primary forms. Unpowered varieties include gliding and parachuting. Powered flight is driven by muscular wings.

To our knowledge, powered flight has evolved at least four times independently. Insects, pterosaurs, birds, and bats each evolved muscle systems capable of moving wings to generate lift.

Beyond powered flight, there are two additional types of natural aerial movement.

The first is parachuting. This is seen in some spiders, which release a string of silk that catches the wind much like a kite, carrying them aloft on ambient air currents.

The second is soaring. This is a specialized form of gliding that utilizes rising or moving air. Under the right conditions, birds with sufficiently large wingspans can ride these invisible highways the same way a surfer rides a wave.

These atmospheric lift zones vary according to topography and weather conditions.

- **Thermal convection**, where warm air rises.

- **Ridge lift**, where wind encounters slopes or cliffs.

- **Wave lift**, created by standing waves of air over mountains.

- **Convergence**, where two air masses collide.

Winged animals can exploit each of these to gain altitude with minimal energy, and with a quiet, graceful mastery of physics that evolution has rendered nearly effortless.[312]

Most of us have seen large birds wheeling high in the sky, soaring on thermals or other atmospheric lift zones. It's a beautiful, almost meditative thing to watch.

Being human, we can't help ourselves. We anthropomorphize. *"They sure seem to be having fun up there."*

But it's not joy that keeps them aloft. It's the clever exploitation of differentials in air speed, temperature, and pressure.

And it's that very exploitation we now apply in theory to *Pelagomorph*'s potential method of flight.

Reports of so-called "UAP" events ordinarily include one shared refrain. The objects move in impossible ways.

Witnesses and commentators aren't wrong. No propulsion system known to us can replicate the behaviors described. Not mechanical or biological.[313]

So, we must look elsewhere. To the unknown.

Clearly, *something* is occurring. With *Pelagomorph* as our biological candidate, it's worth reviewing the leading theories of how a manufactured craft might achieve such feats.

There are a few especially compelling and downright clever solutions on the table.

Some of the most widely discussed and divisive ideas about UAP propulsion originate not in peer-reviewed journals, but in the testimony of Bob Lazar. His story has become the stuff of legend.

If you're reading this book, chances are you're already familiar with Lazar's story. Or at least with the shadow it casts over the broader UAP conversation.

A disclaimer is warranted. Mr. Lazar's accounts, motives, and character are fiercely debated within the community. Many dismiss his claims outright. Others champion his testimony as revelatory.

Such is the case with anyone claiming direct knowledge of technologies or objects beyond scientific consensus.

It's also worth noting that Lazar's narrative directly conflicts with the hypothesis presented in this book. This is true of any account that claims firsthand experience with mechanical craft piloted by nonhuman biological entities.

Future chapters will address that tension more explicitly. For now, it's enough to state that the ideas laid out here may seem equally outlandish. This is why they must be presented carefully, one piece at a time.

We'll get there.

Regardless, the author strongly encourages readers to seek out Bob Lazar's accounts, not just for what they claim, but for what they suggest. His background, path, and perspective are, at the very least, intriguing. In many ways, they are also edifying.

In Bob Lazar's narrative about his time working on and reverse-engineering exotic craft allegedly recovered by the U.S. military, he describes a propulsion method based on gravitational manipulation.

According to Lazar, this was made possible by a then-theoretical material, Element 115, which is now officially named Moscovium.

This claim, depending on how one frames both the science and the man, has served to both buttress and undermine his account. The author will refrain from diving too deeply into the long-running debate over the veracity of Lazar's story. The arguments on both sides are extensive and readily accessible for those who wish to explore them.

What *is* verifiable is that Lazar first mentioned Element 115 in a 1989 interview with journalist George Knapp, broadcast on KLAS-TV in Las Vegas. In that interview, he described Element 115 as a stable isotope capable of amplifying gravity and generating a gravity "wave." He also claimed that it was not naturally found in abundance on Earth and that it served as the core fuel for the alien-built craft he purportedly encountered.

According to Lazar, the amplified gravity field could be shaped and directed to create a gravitational differential. One strong enough to distort space-time itself. This distortion would effectively "fold" space in a specific direction, *pulling the destination toward the craft*, rather than pushing the craft toward the destination. If real, such a method could explain the kind of "impossible" movements observed in UAP encounters. Maneuvers that appear to violate the known laws of physics.[314]

Entire books could be written about Lazar's claims. Even then, they would not capture the full extent of his detailed narrative.

For our purposes, the specifics are less important than the broader framework. When attempting to account for observed behavior that clearly defies our current scientific or logical models, we must be open to entertaining new possibilities. Even speculative ones.

This author makes no assertion of special knowledge regarding Lazar's claims and remains agnostic toward them in general. Still, specific threads deserve cautious attention.

It is entirely possible that a stable form of Moscovium may be synthesized one day. However, to date, there is no indication that such a version would exhibit the gravity-amplifying properties Lazar described.[315]

It is also conceivable that, in the quest to produce stable Moscovium, researchers may have inadvertently created a new isotope of a different element, such as Copernicium. One such isotope, Copernicium-291, has been theoretically modeled to possess a half-life of around 1,200 years.[316]

It remains within the realm of scientific possibility that such an isotope could exhibit unforeseen or even entirely unimagined properties.

Of course, Lazar's theory is only one proposed method for manipulating gravity and space-time in the service of propulsion. Others exist that are equally speculative, but more firmly rooted in theoretical physics.

One such idea is the Alcubierre Drive. This is a hypothetical warp drive that would manipulate space-time itself, as derived from solutions to Einstein's field equations. Within this framework, a craft might attain faster-than-light travel without actually moving through space at superluminal speeds. Instead, it would contract space in front of it and expand space behind it, riding a kind of localized wave.

In theory, such a drive would allow for superluminal travel without violating the known laws of physics.

We also encounter serious challenges here, though. The foremost requirement is a negative energy density. This condition has not yet been observed in any usable form. Additionally, the Alcubierre model would require the emission of enormous quantities of radiation. These levels have not been seen in any of the encounters currently under discussion.[317]

Other alluring suggestions include electromagnetic or plasma-based propulsion, high-frequency oscillation, vibrational thrust, and even hybrid combinations of these and other experimental concepts.[318]

One such theoretical possibility involves metamaterial hulls paired with interior protective field bubbles. This kind of system might dynamically bend or shield space-time around the craft itself.[319]

With all these esoteric and forward-looking models in mind, let us now return to the biological frame.

We will consider how *Pelagomorph noetica* might accomplish similar feats. Not through engineering, but through deep-time evolutionary adaptation within the high-pressure crucible of Earth's hadal zones.

We have already proposed how *Pelagomorph* could survive, and even thrive, through rapid pressure shifts. We have theorized that they might use a combination of internal hydrogel or colloidal matrix, dynamic internal pressure modulation, and a viscoelastic outer membrane.

Now that we've explored both traditional and speculative models of propulsion, the question we turn to is this.

*Can any of these principles be applied biologically?*

And if so, how might *Pelagomorph* mirror or surpass them?

Clearly, the behavior observed in *Pelagomorph* encounters cannot be technically described as "flying." At least not in any traditional sense of the word.

A more accurate description would be 'floating' or 'hovering'.

Several mechanisms for this ability could stem from natural selection and biological origin. For example, if the internal hydrogel or colloidal matrix we previously described were also capable of density tuning, some truly beguiling outcomes would become possible.

By self-regulating the density of this matrix, *Pelagomorph* could achieve buoyancy control in aquatic environments. The same adaptation might also enable hovering or floating in the atmosphere. This could arise from internal manipulation or environmental response.

This is not pure speculation. We already have examples of synthetic hydrogels that can expand, contract, and deform in response to electromagnetic influence.[320]

Another useful property of these materials is their ability to radiate or absorb heat rapidly.[321]

Recall our earlier discussion on soaring. One method that birds use to gain lift is by riding thermal differentials. If *Pelagomorph* were able to create localized heat boundaries around their body, it might reduce nearby air density and generate thermal lift.[322]

Combine that with internal density phase-shifting, and we suddenly have a biologically plausible model for atmospheric hovering.

Let's take it one step further.

In our exploration of deep-sea life and its many strange adaptations, few are as compelling as the ability of certain organisms to generate or sense electromagnetic fields.

Most known examples involve weak field generation.[323] Over vast evolutionary timescales and under extreme pressures, *Pelagomorph* may have learned to exploit EM fields. In doing so, they could exceed anything we witness in the natural world today.

This may manifest not as a constant emission, but in intense pulses. Or in a consistently elevated baseline beyond what familiar biology exhibits.

Such a trait could take many forms. One outcome might be the ability to interact with geomagnetic or ionospheric gradients.[324] This could potentially reduce effective weight, or even generate passive lift.

We can speculate further. If a lifeform like *Pelagomorph* evolved to create a localized electromagnetic pressure differential, they might anchor against or push off the magnetic fields in their environment. Such a mechanism would echo systems proposed for UAP propulsion.[325]

This would produce a biological analog to the Mag-Lev effect, magnetic levitation, without the need for external machinery.[326]

And if such a being emerged from a geologically active trench system riddled with magnetic flux anomalies[327], the conditions for such an adaptation would not just be possible.

They might be inevitable.

Imagine a biological organism evolved with an internal hydrogel or colloidal matrix. One capable of withstanding rapid ascension and extreme pressure differentials. Now add the ability to phase-shift that internal structure, tuning its density from heavy to lighter-than-air. Combine that with the capacity to manipulate thermodynamic and electromagnetic properties in its surrounding atmosphere.

Now we're starting to get somewhere.

Such a specimen would be perfectly at home in the hadal trenches of the ocean floor, yet equally able to move freely through open water and the atmosphere. They could surface and hover with ease.

But even with all that accounted for...

There may yet be other mechanisms of action, tools in their evolutionary arsenal that grant this being an even greater freedom of movement across these environments.

Electrohydrodynamics (EHD) is a well-established principle dating back to the early 18th century. Yet it was only in the 21st century that EHD began to emerge as a viable method of propulsion and lift. This coincided with the advent of lighter, more efficient power supplies.

This technique, also known as ionic air propulsion, creates airflow by accelerating charged particles using an electric field. All without moving parts. Modern drones have begun to experiment with EHD-based levitation.[328] Certain insects are even known to generate weak ion clouds while swarming.[329]

If *Pelagomorph* were capable of emitting charged particles by ionizing the air around their body, they might generate a pressure differential. This ionic wind could be sufficient to lift them silently.

This aligns with many reported UAP phenomena, such as glowing plasma fields, soft buzzing sounds, and electrical interference. A charged ionic envelope might account for all these observed effects.

Granted, EHD requires a dense medium and a high-power input to be effective. But *Pelagomorph* may have evolved workarounds.

Adaptations born in deep time frequently find ways around problems that stump our brightest minds.

Among known oceanic life forms, a standout example of bioelectricity generation is found in eels and torpedo rays. These remarkable creatures are capable of producing hundreds of volts.[330]

Now imagine a deep-time, deep-ocean adaptation in *Pelagomorph*. Long, layered electrolyte sheets running along their outer surface. These would form a biological array capable of generating massive energy pulses in synchronized surges.[331]

A kind of biological capacitor, if you will.

Perhaps, instead of maintaining a constant voltage, *Pelagomorph* charges slowly and then discharges in a sudden, coordinated burst. This might explain why so many UAP reports describe a silent hover, followed by an instantaneous surge into the distance.[332]

Additionally, life that evolved near deep-sea hydrothermal vents has adapted to draw energy not from light but from chemical gradients. This adaptation is powerfully shaped by the inhuman pressures and total darkness of these environments.

At these lightless depths, life relies on chemosynthesis, not photosynthesis. One of the primary mechanisms behind this is the redox reaction. This transfer of electrons between molecules releases energy that chemosynthetic organisms can capture and use as fuel.[333]

Could *Pelagomorph* harness this very process? Might they convert environmental ions into usable charge gradients, which then fuel their bioelectric abilities through chemical potential alone?

It would be an elegant and well-precedented solution.

As for the challenge of using electrohydrodynamic (EHD) propulsion in the less efficient medium of air, we've already touched on some possibilities.

Perhaps *Pelagomorph* can self-generate a localized ionic envelope. This would render the surrounding air briefly conductive, just long enough for EHD to function. Or perhaps they emit a low-energy UV or infrared radiation, subtly heating the nearby air to create a low-pressure semi-plasma zone.[334]

Notably, some UAP reports mention a white halo surrounding the object.[335] That detail aligns closely with this hypothesis.

Finally, *Pelagomorph* might even emit a kind of bioactive mist. A microscopic, conductive cloud such as this would alter the medium itself. It could locally thicken or ionize the air. This would make electrohydrodynamic propulsion not just possible, but silent, seamless, and sublime.

Another possible biological adaptation worth considering is the manipulation of atmospheric pressure waves. If *Pelagomorph* were capable of rapidly vibrating parts of their body, or perhaps their entire form, they could generate localized air displacement.[336] This might produce ripples of pressure similar to miniature sonic booms.

Bob Lazar described a craft manipulating gravity to create a "wave" it could ride. *Pelagomorph* might produce rapid micro-pulses to generate pressure zones. They could then glide along these ephemeral pathways, using their body as a kind of sonic parachute or pressure veil.

This, too, could have its origins in the deep. It would be a trait forged in ancient oceanic conditions, where wave and pressure manipulation might have been key to locomotion or stealth.

Transferred into the atmosphere, the same adaptation might yield improbable aerial abilities. Not through brute propulsion, but through rhythmic, wave-driven finesse.

The final type of adaptation we will explore is the most speculative. It is also by far the most tantalizing.

As loose and wild as we have been in our conjecture to this point, we now venture even further afield. We cast beyond established scientific principles and into the fog-shrouded edge of what might still be possible. But since we've come this far, why not continue a little further?

One of the most puzzling characteristics of *Pelagomorph* encounters is their apparent ability to transcend media boundaries. They appear to move effortlessly between markedly different environments, with no observable disturbance at the boundary.

Objects are seen plunging from the air into water at high speeds, yet no splash is visible.[278] No visible turbulence. Not even on sensitive instrumentation like FLIR.

How is this possible?

Other puzzling phenomena include shimmering transparency, sudden disappearance, and full sensor dropout. Even in direct line of sight and across multiple detection systems.[337]

The most plausible, yet undeniably fantastic explanation? Phase boundary manipulation.

We now consider the possibility that *Pelagomorph* can shift the way they interact not just with water or air, but with the fabric of space-time itself. This is similar to our earlier speculation about membranes capable of adjusting to extreme pressure differentials.

As unbelievable as that may sound at first blush, we do have biological precedent for something very much like it.

Certain deep-sea organisms, such as the glass squid and larval eels, have evolved the ability to change their refractive index to nearly match that of seawater. This renders them effectively invisible, or at least "phased out" of the local environment.[338]

What if *Pelagomorph* could perform a similar trick? Not just in the optical realm. But across refractive, acoustic, or even electromagnetic spectrums. They would become undetectable across multiple modalities.

But there is an even more mind-blowing biological precedent to consider. The utilization of biophotonic crystals and quantum coherence.

These are not theoretical constructs. They are real, *known* biological processes, and they are astounding.

Biophotonic crystals are how certain lifeforms generate color *without* pigment. This includes butterflies, beetles, and some fish species. They instead grow nanoscale photonic structures. These crystalline arrays behave like optical lattices and are finely tuned to absorb all wavelengths *except* those they refract outward.

The result is a display of color that isn't just seen. It is engineered by light itself.

The Blue Morpho butterfly is a stunning example of this. Its wings shimmer with a radiant, electric iridescence that comes not from dye or scale. It instead arises from the precise architectural manipulation of ambient light at the nanostructural level.[339] Look it up at your earliest convenience. You won't be disappointed.

Its color is not a coat. It is an event.

It's not hard to imagine that *Pelagomorph* might also employ similar structures. Not just to dazzle, but to disappear. They might render themselves functionally invisible to the human eye, as well as to the most sensitive instrumentation.

Even more astonishing is the known effect of quantum coherence within biological systems, specifically along enzyme reactions. This process is very relevant to what comes next. It is a phenomenon that is both difficult to accept and widely studied. It's real. It's weird. And it's everywhere.

Enzymes are catalysts. They accelerate chemical reactions by lowering the energy barriers. That essentially brings the right molecules together in the right shape at the right time. In the classical view, an enzyme fits into its target like a lock and key. This helps the reaction proceed by reshaping the participating molecules.

But in the quantum view, something stranger is at play, particularly in reactions involving proton or electron transfers. Here, the particle may not simply leap across a molecular gap. It may quantum tunnel through it.

Even weirder, when these systems are in a state of quantum coherence, the enzyme can exist in a superposition of possibilities. This allows it to simultaneously sample multiple reaction pathways before collapsing into the most efficient one.

Let's pause for a moment to absorb that fully. The enzyme doesn't just work. It *evaluates* all possible ways of working in parallel. And then *chooses*.

This is not a fringe theory. Researchers are studying this process in both human liver metabolism of alcohol and in the way green plants turn light into energy.

To put it plainly, life on Earth evolved to use quantum mechanics in quite specific ways.[340]

Humans are beginning to discover that biology already adheres to the principles of quantum theory, even as we still struggle to comprehend them. And if quantum coherence underlies something as routine as digesting a drink or growing a leaf, what else might it explain?

More importantly, how would this apply to *Pelagomorph?*

It's not a massive leap to imagine that a highly evolved deep-ocean organism could harness this same quantum functionality, extending it far beyond the scale of enzymes.

Perhaps *Pelagomorph* is capable of achieving quantum coherence not only at the molecular level but also across cells or even tissues. This might enable the manipulation of energy, gravity, spatial phase, or *all of these at once.*

What if the incredible behaviors we attribute to "alien technology" are something stranger still?

Quantum biology, at scale.

The author is fully aware that this smorgasbord of speculation may drift well outside the comfort zone of many readers interested in this subject.

Fair enough.

But the mystery remains. Whatever these beings are, the details of their behavior are consistently described across numerous reports. The same seemingly impossible patterns. The same motion. The same vanishing acts.

That consistency demands something more than dismissal. It demands we consider every possibility at hand. Even the ones that may first appear unlikely, or even impossible.

Because we are not dealing with normal anomalies. We are confronting phenomena that utterly defy our current understanding of classical physics, locomotion, and propulsion. They do not match any known behavior, whether biological or engineered.

As we have outlined in this chapter, an organic origin for these behaviors is not only possible. It may be the most probable explanation for what we have observed so far.

We *must,* as Proust urges, *"...behold the universe through the eyes of another, of a hundred others, to behold the hundred universes that each of them beholds..."*

For what of those entities who eschew eyes as we know them?

What of the brain that cogitates in ways that escape our own?

What of the limbs that grasp and feel in ways that human anatomy cannot match?

What of the consciousness that beholds ours, from such a removed vantage that we do not even perceive we are already being beheld?

These questions, and others still stranger, must be asked.

Psychological, mental, or spiritual disposition is incidental. By definition, our lot here on Earth is a human one.

Are we not human, through and through?

And how else can we truly understand our humanness, if not through the eyes of others?

Do we not readily recognize the flaws and foibles of friends and acquaintances, just as they so effortlessly recognize ours? We sometimes muse that the contours of another are easier to discern than the concealed edges of our own being. This principle also holds true here.

Until we are prepared to truly allow ourselves to be seen as a species, we may never truly see ourselves.

Until we accept that something other exists, that beholds us in ways for which we are not prepared, we will never understand that we cannot *be* prepared.

Humans do now, and have always, proceeded by asking questions.

But many times, we ask questions we already have the answers to. Humans tend to ape mastery while merely mastering our own ape-like nature. We pull an idea from the shadows where we long ago secreted it, hold it aloft like a prize, and beat our chests in a self-aggrandizing display of vanity and circular conceit.

We must move beyond this point, bravely and together, against the consternation and puzzlement that come with redefining what we hold sacrosanct.

Knowledge, facts, foundations, and legacies. All will crumble. Not one stone will be left atop another.

*Huzzah.*

Together, we will cheer the apocalypse of certainty itself that awaits us on this path.

It is only through the annihilation of our former, insufficient power of sight that true vision may emerge.

We can follow these waypoints toward consequential truths we tend to overlook. Not because they are hidden, but because we have trained ourselves not to see.

We have only just begun to investigate our strange and wondrous neighbor, *Pelagomorph noetica*.

They have many more wonders to show us. But only if we look.

Only if we *see*, *behold*, *comprehend*, *accept*, and finally, *incorporate* their reality into our own.

# CHAPTER NINE

**Looking Glass in Reverse: De-anthropocentrizing the Familiar**

---

*"One conclusion was forced upon my mind at that time ... It is that our normal waking consciousness, rational consciousness as we call it, is but one special type of consciousness, whilst all about it, parted from it by the filmiest of screens, there lie potential forms of consciousness entirely different. We may go through life without suspecting their existence; but apply the requisite stimulus, and at a touch they are all there in all their completeness ... No account of the universe in its totality can be final which leaves these other forms of consciousness quite discarded."*

*William James, The Varieties of Religious Experience: A Study in Human Nature*

---

**H**umans are awesome.

We possess capabilities and dimensions of being that we are only beginning to grasp. As years become decades and centuries, we will continue to evolve and adapt, as we have done throughout history. If fortune favors, we may even transcend what we now think it means to be human. It's a thrilling prospect, if sometimes shadowed by unease. If we are honest with ourselves as we glance backward at the path that brought us to our current shoreline of reality, we must acknowledge the stumbles, missteps, and outright faceplants along the way. It will fall to the generations ahead to decide which of our choices deserve applause and which are to be condemned. Or worse still, erased entirely, as though they never happened. Is it better to be remembered as a cad or a scoundrel than not to be remembered at all?

There are limits to this, however. What does any current generation truly know of epochs now rendered invisible? Not by selective memory, but by the limits of collective memory itself. We love to recount historical events and debate the motivations or logic behind them. But is there not far more history *unknown* to us than known? Surely. Beyond the tangled trails of human memory, scattered with clues reaching back into the murk of our origins, lies a vaster chronicle. The history of everything *non*human.[341] Is the saga of alligators, horseshoe crabs, or coral somehow less interesting or consequential than ours? Many would say yes. The author disagrees. All histories are entwined, laced together into the grand, shifting tapestry of reality that we now trace our way through.

The intricate, interconnected mystery of life on Earth is humbling to contemplate. It is so ungraspable, perhaps, that the human mind may founder on its shore. And yet, our awesomeness compels us forward. One of the most extraordinary human traits is our ability to climb free of a wrecked vessel and keep swimming toward a distant shore. Just as Paul once did on that fateful voyage to Rome, we press on. This relentless pursuit of what lies just beyond reach is perhaps our greatest strength. It may also be our most perilous vulnerability. Still, we stretch our fingers toward the unknown. We stare unblinking into the face of the arcane and unfamiliar. We ask not just *"What is this?"* but, more crucially, *"What am I, in the face of this thing?"* Friend or foe? Conqueror or conquered? God or dust mote?

Never mind. Because *I am*.

Let us remember not only our long human history upon this Earth, but also the inseparable, unknowable history of the *other*. It swims beside us through the current of consciousness and reality. In knowing one, we glimpse the other. And in knowing the other, we find ourselves staring back in wonder and mystery.

Let us advance together into unmarked territory, across unmapped ground. True adventure begins in the furnace of discovery. Not partial illumination, or safe, sanctioned revelation, but the whole of it. Who among us would be content with the crust of the pie but not the filling? Do we listen only to the minor notes of a melody and plug our ears to the major? Perhaps. But to truly understand, we must be brave enough to taste the whole, to hear the full song of creation.

We are called to deeply inhale the vapors drifting from origins unknown and perceive them as they are, not as we wish them to be. Do not be mistaken. Bravery and foolishness can appear similar. But one is sacred, the other, profane.

We step now into sacred space. Together. We attempt the bravest act a human can muster. To look unflinchingly. To refuse to turn away from that which defies definition, that which will not contort itself to suit the expectations we bring to the encounter.

Why *should* it bend to us? Would *we*?

Could we?

There is only one way to find out.

In the previous chapter, we performed a veritable barrel roll through the possible permutations and manifestations of *Pelagomorph noetica*'s behaviors and capabilities. We have already established that

*Pelagomorph* is no ordinary entity. They are so unusual, in fact, that humanity has failed even to recognize them as a fellow biological being.

Yet they are.

Before we pull back the yoke and climb to even higher altitudes of hypothesis, let's take a brief detour. A lateral drift into new terrain. Not to ask *why* or *how*, but instead *which* and *what*. Is it possible to identify a known taxonomic stratum that could serve as an ancestor of *Pelagomorph*? Can it be narrowed to the closest extant relative?

The author believes so.

We have surveyed a bewitching array of lifeforms across these pages. We have borne witness to the wonders of biological heterogeneity that spring forth from creation itself. In this examination, we have paid special attention to those lifeforms that are not just oceanic, but hadal, those born in the crushing silence of the deep.

It is there, in the chapters behind us, that clues have already surfaced.

By sifting not only through behaviors and adaptations, but by peering beneath the hood into the genetic machinery underlying Earth's biological diversity, we begin to glimpse a handful of candidates emerging. A rare few possess traits so utterly fabulous, so captivatingly *other*, that they merit deeper scrutiny.

Imagine a kind of genetic Olympics, with each class, phylum, or order as an entrant. In such a competition, these would be the most likely medal contenders in their evolutionary events.

First up, we have *Tardigrada*, the indomitable water bear. This author admits great affection for these adorable stalwarts. We have already covered their absolutely enthralling capabilities in detail, so we won't recapitulate here. But just a friendly reminder that tardigrades belong to a very exclusive club of Earth-based organisms that can survive *in space*.[342] You GO, water bear!

Next on the podium, monotremes. Better known to most of us as the echidna and the platypus, these egg-laying mammals are evolutionary outliers in the most delightful and perplexing ways. We have already touched on the truly bizarre platypus. But it's worth restating just how gloriously strange this order is.

For starters, monotremes are mammals that lay eggs *and* produce milk. Somehow, biology permits this anatomical paradox. They also possess spurs on their hind limbs. The platypus variety is venomous. Echidnas use theirs in complex breeding displays. Even more remarkably, they can detect electrical fields via electroreception. This is a sensory skill that few land mammals can claim. And their genome? It's a curious mash-up of reptilian, avian, and mammalian traits.[343]

No doubt about it. Monotremes are unique, charming, and unmistakably distinct.

Next in the parade of evolutionary strangeness are the arthropods.

Certain spiders have spookily complex behavior. Some are even capable of mimicking the specific vibration patterns of trapped insects to lure and ambush *other* spiders.[344] Naughty, naughty.

Butterflies are another arthropod marvel. Their navigational abilities are renowned. But did you know some species retain memory from their caterpillar phase? That's right. During metamorphosis, their bodies dissolve into a cellular soup before reorganizing into something completely new, and *they still remember caterpillar thoughts*.[345] Retaining memory while transforming your entire physical structure? Good on ya, mate.

Ants also deserve a place on the winners' podium. Acting as superorganisms, they form coordinated collectives that mimic a single unified consciousness. Certain species, such as leafcutter ants, even practice agriculture. They cultivate fungi on harvested vegetation, a rudimentary but effective form of bioengineering.[346] Yes, ants have biotech. Don't sleep on them.

We would be remiss not to mention the bombardier beetle as well. This tiny tank co-fires two separate chemicals from its posterior. These mix upon expulsion, creating a rapid-fire jet of superheated gas that reaches temperatures near 100°C (212°F).[347] It fires this chemical like a living machine gun.

The moral of this story? Never mess with a beetle that has a flamethrower in its butt. Words to live by, friends.

Lastly, the mantis shrimp reigns supreme as the undisputed king of arthropod oddness.

Its weaponized appendages are legendary. They strike with such speed that cavitation bubbles form and collapse with shockwave force. This stuns their prey and can even crack aquarium glass. But that's just the warm-up act.

The true marvel lies in its eyes.

Particular species of mantis shrimp can detect circularly polarized light. This is an ability that no other known creature possesses. Circular polarization involves light waves that spiral in a forward direction, unlike linear polarization. The shrimp doesn't just detect this form. It integrates it into a broader perceptual system that includes *all* forms of polarized light. No man-made sensor has ever achieved this. The shrimp's biological optics outperform all current polarized sensor technology.

Even more strikingly, the eyes of the mantis shrimp don't just receive light. They *process* it. Each eye can move independently in all directions, producing two distinct data streams. That's four streams total from two eyes. But wait! There's more. Each eye also features a specialized band of photoreceptors across the middle, known as the midband. This creates a trinocular system within a single eye, granting it depth perception and polarization analysis without requiring the use of the other eye.[348]

To put it plainly, the mantis shrimp does not live in the same visual universe as we do. It perceives dimensions of reality that we can't access. Its world is richer, stranger, and more information-dense than anything we can imagine. We will return to the importance of polarized light in a future chapter. For now, though, one thing is clear.

Arthropods are positively amazing.

Now it's time to turn our gaze to the silver and gold medal champions of evolutionary adaptation, cnidarians (that "c" is silent) and cephalopods.

*Cnidaria*, which includes jellyfish, sea anemones, corals, and hydrae, are among Earth's most ancient and biologically diverse lineages.

This phylum boasts around 10,000 currently recognized species. According to molecular evidence, it also likely dates back at least 740 million years. Fossil records indicate that their appearance occurred around 600 million years ago. But because most cnidarians have soft, gelatinous bodies, they don't preserve well in sedimentary rock. This means their actual historical footprint may be even deeper than we can verify.[349]

Cnidarians are named for their signature evolutionary weapon, the cnidocyte. Also known as a cnidoblast, this specialized cell is a spring-loaded biological stinger. It houses a microscopic harpoon called a cnidocyst, which explosively ejects when triggered, delivering toxin to deter predators or subdue prey. It's one of the most elegant and effective mechanisms in nature. A simple stinging cell, refined over hundreds of millions of years.

Given their breadth of diversity and survival across unfathomable spans of deep time, it's clear that this evolutionary innovation has been an unequivocal success.

Before we explore the truly sensational characteristics found across this phylum, let's take a moment to appreciate the sheer power and sophistication of the cnidarian sting.

Animals like jellyfish and sea anemones are often dismissed by humans as slow-moving drifters, helplessly bobbing along on ocean tides with no control over their path. While it's true they usually *appear* at the mercy of the sea, cnidarians are secretly some of the fastest creatures on the planet.

Not in terms of swimming, of course. But in the mechanics of their venomous harpoon system, they leave virtually all other organisms in the dust.

When triggered by external stimuli, the cnidocyte doesn't just "fire." It *detonates*. The launch speed? A blistering 700 nanoseconds. That's 0.0000007 seconds.

Let's pause to compare that. A standard camera flash lasts about 1/1000 of a second. The cnidarian sting is more than a *thousand times* faster than that.

But speed is only half the story.

The force with which the cnidocyst fires has been measured at an unthinkable 5,410,000 g.[350] That's 5.4 million times the force of Earth's gravity.

For scale:

- Jet pilots typically black out at around 9 g.

- A violent car crash might reach 50 to 100 g.[351]

- This? It's literal biomechanical artillery.

Here's the kicker. It requires no muscles. This entire process is driven by a perfectly balanced combination of hydrostatic pressure and osmotic flow. It's a chemical and mechanical miracle, executed by a creature with no brain and no bones.

It is, in fact, one of the fastest and most powerful cellular processes ever recorded on Earth.

So, right from the start, we can see that these wonderfully perplexing animals defy our expectations on the deepest levels. The speed and force of their sting isn't just a matter of velocity. It feels like a rupture in the very fabric of probability itself.[352]

To the human observer, this entire sequence is unfollowable. There is no wind-up. No blur of motion. No traceable cause. Only instant consequence.

This is not speed in the way we naturally understand it, but something else entirely. It's not a measurement of time, but a *negation* of it.

By the time a reaction could even begin, the event has already ended.

This is not just a weaponization of force. It's a weaponization of time itself.

And as we've already hinted, cnidarians are just getting started.

It's a good time to remind ourselves that what we sometimes casually label "primitive" may only *appear* so from the outside. Many times, the closer we look, the more incongruously magnificent it becomes.

Possibly the most intriguing aspect of cnidarians is their lack of a centralized brain or ganglia. Instead, they operate with a diffuse nerve net. This is a decentralized mesh of neurons arranged around a central axis, not unlike the spokes of a wheel.

This neural net, woven just beneath the epidermis, radiates outward in all directions throughout the body. It is a full-body sensing and response system.

As humans, we might instinctively regard this architecture as primitive or inferior when compared to our own centralized brain and peripheral ganglia. After all, we prize the complexity of our neural command center.

But look a little closer, and you might start to notice something telling. The cnidarian system has its own distinct advantages. They hint at entirely different modes of cognition, perception, and presence.

We will revisit these possibilities again soon.

This distributed neural net, unlike the unidirectional nerve pathways found in humans and other animals, is capable of bidirectional signaling.

In many cnidarians, nerves can fire in both directions. This offers a far more dynamic range of information flow across the organism's body.

Lacking a central brain, their responses are governed instead by localized neural loops. These enable sensation and action to occur simultaneously and autonomously throughout the organism.

In essence, the body *is* the mind. There is no clear divide between perception and response. They are one and the same.

This system is no less advanced than our central nervous system and ganglia. It is instead optimized for different needs.[353] These include:

- **High redundancy** – damage to one region does not paralyze the whole.

- **Fluid motion** – ideal for drifting, reacting, and sensing in a fully aqueous environment.

- **Energy efficiency** – no need to maintain a metabolically expensive, bulky brain.

Rather than concentrating cognition in a single organ, cnidarians embody a more holistic design. Intelligence as a state of being.

A few cnidarians possess sensory structures known as rhopalia. These act as primitive proto-ganglia capable of detecting light and gravity.[354] However, the majority rely solely on the distributed neuronal mesh described above.

We can think of human ganglia as traffic control towers, overseeing and directing the flow of neural impulses with centralized precision. Cnidarians are more akin to a network of autonomous, adaptive traffic lights. Each responds locally to real-time conditions. Decentralized, self-regulating, and endlessly responsive, their system isn't about hierarchy. It's about emergent order from continuous feedback.

As mentioned previously, this phylum is incredibly diverse. It teems with wildly different physical forms and systemic adaptations. One specific group, known as box jellyfish, earns its name from its cube-like shape. Unlike their more languid cousins, these creatures are active swimmers and true hunters. They pursue prey rather than drifting passively into their meals.

Even more unbelievable, certain box jellyfish boast up to 24 eyes. Some of these are exceptionally complex, equipped with retinas, corneas, and lenses! And yet, still no central brain. These animals can track, navigate around obstacles, and detect prey using sophisticated visual input.[355] They do this all while operating on the same distributed neural net we've discussed.

And if all that weren't enough to leave you awestruck by the cnidarians, just wait. Their genomic complexity has even more surprises in store.

Cnidarians are primordial alchemists of multicellular life. Despite their basal lineage, they possess surprisingly complex genomic toolkits. They include HOX genes and WNT signaling pathways, both critical elements that govern body patterning and structure in complex organisms. Even more incredible, they also harbor genes related to nervous system development, despite lacking a centralized brain.[356]

Take the humble hydra. It is a tiny freshwater cnidarian with over 20,000 genes, a number comparable to the human genome.[357] While we humans must resort to prosthetics for lost or damaged body parts, many cnidarians employ a radical alternative. They are capable of stem-cell-powered regeneration.[358] Not only can they regrow body parts, but in many cases, they can rebuild entire bodies from fragments.

Some species take this even further with a bizarre phenomenon known as reverse development. Most jellyfish begin life in a small, simple polyp stage. But a few, most notably *Turritopsis dohrnii*, can revert from adulthood back to that polyp state. This essentially reboots their biological clocks. It is an epic feat that enables a potentially infinite cycle of development from polyp to adult and then adult back to polyp. While they can and do die from injury or predation, these delicate drifters have nevertheless earned the nickname "immortal jellyfish."[359] This regenerative loop is nothing short of wondrous.

Moreover, thanks to mechanisms such as DNA methylation, histone modification, and chromatin remodeling, cnidarians exhibit tremendous epigenetic plasticity. These processes enable them to adjust gene expression dynamically. They respond to environmental shifts such as light, temperature, or toxins. They also adapt to symbiotic relationships (more on this soon) and to regenerative needs.[360] For a lifeform so ancient and regularly overlooked, cnidarians are absolute masters of biological reinvention.

Despite lacking centralized brains, cnidarians possess a highly refined, modular molecular signaling network. Think of it like building with LEGO. They use the same building blocks, endlessly rearranged to create markedly different forms and functions. This system enables elegant biochemical adaptability and responsiveness throughout the body without requiring a central command center.

There's even more. Like the incredibly odd bdelloid rotifers, many cnidarians are capable of horizontal gene transfer. This is the acquisition of foreign DNA from bacteria, algae, or other organisms. In some cases, these borrowed genes have been co-opted into vital functional roles. They support immune response or enable photosynthesis-like symbiosis.[361]

So, while their anatomy and behavior may seem unorthodox, or "primitive" by human standards, nothing could be further from the truth. These are genomic shapeshifters, masters of adaptation and evolutionary trailblazers.

As we move on to our gold medal champion of otherness, let's carry that mindset with us. "Alien" doesn't mean inferior. It might just mean more advanced in ways we haven't yet imagined.

Oh, cephalopods, how do I love thee? Let me count the ways.

These eccentric lifeforms have enjoyed a well-deserved renaissance in the early 21st century. Documentaries, books, and scientific reappraisals seem to appear at every turn. Most of the spotlight has been directed at octopuses, and rightly so. They are particularly spellbinding. But let's not neglect their equally notable cousins, squid and cuttlefish. To fully appreciate why and how cephalopods streak across the finish line in the "Aliens on Earth" Olympics, we must give the whole crew its due.

Unlike cnidarians, cephalopods are not a phylum but a taxonomic class, falling under the phylum Mollusca and subphylum Conchifera. They are not as numerically diverse as the cnidarians, boasting around 800 living species today. But don't let that modest number fool you. It's estimated that over 11,000 extinct species once swam, hunted, and dominated Earth's oceans.[362] Many of these were ecologically critical in their time.

And when it comes to cephalopods' baffling, brain-twisting attributes, it's hard to know where to even begin.

Perhaps the most frequently discussed trait of cephalopods is their unmistakably high intelligence, and rightly so. These creatures display a stunning range of advanced cognitive abilities. Cephalopods are capable of planning ahead and solving complex problems. They also show signs of both short and long-term memory. Some have even demonstrated the ability to recognize individual humans, which is both endearing and eerie.

They learn by trial and error, remembering what works and applying it strategically in future situations. Their spatial awareness and 3D navigational skills are impressively sophisticated, comparable to those of many vertebrates.[363]

Equally engaging, though we'll only touch on it briefly here, is their expert mimicry and camouflage. Cephalopods can instantly shift their skin color, texture, and even body shape to match their environment. They utilize this ability to evade hunters and ambush prey.[364] These formidable adaptations are weapons in the arsenal of predators, as efficient as they are ruthless.

The scope and gradations of cephalopod intelligence are fiercely debated. One thing is abundantly clear, though. Humans have yet to fully grasp the depth of cephalopod awareness. Their admirable brains are only the beginning.

One key reason this debate persists is that cephalopods possess a structurally unconventional nervous system. Its architecture and function challenge every standard metric we use to assess intelligence. From beak to tentacle, their biology defies easy comparison. While humans exhibit some degree of neural distribution, cephalopods are wired in a manner unlike anything else we know. They do have a central brain. But unexpectedly, it is the smallest neuronal cluster in their body, containing around 45-50 million neurons. Flanking it are the optic lobes, which house 120-180 million neurons. This disparity is a clear indication of just how vital vision is to their experience of the world. But the real showstopper is in the arms. Collectively, cephalopods store around 350 million neurons in their limbs *alone*.[365]

As strange as that is, the most thrilling and otherworldly aspect of their nervous system isn't even in their arms or brains. It's in their skin.

Coleoid cephalopods possess a level of camouflage that would make James Bond jealous. If you've ever witnessed it in action, you know it's not just effective. It's *instantaneous*. Picture an octopus gliding from the sandy ocean floor onto a coral reef. It shifts from a smooth tan to a bumpy, mottled orange-and-white in the blink of an eye. That jaw-dropping speed and precision come courtesy of their uniquely wired nervous system.

It all begins with their eyes. They are large, sophisticated structures that in many ways outclass our own. Think of them like a dog's nose compared to ours. It's an entirely different scale of perception. Cephalopods detect extremely high contrast levels and can perceive polarized light. They excel at tracking patterns, brightness, edge detail, and movement. While they are technically colorblind, they possess consummate control over the color and texture of their own skin.[366]

Their secret lies in those proportionally massive optic lobes in their central brain, which process this flood of visual data. But here's the twist. When it comes to skin modulation, that information doesn't travel to the brain. It goes *straight to the skin itself.*

Let's pause to appreciate what that means. Cephalopods' visual and dermal systems are directly linked. They don't just *see* the environment, they *become* it, in real time. Imagine a human walking into a forest, glancing at a tree, and instantly morphing their skin to match its bark. That's the level of biological wizardry we're talking about here.

To underscore the point, up to *two-thirds* of a cephalopod's neurons are found in its arms. This gives rise to a distributed and highly specialized form of visual intelligence. It's one that doesn't begin and end in the brain, but radiates throughout the body.[367]

At the output end of this expressive system are chromatophores. These are tiny sacs of pigment embedded in the skin, each encircled by fine muscle fibers. Think of them like biological pixels. Each can be activated independently, but together they form a massive, dynamic display. The skin itself becomes a living screen.

Even more astoundingly, the skin and arms are equipped with independent neural ganglia. This allows them to process visual information and respond *locally*, without the central brain having to lift a tentacle, so to speak. This is how cephalopods can perform such lightning-fast, precisely targeted camouflage. Their skin doesn't just change. It decides on its own.

And it gets even wilder. Cephalopod skin contains light-sensitive proteins called opsins. These are the very same kind that humans have in our retinas. In essence, their skin sees. This is not a metaphor. It's a literal sensory fabric, responsive to light, pattern, and contrast. Perception and display are seamlessly integrated into a single system.

Cephalopod skin is a smart material in every sense. It's wired, reactive, and strangely intelligent.

The chromatophores in cephalopod skin are primarily responsible for colors like red, orange, brown, yellow, and black. But to achieve their full kaleidoscopic brilliance, these masters of disguise employ two additional types of camouflage cells, iridophores and leucophores.

Iridophores are composed of stacks of microscopic, reflecting plates. These plates create shimmering, iridescent hues of green, blue, silver, and even gold. They do this by manipulating how light reflects off their surfaces. Leucophores, on the other hand, act like adaptive mirrors. They reflect whatever light hits them, making them highly responsive to the surrounding environment.

These three cell types, chromatophores, iridophores, and leucophores, work in concert to produce the dazzling displays of color and texture we see in cephalopods.[368] But color alone isn't the whole story.

Cephalopods can also manipulate tiny papillae in their skin. These small, muscle-controlled projections can expand or retract. They are much like human goosebumps, but with voluntary control. This allows them not only to change color, but also texture and contour. They mimic coral, rocks, sand, or any other natural feature in their surroundings.[369]

Combined, these elements form an uncommonly powerful evolutionary toolkit. It has enabled these soft-bodied, otherwise vulnerable animals not just to survive, but to *thrive* across hundreds of millions of years in Earth's oceans. They don't just blend into the background. They can vanish, shift shapes, and become something else entirely.

As truly impressive as all these characteristics are, we're still not done marveling at the many wonders of cephalopods.

Let's talk circulation. Cephalopods don't just have one heart. They have *three*. First is the systemic heart, which functions like our own. It pumps oxygenated blood throughout the body. Then there are two branchial hearts, one positioned near each gill. These specialize in moving oxygen-depleted blood through the gills, where it can be re-oxygenated. It's an elegant, three-pump system that keeps their bodies running with outstanding efficiency.

And speaking of blood, theirs is *blue*. Literally. That's because cephalopods rely on a copper-based protein, rather than the iron-based hemoglobin found in humans. This protein, hemocyanin, is highly effective in low-oxygen environments, making it ideal for life in the depths of the sea. It delivers oxygen steadily throughout the body and has the added advantage of being less reactive to stress or temperature variations.

There is a tradeoff. This same copper-based blood makes cephalopods sensitive to environmental copper, which can interfere with the function of hemocyanin. Overall, though, it's a small price to pay for an oxygen delivery system that's custom-built for deep-ocean living.[370]

So, let's recap. Three hearts and blue, copper-rich blood. Check and check.

Still don't want an "I 💜 Cephalopods" shirt? Liar.

While the author could easily dedicate several chapters to the cracking array of cephalopod superpowers, we will now (reluctantly) narrow our focus to those most relevant for what lies ahead. That said, we are about to explore what is perhaps the most difficult yet crucial cephalopod adaptation to grasp.

Cephalopods don't just *see* polarized light. They can also *produce* it.

But before we go any further, let's take a moment to clarify precisely what polarized light is and why it's so crucial to this discussion. We've touched on it briefly, but now it's time to give this phenomenon the attention it truly deserves.

Without delving too deeply into the underlying mechanics, it is helpful to remember that light, whether visible or not, is fundamentally an electromagnetic wave. This may seem like a tangent, but it connects directly to the larger themes explored in this book. After all, we opened with a meditation on the nature of reality itself. Why? Because the foundational structure of reality is inseparable from the structures and processes that define all living creatures, including ourselves. If we don't understand where we come from, we can't fully grasp where we are, or where we might be going.

Much of this remains poorly understood, even by the brightest scientific minds alive today. Still, we press on.

To keep things simple, an electric field can generate a magnetic field, and vice versa. These two fields, oscillating at right angles to each other, create a self-sustaining wave. This is a continuous cycle in which electricity gives rise to magnetism, which in turn gives rise to electricity again, and so on. This perpendicular dance gives birth to light, which then travels outward along the axis of propagation, like a radiating ripple across space.

We tend to think of light as a wave. But depending on how we observe it, it can also behave like a particle. This duality is one of the strangest and most mesmerizing truths about light, and it becomes crucial as we continue this discussion. Why? Because these peculiar properties of light may ultimately be the key to identifying what *Pelagomorph* truly is.

As light is produced, it oscillates in a wave pattern, rippling energy through space. The frequency of that oscillation determines where it falls along the electromagnetic spectrum. This range, as measured by modern science, begins with radio waves. These are the longest variety, over a meter in length. The sequence continues through progressively shorter wavelengths, from microwaves and infrared to the narrow slice of *visible light* humans can perceive, then to ultraviolet, X-rays, and finally, gamma rays.

It's important to remember that the light we see is just one small sliver of the universe's radiant output. It is a thin slice in the electromagnetic pie. There may be even broader bands of light and radiation we've yet to detect, much less comprehend.[371]

As if the sheer breadth of this spectrum weren't humbling enough, light waves can also be polarized. In simple terms, non-polarized light vibrates in many directions at once, while polarized light is restricted to a single direction or plane.[372]

Here's an analogy that helps to visualize this.

Non-polarized light is like a pile of cooked spaghetti dumped into a colander. Imagine a tangled mess of twisting, looping strands, pointing every which way.

Polarized light, on the other hand, is like *uncooked* spaghetti still in the box. Imagine perfectly aligned strands, all pointing in the same direction.

So why is this important when discussing cephalopods?

Because they don't just *see* polarized light. They can also *reflect* it.

Let that sink in. These creatures are capable not only of detecting, but also of manipulating a type of light that humans cannot see or naturally emit. That, in and of itself, is arresting.

Possibly the best entry point for understanding this fantastic ability is a strange but true curiosity. Cephalopods are technically colorblind. At least, they are in the way humans perceive color.[373]

Human color vision relies on detecting specific wavelengths of visible light. When white light (sunlight) strikes an object, particular wavelengths are absorbed while others are reflected. A red rose, for example, absorbs every color *except* red, which bounces back to your eye. Inside the retina, two types of photoreceptor cells, rods and cones, convert that light into electrical signals. Rods help with light intensity and contrast, while cones are responsible for color perception.

Both rods and cones contain a protein called opsin. This protein triggers the biochemical cascade that allows the brain to interpret incoming light as something as simple as an orange traffic cone, or a green coffee mug, or a yellow giraffe with brown spots.[374]

A cephalopod, however, sees the world differently.

That same red rose might register to a squid not as "red," but as a 90-degree polarization angle with strong vertical patterning and high contrast. In other words, they don't see *hue*. They instead see angle, alignment, contrast, and light behavior, which is far outside our visible range. These dimensions allow them to distinguish between "colors" using polarization and structural cues rather than wavelength, specific absorption, and reflection.

The sensory structures responsible for this ability are called rhabdomeres. These light-sensitive units are composed of microvilli, which are tiny, finger-like projections. Each bundle of microvilli is oriented in a particular direction. That orientation determines what type of polarized light the rhabdomere can detect. It's like each tiny finger is tuned to one specific angle of light vibration.

This capacity to see at intersecting angles and process overlapping, directional patterns of light is known as orthogonal vision. It's a decidedly different way of parsing the visual world.

The ability to see the angles and directions of light polarization grants cephalopods access to a *hidden universe* we can only guess at. It's reasonable to presume that cephalopods, through this polarized perception, can detect shapes, patterns, and movements of light that are totally invisible to us. Any camouflage used by other ocean-dwelling creatures, if it relies entirely on visible-spectrum coloration, may be utterly ineffective against such perception.

As discussed earlier, cephalopods not only perceive polarized light but can also *reflect* it. This means they may be communicating in private visual channels, sending signals to one another that are completely undetectable to both prey and predator alike. These secret light-based languages would be invisible to human observers, hidden in plain sight.

Where humans rely on brightness, hue, and contrast through rods and cones, cephalopod rhabdomeres provide them with access to a far richer form of visual information. They can detect subtle variations in surface texture, such as tiny shifts in the alignment of sand grains, slight moisture gradients, or micro-ridges in coral. All of these are far too fine for our eyes to register.

In effect, cephalopods don't just *see* objects. They see the interactions *between* objects. They perceive histories, gradients, and tensions in the landscape, like a living map of the seafloor, etched in light. Their orthogonal vision even allows them to detect material composition and assess the angle of a surface by analyzing how light reflects and polarizes against it.[375]

So, where does all of this lead us? What are we to make of these wonderful life forms, these cephalopods and cnidarians, that seem to defy every expectation?

Cnidarians lack a central nervous system, yet they still exhibit hunting, learning, and adaptive behaviors. They wield microscopic venom-harpoons that fire with unimaginable force and speed. Some may even skirt the edge of biological immortality.

Cephalopods, on the other hand, have a distributed nervous system with arms that think independently. Their skin can see and transform almost instantaneously, allowing them to become camouflage virtuosos. They boast three hearts, copper-based blood, and a visual system that bypasses color entirely. It enables them to perceive the world in orthogonal, polarized patterns of light. Waves and angles that are invisible to us.

All of this returns us to the vexing questions we've been asking since the first chapter.

When humans encounter such creatures, our first instincts are to ask...

"How do they think?" "How do they see?" "Why do they behave this way?"

These aren't the wrong questions. But they are limiting ones.

We assume the position of the sole observer.

We forget that the other is observing us as well.

Perhaps a better question is:

"*What is thinking*?"

Is it just human-style cognition? Pattern recognition? Abstract reasoning?

Are there other forms of intelligence that we fail to recognize solely because they do not resemble our own? If they exist, such forms may in fact be equally valid. Perhaps even more sophisticated.

This is not about assigning superiority or inferiority.

The entire ladder collapses here.

We are faltering on the shore of comprehension, faced with true unknown unknowns.

A rose by any other name may not smell as sweet.

It may not look red.

It may not even *be* a rose at all.

It is long past time for us as a species to accept that even life forms as familiar as octopuses, squid, sea anemones, jellyfish, and cuttlefish may be so far removed from our consciousness that they effectively inhabit alternate realities, right here on Earth. Their bodies drift through familiar waters. But their minds? Their perception? Their mode of being?

They may well be elsewhere.

And if these creatures, within reach, within view, within arm's length, can differ from us so profoundly...

Then again, we ask.

*"What of the mind that cogitates along a different axis than Homo sapiens?"*

What of the polarized iridescence that lights up the world in colors we cannot name, sense, or even dream of?

Are those unseen realities any less valid than the wonders we *can* perceive?

No. They are *more* real to the beings who inhabit them.

Would an octopus or squid envy our humanity?

Or might they instead pity the narrow bandwidth of our perceptive reach?

We humans still struggle to identify, let alone decode, language in other life forms.

Is this because language is absent in them?

Or is it because the absence lies within *us*, in our ability to perceive it, to recognize it, to receive what we were never tuned to hear?

This principle applies not only to communication, but to life itself.

We may cite the Fermi Paradox as a lament. *Where is everyone?*

But perhaps its most urgent application isn't outward. It's inward.

It doesn't only haunt the silence of the stars. It whispers to us from beneath our own feet.

From the oceans. From the woods. From the dark crevices of perception where strange minds might wait.

We need not anxiously scan the heavens for visiting intelligences.

The more pressing concern, instead, is how much of our own world have we failed to recognize?

How much of reality, true, present, pervasive reality, do we exclude from our consciousness, purely because it refuses to conform to our five meager senses?

Not only the reality we acknowledge, but also the one that continues to shape every moment. It does so whether we measure it or not. Embrace it or reject it. Accept it or deny it. Digest it or spit it out in confusion.

Perhaps it is precisely *because* these realms are separated from us by the thinnest of membranes that we remain so obstinate toward them.

We mistake the boundaries of our own perception for the boundaries of reality itself.

But this is a critical error.

There has been no final accounting. Only a recounting of what has already been gathered, time and again, like a child stacking the same blocks and calling it a cathedral or a prison. Until we recognize that we are not only the accountants, but also the accounted for, that we too can be summed or negated, multiplied or divided, perceived or ignored... we will remain unready.

Is that why we seem so alone in the cosmos?

Because we choose to be?

Are we truly incapable of contacting a form of consciousness we have yet to conceive of or measure?

Or, more unsettling still, are we entirely capable, yet afraid of what we might discover?

Not just about the other.

But about ourselves?

This, the author suggests, is the true engine humming beneath the so-called Fermi Paradox, a phrase often misattributed to Enrico Fermi and possibly apocryphal.[376] Its lasting power, however, lies not in scientific rigor but in its revelation of our own limitations. It is a perfect example of the self-imposed blinders we so eagerly strap to our own cognition.

We are far more comfortable debating why science fiction has not yet become science fact than we are confronting the possibility that science itself contains mysteries so irreducible and confounding that even our wildest fictions pale in comparison.

This is why we explored, in such detail, the atypical architectures of cnidarians and cephalopods.

Not for novelty. Nor for entertainment.

But because they offer a crucial key, a way to unshackle our assumptions and glimpse what might be possible.

Now armed with that knowledge, we press forward, not only into unseen places but into unrealized potential.

Humans. We really *are* awesome.

And we will prove it, unceasingly.

So, let's do what we do best.

Let's be awesome together. Right now.

Let's pull ourselves up and out of everything we know.

Everything we think.

Everything we are.

This is the only path to realizing the dream of dreamers. Not just those stretching back through our shared past, but also those who exist outside of it.

For dreams do not obey logic.

They do not trace cause to effect.

Within dreams, the impossible becomes commonplace, and the commonplace becomes impossible.

You know this.

You feel it even now, as you read these words.

And the fact that we both know this as you read confirms it as so.

*How?*

Because our dreams are joined.

Not only to each other, but to something else. It is something that evades our conscious perception and escapes the will of thought.

There are, dear reader, other dreams to fill our souls.

And other dreamers to dream them.

Though they drift along currents we dismiss as myth, and trace paths that spiral beyond the reach of our waking imagination, we must follow them.

But we may only do so by preparing ourselves, not just to *be* the dreamer, but also to be *dreamt*.

This is how the paradox is undone.

For when one encounters a riddle that seems unsolvable, the answer is sometimes to let the conundrum *solve you*. Not the other way around.

With all this in mind, let us turn once more to *Pelagomorph* and its observed attributes.

Those deemed "impossible," the author now believes, may finally begin to come into focus.

For our strange neighbor has no use for the definitions we impose upon it, any more than we would for its definition of us.

And yet, here we are.

Both of us.

In this place. Perhaps discarded.

But still waiting to be discovered.

*To be accounted for.*

The requisite stimulus, after all, is not proof or measurement, but awareness.

Awareness that the impossible is not merely possible. It is truly a mirror that steadily reflects back to us the very essence of possibility we pretend to understand.

Does not the so-called "impossible" look upon us with the same disbelief?

If so, then it is only when we *both* reach, both attempting to secure that which cannot be grasped, that we may together seize the treasure that has so long eluded us, darting just beyond the reach of every outstretched hand.

That treasure is contact itself. It is won not through conquest, destiny, or force of will. Instead, it is gained through humility, acceptance, and the gentle wonder that has always marked our most august achievements.

Humans *will* slay these paradoxes, conundrums, and bafflements.

We must.

That seemingly distant pinnacle we dream of? It is, in fact, already within our reach.

The piercing truth we've long failed to perceive lies just beyond the peak itself.

Another climbs as well. Not in competition, but in unity.

Our truest goal was never the summit, but the identity of our fellow traveler ascending the far side of the same metaphorical mountain.

As we are to them.

We will find our rightful place among the stars only when we reckon with our authentic place here, in this world.

We are not emperors or sovereigns of this domain.

We are natives. Countrymen. Kin.

Friends, I implore you...

Lend me your ears, and your hearts.

Feel the truth welling up within you even now.

Is it nobler to seek dominion over what our senses perceive, or to cast off agenda and assumption to embrace the other as it is?

We can only present our true selves when we speak the truth *to* ourselves.

Only when we reach out in joy, astonished by the joy that meets us in return, will we truly arrive in the station we were born to occupy.

Despair is born of ignorance.

Delight of innocence.

Another now joins our bitter struggle, and, in doing so, transforms it, through simple presence, into a tender journey.

We need only surrender our ignorance, worn and worried like a rotten tooth, extracted in a moment far gentler than its persistent ache ever portended. Free of that pain, we will behold our glorious destiny. Not as wardens of a menagerie, but as siblings in a garden.

Our sisters and brothers need us desperately.

We need them too.

Do you not, even now, feel a pang in your soul?

A cry for those who toil just beyond the veil of this looming horizon?

The embrace has already begun.

We are hurtling toward reunion with ever-increasing speed.

Let us meet it, together, with senses and intentions thrown wide in alacrity and wonder, welcoming the discarded forms of consciousness we once failed to recognize.

For in embracing the other, we finally return to ourselves.

And what could be more human than that?

# CHAPTER TEN

## Asymptote Fidelity: Apparitions in Evolutionary Syntax

*"Pragmatism avails a savior far more than aestheticism."*

*Ted Chiang, Stories of Your Life and Others*

In our quest to arrive at a destination informed by logic and clarity, it has been necessary to repeatedly choose the path less traveled. There's a reason that clichés embed themselves so firmly in the fertile soil of our shared lexicon. They persist not because they are lazy, but because they are reflexive. Truisms speak to us in shorthand, across time and experience. They hum with a frequency that bypasses the intellect and goes straight to something older. We've all been presented with the familiar choice of the low road or the high road, the rocky path or the smooth one, the dense and forbidding woods or the poppy-strewn field, inviting us forward with butterflies and sunlight.

But the deeper we tread along the circuitous tracks that define our human existence, the more frequently we find that choosing which path is *truly* less traveled, less obvious, less damaging, or less corrosive to progress can be confounding. It behooves us, both as individuals and as a species, to examine our actions carefully before taking a step forward. For once chosen, the path we reject may disappear behind us. What's left behind is not always a door we can reopen.

History reveals this not just in our own experience but also in that of the world around us. How many times has a path been chosen for us by those who came before, pressing onward and leaving us to contend with the consequences? A person who has not eaten in days may find nourishment not through personal insight, but instead by witnessing what another creature consumes. Hidden spaces or escape routes, sometimes the difference between life and death, are sometimes only revealed through observing the movement of others. Even faint, lingering evidence of past action, long buried by time, can reignite the scholar's curiosity or illuminate a pilgrim's search.

To this list we must add one more. Never underestimate practicality. At times, the correct course of action reveals itself not through brilliance but through the obvious, if we are humble enough to accept it. Many of the most profound realizations have come not through discovery but from remembering. However, we must acknowledge that what has been forgotten can, at times, be far more daunting than acquiring what has yet to be known.

Time and again throughout our shared history, humans have planted flags of conquest upon the hills of supremacy, finality, fundamentality, or unassailable centrality. We then watch in vain as these markers are rendered quaint, displaced, or altogether erased by oncoming waves of knowledge that redefine the landscape entirely. What once stood as a fortress of certainty is revealed, in time, to be a crumbling monument to hubris or a relic of superstition.

Were we wrong to thrust those banners so hastily into the yielding crust of belief? Perhaps. Perhaps not. It is not the author's place to pass judgment. The ideas contained within these very pages may themselves be met with skepticism or even ridicule. So be it. Like any messenger emerging from an unfamiliar corridor, clutching some strange and galvanic treasure, there is no boast in the telling. There is only awe. Only gratitude for the glimpse. Nothing here is presented in the spirit of authority. At best, the hope is that it may be found valuable, or at the very least, propitiously sincere.

Admittedly, this aim is lofty. What is offered here is not a conclusion, but a re-centering. It is intended as a fresh calibration of human thought regarding our stance not only within Earth's biosphere, but within the wider arena of the universe itself. The same gravitational pull of thought that once beckoned Galileo, Semmelweis, Wegener, and Heisenberg whispers to us still.[377] It murmurs at the edge of our perception.

*"Look! The very stone your foot may strike in affliction may also be the marker by which your course is redirected to a more fruitful path. The beast you dread encountering along the way and whose presence you scheme to avoid may yet be the guide who leads you to triumphs still unimaginable."*

Indeed, it may be only when we are well along our way that we begin to grasp the true significance of where we have been headed all along.

Even as these words are being written, new discoveries continue to emerge from the very fields we have touched upon. With each passing day, the foundational stones of our mutually agreed-upon understanding of life are not just being rearranged. They are, in some cases, removed altogether.

Among the most sacred of these assumptions is the idea that all complex life on Earth requires oxygen to survive and reproduce.[378] Yet even this pedagogical cornerstone is no longer immune to reevaluation. We now have a stunning exception in *Henneguya salminicola*, a microscopic parasite of salmon. This organism has simplified its biological structure to such a degree that it no longer retains mitochondrial DNA. As a result, it lacks the machinery required for aerobic respiration. It does not use oxygen *at all*.[379]

This discovery is more than improbable. It is paradigm-shifting. It reminds us that evolution, far from being a linear ascent toward greater complexity, can proceed by subtraction. It may delete components once thought essential in pursuit of elegance, efficiency, and environmental adaptation. Notably, this loss in *Henneguya* is not ancestral, or preserved from some primitive origin, but rather a secondary simplification. It is the result of a later adaptation, tailored to the demands of a new environment.

This arrangement is well worth remembering as we continue our exploration of *Pelagomorph*. Evolution is not directional. Nor is it loyal to the narratives we impose upon it. It is elastic, amorphous, and persistently inventive, ruthlessly practical in its pursuit of function over form.[380] Life, in all its iterations, remains miraculous not for its conformity to our expectations, but for the spectacular ways it defies them.

So, what of *Pelagomorph noetica*?

We've come this far together, and still you remain, pressing onward through these shifting permutations of thought and conjecture. Perhaps you already sense where we are headed. Perhaps not. Either way, the time has come to acknowledge our traveling companion fully and lay bare the breadth of this supposition, out in the open, where it belongs.

When we read the witness accounts and sensor data surrounding object sightings, whether from Aguadilla, the *USS Omaha*, the Nimitz encounters, or countless other instances, we marvel not only at their details but also at their implications.[381] The described behavior of these objects consistently defies our expectations and transgresses the very boundaries of what we believe to be physically possible. We experience this wonderment regardless of how much faith we place in the veracity of the reports. The behavior itself, a kind of effortless transgression of reality, strikes us as simultaneously impossible and eerily familiar.

Why is this?

No matter our scientific training or level of technical literacy, we feel a strange convergence in our gut. It is an impossible bloom of both disbelief *and* recognition. We are confounded, yet smitten. It is this numinous pairing, fascination connected to longing, that lies at the heart of what we seek in these pages.

Perhaps there is something atavistic, embedded within our psyche and DNA, that stirs in response to these events. Though we fail to reckon with the full nature of these spooky apparitions, we likewise hesitate to define them as wholly "other." They feel like emissaries from a place just beyond knowing. And yet, paradoxically, they seem to originate from somewhere deep within us. Somewhere, we have forgotten.

We are like amnesiacs caught in a dream of perpetual déjà vu. Our awareness hovers, almost but not quite able to name something we've only just rediscovered in a jumbled drawer of lost ephemera. Fingers close around it, expectantly, hopefully. Still, it slips away, skating just beyond our capacity to incorporate it. And yet it remains, warm in our palm, impossibly familiar.

This, dear reader, is *Pelagomorph noetica*.

Our neighbor. Our cousin. Ascending the far side of the same mountain, intent on meeting us in mirrored amazement. Surely they regard us in kind. How could they not?

As we will explore together in this chapter, our mysterious companion is something more than sentient. They are, by all meaningful measures, *sapient*. A symbiotic biological union born of ancient lineages, descendants of Cnidaria and Cephalopoda, intertwined in a form of life we are only now beginning to actually name. Could others like them exist? Possibly. But it is the author's belief that this distinct evolutionary branch has developed sapience beyond the reach and beneath the notice of human perception.

And this, perhaps, is no accident.

This deliberate evasion, this elegant obscurity, seems to be a chosen path. It appears to be a willful avoidance shaped by pressures and motivations we may never fully comprehend. Not now. Perhaps not ever. And yet, even without direct input from *Pelagomorph* themselves, we can infer much. Chief among these inferences is that they seem to have had no difficulty remaining hidden. They keep to their own domain, undisturbed. Is this distance entirely of our own choosing? Or does the choice lie elsewhere?

Is our collective lack of deep-ocean exploration, our failure to penetrate 99.999% of Earth's aquatic abyss, entirely voluntary?[40]

Or has there always been something pushing us back?

How else do we explain our relentless obsession with overturning stones on distant planets, while leaving this neglected and unexplored terrain here at home untouched? How do we justify this selective blindness, this willful hesitation, when the unknown lies so close?

Cephalopods are renowned for their secrecy and sophisticated avoidance strategies.[382]

Consider, for instance, the elusive colossal squid. Our knowledge of its full adult size comes not from direct observation, but from the beaks recovered within the stomachs of its mortal enemy, the sperm whale.

These massive squid can exceed 1,000 pounds in weight and stretch up to 46 feet in length, tentacles included. And yet, no human has ever witnessed a living adult in the wild.[383]

Not once.

Other species, such as *Gonatus antarcticus*, have only recently been observed alive. For decades, they were known solely through remnants discovered in the bellies of predators or as deceased specimens tangled in the nets of deep-sea fishermen.[384]

Perhaps most telling of all, the genome of cephalopods, especially the octopus, reveals a superb level of complexity. Within its DNA lies an extensive array of genes found in no other animal on Earth. This isn't just unusual. It's paradigm-breaking. It points to an evolutionary history still shrouded in mystery, a story we have barely begun to decipher.[385]

And yet, let us be clear. We are not describing a cephalopod.

We are speaking of something entirely other.

This book proposes that, at some point in evolutionary history, in the darkest, most inaccessible regions of the ocean, two ancestral lineages entered a radical symbiotic partnership. One was derived from cnidarians, while the other was derived from cephalopods. Over deep evolutionary time, this partnership did not just persist. It evolved into a hybrid lifeform of marvelous complexity. It is the one we now call *Pelagomorph noetica*.

Shaped by the intense pressures and unique forces of its abyssal environment over hundreds of millions of years, this lifeform developed abilities and characteristics so advanced, so foreign to human expectation, that their manifestations are reflexively classified as "UAPs" or "UFOs."

Given *Pelagomorph*'s formidable potential, this is hardly surprising.

Their movements through space-time defy familiar classification. When witnessed, they appear to us as something mechanical, artificial, crafted.

As we will explore in the chapters ahead, this illusion may arise not from error, but from the limits of our perception.

We have already touched upon some of the possible features and systems that might underpin their apparent capabilities. These include internal mass redistribution, tunable hydrogel or colloidal matrices, extreme morphological plasticity, reactive dermal architectures, and modulated electromagnetic fields. But we are not done.

We now turn to the even more startling phenomena of splitting and merging behavior, time distortion, deliberate avoidance strategies, and sophisticated forms of bioluminescence.

Some of these may seem implausible, perhaps even absurd.

However, when we consider the demonstrated abilities of cephalopods and cnidarians, and then imagine what a continuous, uninterrupted arc of symbiotic evolution might yield over half a billion years of development, the implausible begins to take on a haunting familiarity.

Throughout this exploration, we must remember the existing, real-world mastery that octopuses and other cephalopods already exhibit. We have already discussed the intentional reshaping and presentation of their bodies in ways specifically designed to distract, mislead, or frighten potential observers. These are not accidents of biology. They are intentional acts of survival.[386]

*Pelagomorph noetica*, if they exist, do not conform to our notions of visibility, boundary, or locomotion. Why should they? They did not evolve under the constraints of our expectations. Nor did they grow beneath our sun, in our air, or within our frame of reference.

Their trajectory, entirely practical yet radically divergent, was forged in a world we have never entered. It is a domain so removed from our upright, oxygen-fed, sunlit perspective that we must abandon pretense and admit that, whatever they are, they may not have evolved to be seen by us at all.

One type of cnidarian we have touched on previously is the siphonophore. Perhaps the best known of these is the Portuguese man o' war. It is a complex marine organism that, at first glance, appears to be a singular life form. In fact, it is a cooperative amalgam, a colonial assembly of smaller, highly specialized organisms known as zooids. These zooids cannot survive independently. Yet together they form a single, integrated system which functions as if they were one being.

Siphonophores are elaborate in both structure and form. They are capable of assuming various shapes depending on their species and environmental context. While no known siphonophore can intentionally split into subgroups and later reunite, they are extraordinarily fragile. This fragility offers a telling clue. Some species reproduce or disperse through detachable subunits known as *eudoxids*. These fragments drift free for a time before giving rise to new colonies. It's a living system composed of specialized, semi-autonomous parts, capable of partial independence and self-renewal.[387]

Siphonophores are not alone in this regenerative prowess.

Numerous marine organisms display fanciful capabilities in this domain. Some species of starfish, for example, can regrow an entire body from a single severed arm.[388] Marine worms, sponges, corals, and sea anemones all exhibit forms of regenerative replication. Each has its own peculiar strategy. However, without a doubt, the true champion of biological reassembly is the humble hydra.

These tiny, solitary carnivores possess one of the most astounding regenerative abilities in the animal kingdom. Even when a hydra is torn into numerous pieces, under the right conditions, those fragments reassemble into a functioning whole. The only absolute prerequisite? That at least a small portion of the hydra's head region remains intact. If even 20 cells from this region survive, they can serve as an organizational nexus that guides the rest of the disaggregated cells to reform into a coherent, living structure.[389]

This is not science fiction. This is nature's fact.

When we consider possible mechanisms underlying *Pelagomorph*'s reported "splitting" behavior, the hydra's regenerative abilities present one of the more arresting biological precedents. It's worth entertaining the possibility of a hybridized system. One that combines the hydra's cellular reassembly with the distributed neural architecture of cephalopods. Such a configuration might, in theory, allow the organism to split and rejoin physically. It could conceivably also enable autonomous, continuous neurological function within each subdivided portion, before, during, and after separation.

It's an idea that is both amusing and charming. Through the imaginative freedom of speculative biology, we will consider other expressions of this phenomenon. These events are often observed but still poorly understood.

The first possibility we will examine is a form of distributed neural swarm intelligence. Imagine starting with the cephalopod's decentralized neurology, each arm capable of independent processing. Now combine that with the diffuse nerve nets of cnidarians, and then place the hybrid in the Darwinian forge of a zero-light, high-pressure abyss for hundreds of millions of years. Could the result be a biological swarm consciousness?

One could imagine such a being comprised of semi-autonomous "nodes." Each would possess limited awareness but be linked into a cohesive and responsive whole capable of separating and rejoining as needed. This would go well beyond a siphonophore. It would instead be a kind of super-organism, perhaps even stranger than any we have yet observed.

A conceptual analog appears in Vernor Vinge's brilliant novel *A Fire Upon the Deep*, part of his *Zones of Thought* series. In it, we encounter the Tines. They are an alien species composed of packs of dog-like creatures. Each pack constitutes a single conscious being. Individual members of the pack can act independently, but only attain full cognition when joined with the others. This type of biologically distributed mind is a powerful narrative device. It is also perhaps an even more plausible evolutionary structure than we have yet considered.

In such an arrangement, what looks like "splitting" may not be separation at all. It may instead be a pattern shift. This might be an adaptive modulation used for communication, defense, or even as a mechanism for moving between media, such as water and air.

Another variation on this cluster-based design imagines the deep ocean's pressures selecting for a fractal form. Here, the organism is not modular in the usual sense, meaning it lacks dissimilar, specialized parts. Instead, it would be a body made of repeating, self-similar units. Smaller replicas of the whole.

If this were the case, the observed "splitting" might not be segmentation or division at all. Instead, it could be multiple instantiations of the same entity. Each would be functionally complete and perhaps even capable of autonomous action. This would be a different kind of unity. It would be iterative, resilient, and strange.

While this may seem speculative, nature offers us revealing hints. Octopuses occasionally express odd mutations that result in bifurcated or even trifurcated arms.[390] While not truly fractal, such expressions reflect the inherent plasticity of cephalopod morphology. They also point to underlying genomic oddities that defy typical vertebrate expectations.

One of the most consequential of these anomalies is their capacity for robust RNA editing. This rare biological trait enables cephalopods to rapidly and precisely alter protein synthesis without requiring changes to their genetic code or DNA.[391] It also allows for far more agile environmental adaptation than most multicellular organisms can achieve.

How might such a capacity play out over hundreds of millions of years in an unbroken, pressurized, lightless evolutionary corridor? Could a life form emerge whose physical form loops back on itself in circular patterns? Nature, after all, is an ever-experimental sculptor. Could evolution select for a morphology that is not linear or singular, but iterative, patterned, and self-reflective?

We are left with the tantalizing realization again. There is much in cephalopod genetics we do not yet understand. And in that blind spot, possibility waits.

The idea of a fractally constructed descendant of a symbiotic cnidarian-cephalopod hybrid is undeniably bemusing. However, a more plausible explanation for the observed "splitting" behavior may lie in the realm of advanced camouflage.

In this interpretation, *Pelagomorph* may not be dividing their physical form at all. Instead, they may possess the ability to manipulate the visual signature of their outer membrane in ways that create the *illusion* of separation and rejoining. One possibility involves a localized cloaking effect. Perhaps the central body becomes temporarily transparent or light-refractive, giving the impression that the organism has split in two.

But this deception could be far more sophisticated.

Imagine a coordinated extension of multiple appendages, each tipped with bioluminescent structures that radiate outward from a camouflaged core. If the intervening arm structures were visually erased via active skin modulation, the result would be a radial display of luminous "nodes" hovering in apparent separation.

Cephalopods already possess the building blocks for such a feat. Their skin contains chromatophores for pigmentation, iridophores for iridescent reflectivity, and leucophores for reflecting ambient light. *Pelagomorph* may possess similar structures or a similar evolutionary refinement. It is entirely conceivable that they use not just passive camouflage but also "projective mimicry." This would cast light and visual cues outward in intentional patterns.

This wouldn't merely serve as concealment. It would be communication, misdirection, or perhaps even ritual.

Considering the known capabilities of cnidarians and cephalopods, such a strategy seems less far-fetched than our initial astonishment might suggest. Indeed, it may be the most biologically reasonable path to producing the kind of optical phenomena described in consistent witness reports of "UAP." Far from science fiction, it could represent the apex of evolutionary adaptation. This kind of cloaking strategy would take millions of years to develop.

As we have previously explored, quantum mechanics plays a demonstrable role in biological processes. While the specifics are still debated, this mechanism is supported by compelling evidence.

Quantum effects appear to be involved in phenomena such as photosynthesis, avian navigation via magnetic fields, and even specific enzymatic reactions.[392] If such strange and delicate wonders are already embedded in nature's playbook, then it becomes conceivable that biology could harness even more esoteric quantum functions under the right conditions.

This possibility becomes even more tantalizing when we consider the bizarre and still not fully understood genetic structures of both cnidarians and cephalopods. These creatures already demonstrate marked deviations from biological norms. In light of growing evidence that biology utilizes aspects of quantum entanglement, we now pose a further question. Could *Pelagomorph* operate within a regime of biological quantum entanglement?

If so, this might offer an elegant explanation for the oft-reported "splitting" behavior. Rather than a literal division of form, what we witness may be the result of quantum desynchronization. This would be a temporary displacement of individual entities within phase space. In this view, "splitting" is the visual artifact of entangled segments becoming momentarily out of sync. Their subsequent "rejoining" would not be a physical merging. Instead, it might be a quantum rephasing. In that moment, the organism locks back into coherence as a singular, unified perception.

Yes, this is speculative. But no more so than prevailing interpretations that attribute these phenomena to hyper-advanced mechanical craft of unknown origin. Which of these explanations is more plausible?

Do these sightings represent artificial aerospace vehicles traversing Earth's skies from another planet or dimension? Are they manufactured craft navigating complex atmospheric and oceanic mediums in silence, with no signs of propulsion or material fatigue?

Or are we encountering an as-yet-unclassified *biological* entity? Could it be one that evolved right here, in the unexplored pressures and darkness of Earth's deep oceans?

If we allow ourselves to step outside the gravitational pull of popular narratives and abandon the inherited presumptions of the modern UFO mythology, a bracing clarity emerges. Many of the behaviors described, such as splitting, merging, disappearing, and responding with apparent intelligence, do not resemble machines. They resemble life.

The reader is invited to decide for themselves. But the author asks gently... which path, though less trodden, leads more clearly toward sense?

Another highly speculative, but tantalizing possibility involves a theoretical enhancement of the reflectin proteins found in cephalopod iridophores. As discussed earlier, these sophisticated cells manipulate light via stacks of specialized, plate-like structures. Those structures generate vivid color displays through constructive interference. Now imagine a far-future iteration of this system, refined over hundreds of millions of years through natural selection. Might *Pelagomorph* have evolved the ability to use such cells not only for color modulation, but as a kind of biophotonic projection system?

If so, we could be witnessing the biological equivalent of optical projection tomography. This technique is used in human science to create 3D images from scattered light.[393] In *Pelagomorph*'s case, this might manifest not as a literal division of the organism but instead as the illusion of separation. Imagine a shimmering mirage of twin selves, generated to perplex, to shield, or for purposes yet unknown.

This false doubling could serve as a camouflage tactic or defensive shroud. Alternatively, it might function as a decoy signal, diverting attention away from the entity's actual physical location. In such a case, what FLIR sensors or optical instruments capture might not be the creature itself, but a projected phantom. What we perceive would be a misleading image. It may be born of bioluminescence, structural coloration, or electromagnetic field distortions emitted intentionally or incidentally by the organism.

Whether this projection is a deliberate act of deception or a passive by-product of other metabolic or neurological functions remains unclear. But in either case, it powerfully challenges our assumptions about what we are observing when we point our instruments toward the strange.

This brings us to a final, particularly compelling avenue of speculation regarding the "splitting" behavior observed in conjunction with *Pelagomorph* encounters. It is electromagnetic field manipulation. If this life form has, through evolutionary pressure, developed the ability to generate or distort localized EM fields, it opens several revealing possibilities.

For instance, a directed electromagnetic disruption could scramble or disable nearby detection systems, such as radar and sonar, or even the fragile instrument of human visual perception itself. Such distortion might not just jam sensors. It could produce phantom echoes, perceptual misalignments, or apparent duplications. These are precisely the kind of visual anomalies regularly reported in UAP encounters.

While cephalopods have not demonstrated any confirmed use of EM fields, many cnidarians are sensitive to electrical and mechanical stimuli and use them for orientation, navigation, or detecting ocean currents.[394]

Building on this, it is not a great leap to imagine a symbiotic life form, shaped by the immense pressure and electromagnetic turbulence of a hadal trench, evolving internal structures that interact with such fields more directly.

Perhaps *Pelagomorph* developed a type of piezoelectric matrix, converting physical pressure or internal motion into electromagnetic energy. Or perhaps they possess field-sensitive membrane structures that enable them to sense gravitational and magnetic flux, just as other creatures do, by feeling currents or shifting winds. In an atmospheric context, such capabilities might produce luminous plasma-like effects, disrupt nearby instruments, or even facilitate non-inertial motion. This could be especially pronounced along geophysical "ley lines" or zones of natural magnetic anomaly.

Should those electromagnetic field distortions be strong enough, they might also project or fracture the creature's image. This might create the illusion of splitting, doubling, or phasing in and out of view. Not through magic or machinery, but through a mastery of fields and perception which we are only beginning to imagine.

Speculation along these lines is driven less by whimsy and more by practicality. The reported visual anomalies associated with

*Pelagomorph* are too consistent and too repeatable to dismiss outright. We know, at minimum, that this life form can present the appearance of splitting or duplication. Whether the mechanism is camouflage, projection, or something more exotic remains to be seen. Until more empirical data is available, we must remain open to multiple hypotheses.

And yet, visual trickery is only part of the story.

We now shift our focus to another recurrent and distressing element of *Pelagomorph* encounters, the distortion of time itself. Often referred to as "lost time," this phenomenon goes beyond simple amnesia or bewilderment. It includes reports of time dilation, temporal disorientation, and disruptions in the subjective experience of chronology.

In some cases, witnesses describe hours passing in what feels like moments, or vice versa. Others report sequences of events that seem scrambled, nonlinear, or entirely unmoored from standard cause-and-effect logic. These are not isolated incidents. Nor are they always confined to close contact experiences. Even distant observers have recounted strange temporal effects. This strongly suggests that whatever mechanism is at work may not be limited by proximity.

Reports of time distortion, so-called "lost time," are so consistently and widely described across encounters that we would be remiss not to attempt some form of explanation.[395] While many of the ideas we are about to explore fall squarely within the realm of speculation, none are beyond reason. In fact, some may be grounded in well-established principles, albeit applied in unfamiliar ways.

A leading candidate in this regard is the manipulation of electromagnetic (EM) fields. As with earlier discussions of *Pelagomorph*'s visual and locomotor phenomena, EM field distortion remains a plausible and unifying hypothesis. It may explain not only splitting and rejoining effects but also apparent changes in observers' perceptions, memory disruption, and warping of temporal awareness.

Could it be that such field manipulation creates a localized region of distorted space-time around the organism? If so, the resulting time effects might be directed. They could be a defense mechanism, a navigational aid, or even a form of intentional obfuscation. On the other hand, these anomalies may be incidental by-products of unrelated biological processes, like sonar to a bat or the humming of power lines to human ears. They are perceptible, but not deliberate.

Whether intentional or emergent, this kind of phenomenon suggests a biology that is not only advanced but also incommensurate in its relationship to human perception of space, time, and consciousness.

Another potential culprit could be a form of neurochemical interference delivered via an aerosol mist or particulate secretion. This would be aligned with known cnidarian capabilities. Given the considerable potency and complexity of venoms found within cnidarian nematocysts, it's not difficult to envision an evolutionary pathway in which deep-sea environmental pressures selected for a chemical defense mechanism designed to unsettle, disable, or disorient nearby organisms.[396]

In aquatic environments, these substances are injected or diffused through contact. But if carried into the airstream, during flight or at the surface, the mechanism could become aerosolized. In such a scenario, even brief exposure to the field or mist surrounding *Pelagomorph* could result in perceptual and cognitive distortion in humans.

Some existing cnidarian venoms already demonstrate the ability to induce confusion, panic, and neurological disruption. A striking example is Irukandji syndrome, a condition caused by the sting of a specific type of box jellyfish. Victims report intense physical symptoms, but also a profound and inexplicable sense of "impending doom." This psychic state is so reliably induced that it has become a diagnostic feature.

The precise biochemical pathways remain under study. However, it is known that these toxins interfere with neuronal communication, sometimes enhancing excitability to the point of triggering seizures.[397] It is not unreasonable to imagine a compound evolved along similar lines, perhaps secreted through skin membranes or defensive structures. Released in vapor form, this could trigger pervasive disorientation, even a fractured or altered sense of time.

Another salient possibility for this reported anomaly is the overlapping of neural fields, whether as a passive byproduct or an active interaction. If Pelagomorph operates through a distributed neural network or field-based cognition system, this could place it fundamentally out of phase with human neurobiology. The clash or interference between these systems may result in a range of unintended consequences.

Should the neural field of Pelagomorph fall outside the frequency, architecture, or harmonics of human cognition, even brief proximity could create unpredictable disruptions. Such disruptions could manifest as a disjointed perception of time, paralysis, altered consciousness, or surges of sensation and emotion. In their most severe form, they might leave behind dissociation, trauma-like symptoms, or even permanent shifts in personality and worldview.

Some experiencers report sensations of telepathic communication, intrusive thoughts, or a perceived manipulation of the mind itself. Others describe developing extreme electromagnetic sensitivity or aversion to electrical fields following close encounters.[398]

For anyone familiar with the broader literature surrounding UAP-related experiences, these reactions are immediately recognizable. They recur with unequivocal consistency across disparate cases. Frequently, they are accompanied by the same eerie sense of mental interference. This is reported as being observed, known, or altered. Within the framework presented here, such effects may be reframed not as psychological aberrations but as the consequence of interacting with a biologically and cognitively divergent life form.

As previously postulated, if *Pelagomorph* is manipulating or navigating quantum fields or processes, the results could extend far beyond visual distortion. Such a being might induce not just spatial anomalies but acute temporal dislocations. Should *Pelagomorph* be capable of achieving quantum coherence, or have evolved to exist partially within a liminal quantum state, they would not be fully collapsed into the classical space-time framework we inhabit. In such a case, they would hover near the edge of physical realization, only intermittently intersecting our consensus reality. They would occupy a zone of conditional being, sometimes present, sometimes phased beyond reach.

This would grant *Pelagomorph* a radically different relationship to locality, duration, and continuity. Humans are naturally tethered to the stable flow of time and spatial permanence. When interaction with *Pelagomorph* occurs, our brains might fail to process the encounter. One helpful analogy is that of vertigo. Housed in the inner ear, the human vestibular system maintains balance by synchronizing internal perception with external motion. When this system falls out of sync with visual input, the result is disorientation, nausea, dizziness, and a glitch in embodied experience.[399]

Now, replace motion with time.

A human near an entity that is partially out of quantum phase might experience a similar mismatch. Not vestibular, but temporal. The result would be a kind of temporal vertigo.

This neurological misalignment could produce memory gaps, perceived "lost time," time dilation, scrambled event sequencing, dissociation, or the disconcerting sensation of jumping across moments, much like skipping stones across a pond.

One way to conceptualize this is to imagine time as a stack of translucent film slides. Each frame represents a discrete moment in a linear sequence. Humans experience time by flipping through these slides one at a time. Sequential moments are revealed in succession. A quantum-entangled life form such as *Pelagomorph* might instead move diagonally through the stack, touching multiple frames at once or out of order. They might intersect briefly with the moment you occupy before veering away again.

Human brains are built to perceive time as a linear progression. Faced with a nonlinear intrusion, a frame that doesn't belong or once did but no longer does, they flounder. The result could be as subtle as a missing second or as shattering as the realization that the order of moments has reversed, split, or doubled back.

At first glance, these suppositions may seem fantastical or outlandish to the reader. They strain against the confines of the systems and processes we expect to encounter in the natural world, at least as we understand it. But that is precisely the point.

The author has taken care throughout this work to elaborate on some of the most extreme biological and evolutionary outcomes currently observed in nature. When first discovered, many of these seemed inexplicable. They bordered on the unbelievable. And yet, they exist. Their reality is not speculative. Their behaviors are not theoretical. They are facts of life on Earth.

At the risk of repetition, we must keep in mind two facts that serve as pillars for everything presented here:

1. The mute vastness of the deep ocean remains unexplored. Its secrets are sealed beneath pressure, darkness, and the passage of time.

2. The duration of uninterrupted evolutionary development in those deep regions decisively exceeds any span of terrestrial observation or scientific scrutiny.

These environments have progressed along a timeline wholly separate from ours. They remain insulated from surface disruptions and are governed by selective pressures that we are only beginning to fathom.

And as we have seen repeatedly, from the strange genome of the octopus to the oxygen-independent metabolism of the myxozoan *Henneguya salminicola*, our expectations have no bearing whatsoever on which genetic solutions nature selects. Evolution answers to its environment, not to our imagination.

That alone should unsettle our sense of finality, opening the door just a little wider to what comes next.

The biological paths that evolution takes may appear inscrutable to us. But we must always return to one essential truth. At its core, evolution is a practical process. No matter how bizarre or idiosyncratic an adaptation may seem, all are shaped by the pressures of survival. None adheres to any allegiance of form or familiarity.

Nature does not abhor aesthetic principles. But neither does it cater to human expectations of symmetry, order, or beauty. Its selections are indifferent to our sensibilities. The genome follows the gradient of advantage, not the guideposts of human design.

And yet, within that raw practicality lies something profound. It is a beauty far deeper than ornament or arrangement. Whether one is rooted in scientific materialism or guided by sacred faith, the outcome remains the same. Life is a *miracle*. Creation is *wonder*. The disagreement lies only in the name of the creator.

At the threshold of grasping a sapient life form that may have evolved in parallel, hidden beneath the waves and beyond our awareness, we must cling to an enigmatic truth. Just as we cannot fully comprehend its nature, neither can it entirely grasp the human condition. We are co-inhabitants of a shared universe, each shaped by different pressures, histories, and vantage points. But wonder remains the common language between us.

This is perhaps the most beautiful and sublime reality of all. Whether you kneel at the altar of reason and scientific method, or bow at the pulpit of faith and divine doctrine, are we not all genuflecting upon the same Earth? We tug the same thread from opposite ends, expecting ancient wisdom or ineffable mystery to greet us. Instead, we find each other. Both faces, ours and theirs, stare back in wonderment. Like strangers glimpsing one another in a mirror, we each expect a reflection and find instead the unfamiliar made flesh.

When a phantom appears in place of the familiar, or the path delivers an old friend where none was expected, we may be disconcerted or even stunned. Humans are creatures of pattern and prediction. We rely on anticipated outcomes to guide us through a world prone to resisting harmony. However, as we have explored throughout these pages, both genuine growth and true awakening are born of disruption. It demands that we loosen our grip on the comforting notions we have clutched against our chests like talismans, afraid to let go.

These tchotchkes of the soul, however cherished, are not lifeboats. They are anchors. It is only when we release them and surrender to what appears to be imminent annihilation that we begin to rise up and over the jagged cliffs of certainty. We climb beyond shadowed crevasses we once feared would consume us, into a sky not yet named. We soar at last toward a dawn still unimagined.

KEATON RYON

# CHAPTER ELEVEN

### Epistemological Scotoma: Abnegation of Axiomatic Fallacies

---

*"We have no need of other worlds. We need mirrors.*
*We don't know what to do with other worlds. A single*
*world, our own, suffices us; but we can't accept it for*
*what it is."*

*Stanisław Lem, Solaris*

---

O ne of the most frequently reported features of *Pelagomorph* encounters is some form of light display or luminescence. These manifestations vary widely. They are usually described as patterned, flashing, or pulsing fluorescence. Color schemes range from monochromatic glows to vibrant, shifting arrays.[400] When drawing parallels between these reports and a hypothetical sapient descendant of cnidarians or cephalopods, this detail is among the easiest to account for. Bioluminescence is commonplace in the lightless, high-pressure zones of the deep ocean. Still, despite the apparent similarities between these groups, caution is warranted. The specifics matter.

Throughout the process of gathering and analyzing the accounts presented here, many cherished assumptions have been reduced to ash upon the hearths of our hearts. Sacred cows have been sacrificed wholesale. Truths once held self-evident have slipped through our fingers like fine dust. These are not failures. They are revelations.

Sturdier truths emerge, born from fracture. Icons once venerated are transformed, resplendent in their refusal to be worshiped. These wonders are hidden not in the heavens but in the overlooked, mud-caked, and marbled world beneath our feet. This planet we inhabit teems with endless variation. It is limitless not only because of what it contains, but also because of what we cannot see. The sensory tools that humans wield are tragically unfit for complete revelation. Even what we *can* perceive may overflow our empirical cups, flooding the mirror with more than just our reflection.

Bioluminescence is not unique to ocean life. Nor is it confined to aquatic environments. As we have previously explored, other branches of life have evolved this chimerical biological adaptation. Among land-dwellers, it's primarily the arthropods who steal the show. They are joined by a handful of annelids, several varieties of fungi, and even a genus of freshwater snails that produce glowing slime.[401] But the undisputed champions of this luminous trait are denizens of the deep. The sheer variety of ocean life capable of glowing is seemingly unbounded.

It is estimated that between 80% and 95% of species in deep-sea environments exhibit some form of bioluminescence. This includes echinoderms, tunicates, crustaceans, clams, sea slugs, and sea snails.[402] But the two reigning dynasties in this glowing pantheon are, unsurprisingly, our old friends the cnidarians and cephalopods. They are not just masters of bioluminescent display. They are also maestros of variety, elegance, and intent. These two endlessly inventive branches of life have evolved this trait in nearly every imaginable shape, pattern, and function. Given their equally unusual genetic toolkits, this comes as no real shock. Their abilities in DNA and RNA editing are unparalleled. And just to seal the point, *over* 97% of cnidarians are known to glow.[403] That's not a quirk. It's a calling card.

As discussed in a previous chapter, the vampire squid is an utterly captivating organism. Neither an authentic squid nor an octopus, it belongs to its own eponymous order, family, genus, and species. Its body is cloaked in photophores, specialized light-emitting organs that blanket nearly its entire form. *Vampyroteuthis infernalis* can produce bursts of light to disorient nearby creatures.[404] This is a defensive adaptation that invites comparison to *Pelagomorph*'s own reported displays. Could this ability to project rhythmic or sequenced bursts of light serve purposes beyond simple distraction? Stroboscopic or pulsed light has been shown to alter human perception of time itself, distorting our sense of temporal duration and flow.[405]

Beyond this, the vampire squid also possesses larger, more complex photophores clustered at the ends of its arm tips and at the base of its distinctive, ear-like fins. When threatened, it draws its dark, cape-like arms up over its body, tucking the glowing appendages into a ring around its head, rendering it momentarily unrecognizable. This visual reconfiguration closely resembles the shifting light arrangements reported in many UAP encounters.

But its cleverest trick of all may be ejecting a sticky cloud of bioluminescent mucus when disturbed. Lacking the traditional ink sac found in other cephalopods, the vampire squid instead employs this alternative defense. This glowing, viscous substance not only disorients but also clings to the predator, marking it in turn.[406] For any hunter, this is a dangerous liability in the abyss. Translated into the *Pelagomorph* framework, such an adaptation could plausibly explain the customarily reported "splitting" or dispersal behavior. Light structures breaking apart, separating, or duplicating are routinely observed. A tactic of evasion, yes, but perhaps also a sleight of light, meant to perplex, camouflage, or even communicate under threat.

Another elaborate adaptation of bioluminescence is demonstrated by the bobtail squid, which is a cephalopod closely related to the cuttlefish. These remarkable creatures have evolved a symbiotic relationship with bioluminescent bacteria, which inhabit a specialized light organ within the squid's mantle. In an innovative feat of coevolution, the squid nourishes these bacteria with a solution of sugars and amino acids. In return, the bacteria emit a glow calibrated to match the light striking the top of the squid's body. That glow effectively erases its silhouette when viewed from below.

This counter-illumination renders the squid nearly invisible to predators as it moves through the water column. The system is spectacularly sophisticated. Within the squid's light organ are filters that balance ambient light, a lens that softens the bacterial luminescence, and a reflector that channels it downward into the dark.[407] All this complexity in a creature that measures just 1 to 8 centimeters (roughly 1/3 to 3 inches) in length.

The elegance of this symbiotic camouflage, which manipulates light to reduce a visible profile from specific angles, is highly relevant to our discussion of *Pelagomorph*. It hints at the myriad and subtle ways an organism could co-opt environmental light, microbial allies, and its own anatomy to sculpt invisibility, engage in mimicry, or manifest dazzling displays of misdirection. If a tiny squid can master such precision, imagine what evolutionary forces might grant to something born of deeper waters and stranger destinies.

Turning to cnidarian bioluminescence, we find no less ingenuity, only a different kind of strangeness. *Atolla wyvillei*, a deep-sea jellyfish sometimes referred to as the "alarm jellyfish," has evolved a dramatic defensive strategy reminiscent of *Vampyroteuthis infernalis*. When threatened, the *Atolla* unleashes a hypnotic sequence of bright blue concentric flashes. This underwater burglar alarm is meant to attract much larger predators. The apparent hope is that its attacker will soon become someone else's prey.[408]

Descriptions of *Atolla* lean into the otherworldly. Terms like "UFO" and "alien" frequently crop up. This is a testament to how far its morphology and behavior drift from terrestrial expectations. In earlier sections, we speculated that *Pelagomorph* might deploy variable-frequency light patterns to affect human perception, particularly our sense of time. In this context, the *Atolla*'s pulsing display stands as one of the most conspicuous natural analogues we have yet found. If *Pelagomorph* could synchronize flashes to match human Alpha or Theta brain waves, the result might be more than radiant. It could subtly unravel our perception of linear time itself.[409]

Another quiet champion of bioluminescence is the mighty coral. Incredibly, coral is believed to have developed this ability as far back as 540 million years ago. This predates any known competitor by a sizable margin. In addition to their ancient lineage, corals possess the mesmerizing ability to glow in slow, undulating waves. Light travels across their bodies like a living aurora. But this brilliance isn't merely ornamental. Corals can absorb harmful wavelengths of light, such as ultraviolet and high-energy blue, and then re-emit them as softer hues of pink, violet, or purple. This transformation serves a vital purpose. It shields the organism from photonic harm.[410] It is bioluminescence not as spectacle, but as buffer, a shimmering defense written into the code of survival itself.

The breadth of these luminous displays is resplendent. It is marvelous not only in form, but in the sheer variety of expression. Yet rather than dwell solely on the taxonomies of bioluminescence in deep-ocean life, it is also compelling to note the uncanny parallels between these biological patterns and those generally reported in UAP encounters. Correlation is not causation. The resemblance alone proves nothing. But when studied side by side, the echoes become difficult to ignore.

Consider the gentle swaying, the rhythmic blinking, the looping, coordinated dances of light, circling, weaving, flaring, vanishing. Such motion could easily describe a cephalopod or cnidarian drifting through the midnight zone, or a strange set of orbs pulsing silently above a desert sky. If we allow ourselves to lean into that suggestive symmetry, to imagine what *Pelagomorph noetica* might be equipped with, the possibilities don't just expand. They ignite.

Given the mechanisms of bioluminescence observed in ocean life, it's reasonable to infer that *Pelagomorph* might employ some of these same methods. One of the most conventional is the enzymatic reaction in which luciferase catalyzes the oxidation of luciferin.[411] But if we introduce cofactors like calcium and magnesium into this equation, the potential complexity and capability expand accordingly.

In some jellyfish, calcium acts as the trigger for light emission. Control over calcium concentrations within the cellular structure allows for precise timing and intensity. If *Pelagomorph* evolved the ability to generate localized calcium pulses, they could selectively illuminate different body regions. This might create rhythmic flashes or even complex visual sequences. Magnesium, a stabilizing agent in enzymatic processes, adds another layer of finesse. It supports ATP stability, buffers pH levels, and helps maintain structural integrity.[412] This could enhance the longevity and reliability of a light display. An internal magnesium reservoir could enable *Pelagomorph* to emit a sustained, hovering glow without draining the energy needed for other functions, such as motion or orientation. This also aligns with UAP sightings. These typically include descriptions of glowing forms that pulse, hover, or remain steady over extended periods.

Taking the idea even further, we might imagine a tunable system of ion-driven luminescence. This may involve the real-time modulation of calcium, magnesium, sodium, potassium, or chloride concentrations. If *Pelagomorph* could selectively flood targeted tissues with ion gradients, the resulting interplay could yield coordinated, shifting hues across full-spectrum displays that pulse with biological intent. These are precisely the kind of light displays reported in countless UAP encounters.

Another possible adaptation that could grant *Pelagomorph* the luminous signature reported in UAP encounters is an array of embedded bioferromagnetic nodules positioned in or near their light-emitting tissues. Such a structure could allow internal or ambient electromagnetic fields to actively shape the refraction or direction of light. We already know that many animals naturally produce magnetite crystals.[413] If evolution favored the placement of magnetite clusters near *Pelagomorph*'s luminous organs, and if these were surrounded by polarized proteins or biophotonic crystals, then EM fields could influence them in meaningful ways.

These interactions might include shifts in orientation, changes in tissue tension, or subtle alterations in refractive geometry. Picture a transparent, flexible sheet stretched across an array of tiny magnetic marbles. As the magnetic field shifts, the marbles spin and nudge, warping the sheet like liquid glass. Shine light through it, and the resulting output shimmers, bends, and dances, even if the light source itself remains steady. Now imagine *Pelagomorph* doing this across their entire body, with bioluminescent cells providing the internal glow and their magnetite-laced skin modulating the light in real time.

The result would be a dynamic skin that pulses and shifts. Their inner luminescence would refract as though it breathed with radiant intent, shimmering like liquid metal or drifting like neon fog.

Starting to sound familiar?

An anatomical design like this might yield a surface alive with shifting light. Radar and FLIR would return little more than noise, with night vision revealing a faint halo dissolving at the edges and leaving ghostly afterimages. To those who witness it, the apparition may appear as a plasma orb, an energy field that seems aware.

Should these magnetite structures be capable of re-aligning with fine control, *Pelagomorph* could conjure visual effects bordering on the magical. They might refract light away from the observer, channel infrared by dimming one side while glowing on another, or craft angular camouflage so precise that they would vanish except from chosen angles.

The military would salivate over technology like this. It's not just stealth. It's Stealth Technology 2.0.

The luciferin-luciferase enzymatic reaction, augmented by cofactors and influenced by magnetite or EM field arrays, would be just one possible method by which *Pelagomorph* might produce the kinds of glittering light displays so often reported. Another possibility is that they evolved a symbiotic relationship with bioluminescent bacteria, similar to the one found in the bobtail squid. This would allow for a lower-intensity, yet long-duration glow, with the potential for tunable brightness and hue.

Or perhaps *Pelagomorph* employs a more active, modular skin system akin to that of cephalopods. If they also wielded chromatophores, iridophores, and leucophores, this would grant them precise control over both the wavelengths of light they emit and those they reflect. Just as cofactors might enhance enzymatic bioluminescence, so too could this modular skin be layered atop or woven into the same system, creating an intricate and responsive display matrix.

Taken together, these systems suggest that *Pelagomorph* may not just glow, but *perform*. To imagine how such a display could function in real time, we might turn again to cephalopods, whose layered dermal structures offer an enthralling biological precedent.

Imagine specialized dermal glands, where the luciferin-luciferase reaction is catalyzed, functioning as light-emitting nodes, each akin to a single LED. We can then begin to imagine the skin of *Pelagomorph* as a kind of living, luminescent screen. These nodes could respond to various inputs, including neural commands, electromagnetic field exposure, environmental cues, or even social signals such as threat or curiosity. The result could be pulses, glows, or steady illumination. This would form the base layer.

Overlaying this would be a modular array of chromatophores, pigment sacs that expand or contract. In *Pelagomorph*, these could act like color gels over stage lights, filtering the wavelengths emitted. They might also modify light through pigment density, chemical tuning, or even polarity shifts that interact with polarized light. This becomes the intermediary layer.

Above that lies a third tier, the iridophores and leucophores. Iridophores are stacked, crystalline, mirror-like cells that are capable of reflecting specific wavelengths, depending on their angle, layer spacing, and neurological control. Leucophores scatter light, creating a neutral "canvas" upon which other effects can play. In *Pelagomorph*, this final layer could amplify internal luminescence, redirect it across the surface, polarize it selectively, or generate complex waveforms that ripple like living plasma.

Collectively, this entire surface would not just shimmer, it would *paint with light*. This effect would not only act upon *Pelagomorph* themselves, but also into the very medium surrounding them.

A stable internal light source, combined with modular control layers, could enable real-time changes in color, pattern, brightness, and polarization. In the air, such displays would register in the human visual cortex as pulsing plasma spheres or shimmering orbs. They would appear as objects that split apart, fade, or rearrange their shape with precision.

If *Pelagomorph* could cross-link this dermal system with electromagnetic field emissions, the effect would scale even further. They could coordinate ultraviolet and infrared pulses with visible light, alter radar or thermal signatures, and essentially *sing* across the entire electromagnetic spectrum, from human-visible wavelengths to machine-detectable bands.

Interestingly, this is the very behavior we associate with UAP phenomena.

It is no wonder that so many witnesses interpret these encounters as mechanical, engineered, or extraterrestrial in origin.

Faced with such controlled, radiant precision, what else could we assume?

When humans see something hovering and maneuvering in ways that defy physics, while radiating light in structured, coordinated patterns, we instinctively assume this is a machine, craft, or technological marvel beyond our current reach. Rarely, in that moment, would it occur to us that what we are witnessing might be a *life form*. Let alone, one that evolved in conditions so unassimilable and extreme that they lie beyond the edges of our conceptual maps.

The evolutionary pressures that shaped *Pelagomorph* were applied over epochs of deep time. These are not only unimaginable but perhaps even unknowable to the human mind. This may be why so many UAP witnesses describe their experience not just as strange, but *unsettling*.[414] Even when the sighting is interpreted as mechanical or manufactured, there remains a lingering impression of something unspeakable, an element that witnesses can't quite name. It is a presence that feels... *other*.

That otherness is *Pelagomorph*'s signature. Not just more advanced, or more "intelligent," but removed entirely from the scaffolding of human experience. They seem to exist, not on the next rung of our evolutionary ladder, but on a different ladder altogether. Their senses, locomotion, modes of perception, and communication are not just beyond our own. They pass *through* them, bending around what we know and illuminating shadows we don't even recognize as blind spots.

Perhaps the most puzzling and cryptic of *Pelagomorph*'s reported behaviors is their ability to execute extreme directional shifts, defy inertia, and even appear to *teleport* from one point in space to another. These are not one-off anomalies. They have been documented repeatedly by both eyewitnesses and sophisticated instrumentation.[415] And yet, despite years of scrutiny, no consensus has formed around how such feats are accomplished.

Once again, this author suggests that a biological explanation may be more plausible than a mechanical one.

One of the primary obstacles facing any living occupant of a high-speed craft, particularly one performing sharp turns or instant accelerations, is g-force. Our standard, of course, is human biological structure and hardiness. Professional fighter pilots, aided by pressurized suits and rigorous training, can endure up to 9 or 10 g. That's about nine times the gravitational force you experience standing on Earth. Positive G-force, as in a roller coaster's downward plunge, pulls blood away from the brain. Negative G-force, like cresting a steep rise, forces blood back toward the head, disorienting or even injuring the pilot. Both are biologically stressful, and both set hard limits on what we consider survivable motion.[416]

Then there's the "Tic Tac" incident.

During the 2004 Nimitz encounter, trained U.S. Navy pilots observed a white, oblong object drop from 28,000 feet to sea level in *under one second*.[417] This is an acceleration that defies not only engineering but also biology. To risk understatement, no known organism, machine, or material structure could survive such a descent. Not without being liquefied.

And yet, something did.

When humans witness something that moves in ways that defy our understanding of physics, we instinctively try to place ourselves, or someone like us, at its controls. Our minds project a pilot, a cockpit, a machine. When the behavior of that "craft" doesn't align with our sense of what is physically possible, we begin to doubt our own perception. This is not irrational. It's protective. The human brain is assiduously invested in the continuity of time, space, and causality. When that framework is disrupted, the psyche wobbles.

If you have ever experienced an earthquake, even a mild one, you have felt a shadow of this effect. The author was living in West Los Angeles during the devastating 1994 Northridge earthquake.[418] Two impressions remain crystal clear. The first is how viscerally disturbing the event was. The second was how quickly most longtime residents seemed to move on, mentally unfazed. For the author, the anxiety lingered. It surfaced as nightmares, obsessive thoughts about aftershocks, and a strange, almost manic vigilance. This was recognized, even then, as an overreaction. And yet it would not subside. Only a multi-day trip to a remote, quiet town in central California restored some semblance of normalcy.

This pervasive unease came from one simple, inescapable truth. *The ground moved*.

For someone who had lived their entire life assuming the Earth's solidity, this violated a silent contract with reality. When something that should not move does just that, it shakes more than the floorboards. It fractures the lens through which we view the world.

Witness accounts of *Pelagomorph* encounters suggest a consistently similar psychological response. Those who report them are not defined by any demographic or psychological profile. They are not predisposed to delusion. Nor do they share a standard belief system, background, or worldview. What they *do* share is an enduring echo, a quiet, sometimes disturbing dissonance that lingers after the experience.[419] It is not only that they saw something strange. It is that what they saw does not align with human expectations of space, time, or causality. Once that expectation is broken, it cannot be easily repaired.

In weighing *Pelagomorph*'s extraordinary capabilities, it is worth remembering that they do not break the laws of physics. They bend them in ways not yet understood. What appears impossible may only be unfamiliar. Their movements suggest an exploitation of locomotion, momentum, and space-time positioning through means that remain beyond present human comprehension.

Consider their proposed anatomy. This is a life form shaped under unrelenting oceanic pressure in a realm devoid of light. They possess no internal skeleton, as humans do. Instead, they likely rely on a muscular hydrostat, a structure of fluid-filled tissue and controlled compression, much like the mechanics of an octopus arm or even the human tongue. In an environment where external force is challenging to generate, precise internal regulation becomes not just useful, but essential.

*Pelagomorph* may function as a kind of biological gyroscope, their body composed of layered hydrogel-based colloids, electromagnetically sensitive contractile fibers, and internal fluid masses or vacuoles. These structures could shift dynamically, redistributing mass in microseconds.

A push of internal weight to one side distorts the outer form, redirecting momentum without relying on external thrust. This could explain the sudden, silent shifts in direction reported in UAP sightings, changes with no deceleration, wake, or propulsion trail.

To us, it might appear as a 90-degree turn at Mach 10. To *Pelagomorph*, it is a simple recalibration, a redirection of mass, or a pivot within their own fluid matrix.

We are wired to think of motion as something that occurs through space, such as a plane in flight or a car on a road. But *Pelagomorph* may operate differently. Think of them instead as a soft-bodied, gyroscopic swimmer in four-dimensional space. Not moving *through* reality, but pivoting *within* it.

No engines. No flame. Just a living compass, turning their center to face a new direction.

Our hypothesis, that *Pelagomorph* is composed of heavy, yet fluidic internal masses capable of shifting or pulsing through their body, suggests a structure not unlike a biological ballast system. Imagine a central "ballast blob" overlaid with a flexible, muscle-like sheath. This outer layer could squeeze or contract in precise patterns, redistributing internal weight and directing motion.

Within this system, we might also envision specialized compartments, or vacuoles, filled with unusual substances. These might include supercritical fluids, pressure-stable colloids, or magnetically responsive particles that can be repositioned upon triggering. Such adaptations could serve not only as mechanisms for movement but also as pressure-management systems. This would allow *Pelagomorph* to survive the punishing depths of the ocean and then ascend rapidly into the atmosphere without catastrophic decompression.

This brings us back to one of the most dramatic observations ever recorded.

As mentioned previously, during the 2004 *USS Nimitz* encounter, pilots and onboard instruments tracked an object dropping from 28,000 feet to sea level in under one second. No fireball. No sonic boom. No heat signature. If a human craft were to attempt such a maneuver, it would experience over 5,000 g. This is an instantly fatal force. Even our fastest jets, at full speed, require several seconds to vanish over the horizon.

Commander David Fravor reported that the object later transitioned from a hover to complete disappearance in the blink of an eye. No shockwave or thermal bloom. No atmospheric disruption at all.[420]

We ask ourselves, how? What is it? What kind of technology could allow this? But these are the wrong questions. The answers we seek lie elsewhere.

This, then, is the heart of the conundrum we face. We must ask not just what *Pelagomorph* is, or how their abilities are possible, but who is behind them.

Humans tend to assume a privileged vantage point, a self-appointed catbird seat from which we define intelligence, self-awareness, and sapience. But this illusion is ours alone. Reality has not consented to it.

Two things become clear when we examine reports, whether from Navy Lt. Graham Bethune near Keflavik, JAL Captain Kenju Terauchi north of Anchorage, or the *USS Nimitz* encounters as recorded by Kevin Day and Commander David Fravor.[421]

1. *Pelagomorph* **is superior to** *Homo sapiens* **in every measurable and observable metric.**

2. **They have not, to date, acted aggressively or with hostility toward us.**

In fact, given their evident command of the physical domain, including an effortless traversal of space, air, and gravity, we can speculate beyond mere passivity. They appear deliberately non-confrontational. In encounter after encounter, *Pelagomorph* exhibits not only restraint, but something more nuanced. We witness a kind of awareness, wisdom, and quiet sagacity that might best be described as parental.

This pairing, ascendancy coupled with benevolence, sits at the very center of this thesis. And yet, when people reflect on these encounters, that thread is too often missed. We focus on the movement, the speed, the shockwave that never came. But movement is not the headline.

The real story is the unwavering precision without aggression. It is the ability to act, and the repeated decision not to. What we are witnessing may be less about dominance and more about communication.

One crucial detail is usually omitted from reports of the Nimitz encounter. The author has reviewed several firsthand accounts from these brave individuals, and the witnesses *do* describe it. The omission lies instead in how others retell the story.

Media outlets and interviewers tend to gravitate toward the most digestible, well-worn fragments. They focus on the visual spectacle, radar anomalies, and Top Gun drama. In doing so, they gloss over what doesn't neatly fit within our most basic expectations. This tendency is not malicious. It's human. We filter reality through the lens of what we are prepared to accept.

But this is precisely why we continue to miss what may be the most critical signal of all.

It's not unlike the way we respond to a cat dropping a dead mouse on our doorstep. From the cat's perspective, this is a gift, an offering, an act of recognition. And what do we do? We recoil in disgust. We misinterpret the gesture, unable to see it for what it is, an overture of connection across a cognitive gulf. One species reaching toward another in the only language it knows.

To the cat, this is love. To us, it's a mess.

Are cats aloof, inexplicable creatures? Or are we too self-involved to understand the sincerity of their gestures? What do we offer in return? Some readers may nod solemnly here, pledging to slay their cat's enemies in kind. Others might shrug and mutter, "It's just a cat."

And this is the question, no, the *wrong* question, we keep asking of *Pelagomorph.* If we can't comprehend the meaning in a feline's rodent gift, how prepared are we to recognize the pure, perhaps unfathomable motivations of a nonhuman intelligence?

The popular phrase "shit happens" contains an important, embedded truth. It *will* happen, whether we are ready or not.

*Pelagomorph* are happening to us.

How *we* react may not matter nearly as much as how they choose to respond to us.

Here's the detail that is overlooked. Not obscured, just missed.

During Combat Air Patrol (CAP) training missions, pilots are assigned a predetermined rendezvous location known as a CAP point. This fallback position serves as a silent contingency. If the mission is interrupted or something unexpected occurs, all aircraft are to regroup at this location. It eliminates the need for midair communication or confusion under pressure.

The location of a CAP point is kept secret. It is not broadcast or discussed over open channels. There would be no reason for any pilot, controller, or observer to reference it publicly.

After Commander Fravor's direct encounter, his aircraft was low on fuel. He broke off and began returning to the *USS Nimitz*. The object they previously pursued had vanished, reportedly accelerating "like it was shot out of a rifle." Multiple similar objects were also reported as present in the area at the time.

Then came a call from the *USS Princeton*. The object was now *at the CAP point*. This location was 60 miles from the pilots' current position.[422]

The odds of this being a coincidence are vanishingly small. Especially given the object's earlier behavior, which included mirroring pilot maneuvers with crystal clear precision.

So, what does this mean?

To the author, this suggests one of two possibilities.

1. A complete, cosmic coincidence.

2. A deliberate action. A reply. A rendezvous.

If we accept the latter, the implications are provocative. The gesture, at once simple, precise, and deliberate, appears to convey several messages simultaneously.

- "I am resisting your aggression without returning it."

- "I understand your intentions, not just operationally, but personally."

- "I hold superior command over the shared physical space we occupy."

- "By arriving at your secret rendezvous point, known only to you, I signal a desire not to oppose your plan, but to align with it."

If this was communication, it was not a message sent in words.

It was understanding, enacted.

Imagine a person arriving in a foreign land where they do not speak the language. The customs are unfamiliar. They drift at the edges, unseen, not understood by those around them. One day, they find children playing a complex physical game. The rules are unspoken but clearly known to the participants. The visitor watches from the edges, quietly decoding. Slowly, the logic of the game begins to reveal itself. Its rhythms. Its goals. Its secret grammar.

Then, at last, the moment comes. The outsider steps in, tentatively at first, then with growing confidence. When the next point is scored, the winning team breaks into a choreographed celebration. It's a gesture they have done many times before. This time, the newcomer joins, mirroring the movement exactly, showing not only comprehension but intent. I see you. I understand. I want to be part of this.

This, perhaps, approximates how *Pelagomorph* may view us.

It is also why the author believes humanity must begin a focused, deliberate effort to bridge the gap while there is still time to do so with grace. The answers we seek will not come in forms we recognize. Nor will they match the questions we think we are asking.

We must be prepared to answer in turn.

*Pelagomorph* may also be seeking understanding of our intentions, our patterns, our story. The opening of this door is not one-way. Ingress and egress occur together. With every question we ask, we offer a reflection of ourselves. We are not just reaching into *Pelagomorph's* world. We are inviting them into ours.

Only when both hands hold the mirror can we begin to understand one another, not as anomalies, but as cohabitants.

Not as intruders, but as kin.

Not as enigmas, but as shared phenomena of the same Earth.

# CHAPTER TWELVE

**Inertial Decoupling: The Heresy of Lift and the Fall of Ordinance**

*"You cannot buy the revolution. You cannot make the revolution. You can only be the revolution. It is in your spirit, or it is nowhere."*

*Ursula K. Le Guin, The Dispossessed*

Throughout this book, we've imagined new possibilities. They are not ideas born of humankind's shared experience, but plucked instead from the obstinate mist of what has yet to be reckoned with.

Concepts like these are usually met with scorn or derision, and rightly so. Fierce debate should always accompany ideas that dare to step beyond convention or challenge the sacrosanct. And what belief could be more sacred to our species than the one that humankind stands at the apogee of thought, evolution, and cosmic relevance?

We have boldly claimed that seat, just as any creature might. The confidence with which we assert this position holds a spectrum of meanings. Some feel justified by intellect. Others feel divinely appointed. Still others may be unsettled by the possibility that something else, older, stranger, or more advanced, might loom above even our loftiest ideals.

What if we are not the observers, but the observed? Are we coolly regarded in the same way we study the daily rhythms of, say, a meerkat colony?

Humans are not only prone to anthropomorphizing the unfamiliar. When confronted by life that is *too* discordant or unorthodox, we may respond with something subtler and more defensive, anthropominimizing. We shrink the other into something more palatable or less threatening. It is a trick of the mind to protect the sanctity of the self.

In this chapter, we will examine the observed behaviors of *Pelagomorph* in greater detail and propose specific biological mechanisms that might underlie them.

The author claims no special access to these shockingly agile and capable beings. But one thing seems clear. They are evolutionarily honed to move as sovereign presences within their world.

We have already touched on some of their more fanciful traits. Now, we delve deeper into how those mechanisms might in fact function. While we cannot claim certainty about the forces at play, we *can* observe behavior. From those observations, we can construct hypotheses. What framework best fits the patterns we see? What theory most closely aligns with the data, however limited, that our senses and instruments report?

As stated before, a biological origin remains the most plausible explanation. Not just for the aerial maneuvers observed, but for *Pelagomorph*'s behavior in total. Their conduct does not suggest cold calculation or utilitarianism. It feels alive, responsive, and purposeful.

Much like past reckonings, when humans were forced to abandon geocentrism or discard miasma theory in favor of germ theory, we now face a new epistemological frontier.[423] We stand witness to events that lie outside our current models of understanding. We cannot yet define them. We struggle even to measure them.

And so, we lay new foundations.

But the question remains. Will we search only inward for the tools and blueprints we need? Or will we dare to imagine that those who have already climbed farther up the evolutionary ladder might offer us a hand, not in dominance, but in the furtherance of understanding?

In the landmark film *Schindler's List*, a pivotal conversation takes place between Oskar Schindler and the SS officer Amon Göth. It's a moment that serves as a microcosm not only of the film's narrative, but of human existence itself.

Göth, overseeing a concentration camp, passes idle hours on his balcony, rifle in hand, calmly murdering Jewish prisoners as they struggle to carry on their daily patterns. In private, Schindler tries to reframe Göth's distorted vision of the world. He hopes to plant a seed of rhetoric, however fragile, in the barren soil of Göth's conscience.

In his drunken haze, Göth insists that control is power, holding up Schindler's sobriety as an example. "You're not a drunk," he says, hinting that Schindler's restraint is a kind of mastery he cannot claim for himself.

But Schindler challenges him.

"Is that why they fear us?" he asks of the prisoners.

Göth responds flatly: "We have the power to fucking kill. That's why they fear us."

Schindler corrects him quietly. The fear comes not from punishment but from its randomness. The Jewish prisoners do not only fear death. They fear the capriciousness of its arrival, the knowledge that it may come for no reason at all.

Ordinarily, if a crime is committed, it is met with punishment. Schindler makes this plain. "That's not power," he says. "That's justice." Consequence follows cause. Retribution earned through action.

Real power, he suggests, takes another form.

"Power," Schindler tells Göth, "is when we have every justification to kill, and we *don't*."

Göth sneers. "You think *that's* power?"

Schindler leans closer, steady in his persistence. He offers Göth a vision framed not by moral light but by ego's shadow.

He speaks of "the Emperor."

He names none in particular. He does not need to. It is the idea alone that matters.

Schindler imagines a guilty man collapsing to his knees before the Emperor, begging for his life. Death is expected. It is deserved. Yet the Emperor pardons him. Not out of duty, not from innocence, but solely because he *can*.

"This worthless man," Schindler says, with the faintest smile, "he *lets him go*."

Göth continues to scoff. "I think you *are* drunk."

Schindler doesn't blink. "That's power, Amon," he says. "*That* is power."

It is a quiet appeal, veiled in flattery. A rhetorical gambit meant to draw Göth toward a higher ideal, one Schindler hopes might echo in the chambers of his narcissism. It is a plea for transformation, cast not as morality but as dominance transfigured.

But Göth cannot follow.

Later, alone in his room, he faces the mirror and touches his own reflection. "I pardon you." he whispers.

But the words are empty.

The gesture is performative. To pardon, to forgive oneself, demands a reckoning with the past. Göth cannot endure it. The burden of his actions is too great. His soul recoils.

He turns away.

In the end, Göth rejects Schindler's vision of power. It is not that he disputes it, but that he cannot bear the cost of assent. He clings to fear over responsibility, denial over transformation.[424]

Even when the door stands open, and the better room lies before him, he will not cross the threshold, perhaps convinced he is either unworthy or unable to occupy it.

The author's interpretation of this scene is not the only one. It's possible that Göth does not struggle, but is instead without conscience, vacant instead of tormented. For our purposes, what matters is not the distinction but what the moment reveals.

The point is not Göth's absence of integrity, but his refusal to face himself. To step toward truth would mean setting that self aside and granting a pardon he had earned through reckoning. Confronted with the door to genuine power, he recoils, clinging instead to control.

The analogy may seem exaggerated, even extreme. Yet if we examine humanity's historic relationship with the natural world, the parallels become unmistakable.

Are we in fact *so* different from a prison guard, drunk on his own imagined dominion? Do we not liquidate and erase at will, often not out of necessity, but convenience or passing whim? Even when we justify our destruction in the name of progress, do we ever truly account for the cost?

We imagine ourselves as the stewards of Earth, the authors of its future. But this is an illusion. We are *of* the world, not above it. We exert no authentic control. Our behavior betrays this. It is marked by volatility, rash reactions, and an inability to align with the long arc of consequence.

Are our contributions truly for the world's benefit, or only our own? When we defile the environment, do we only damage the world, or do we also harm ourselves?

What is the real cost of the progress we claim as our triumph? And who, ultimately, pays the price?

We may indulge in fantasies of authority, believing we are the ones steering the course of fate. But fate is indifferent to our imagination. The outcomes are imminent, regardless. The only thing we have a say in is whether we meet them with grace or delusion.

The extinction events that have echoed throughout Earth's history make one thing abundantly clear. We are not in control.

We are passengers, nothing more.

We float as fellow travelers aboard the ark, drifting alongside our neighbors. Though we may busy ourselves with routines or illusions of purpose, we all await the same unknown horizon. Still, humans fancy ourselves captains of the vessel, despite having no knowledge of where the wheelhouse lies, or whether it exists at all. There is comfort in this delusion, a strange reassurance in pretending to steer. Let's abandon this hollow pantomime and instead take hold of something real.

Look into the mirror, not with judgment, but with courage. We can pardon ourselves for the deception, the grasping, the fantasy of control. And then, let us turn away from that reflection. What might we see, once the scales fall from our eyes? What becomes visible when we finally stop mistaking our image for the world itself?

That is where we now turn.

To understand *Pelagomorph* is to confront an entirely new vision. Not just of life as we know it, but of life as it *might* be. Their observed behaviors and capabilities defy every model we've built from the narrow lens of human experience. But they do not defy nature. In fact, they echo what nature has been quietly demonstrating all along.

Together, we have explored how quantum phenomena are not confined to particle physics or theoretical models. They are embedded in the fabric of biology itself. Across the living world, quantum effects shape the essential functions of life. These include photosynthesis, enzyme catalysis, avian navigation, olfaction, and even DNA repair.[425]

Photosynthesis is a compelling example. This process, by which green plants convert sunlight into energy, is nothing short of miraculous.  When photons strike plant cells, they excite the molecules within them into action. But instead of marching down a single predetermined pathway, these molecules do something unbelievable.

They explore *all possible paths at once.*

This is called quantum superposition. It is a perplexing but real process of nature. In this state, a single particle exists in *multiple places simultaneously*, until one path is ultimately "chosen" and collapsed into. This is not a metaphor. It's physics.[426]

Imagine you are standing before 100,000 identical closed doors. You're starving, and behind one of them is a piping hot pan of your favorite food. If you had to open each door one at a time, how long would it take? Is your meal behind door 999 or door 99,999? There's no way to tell.

What if, using the same energy required to open just *one* door, you could open *all* of them *simultaneously*?

You would instantly know exactly where that lasagna was hiding, and you'd be eating it before someone else even turned the first knob.

This is what happens during photosynthesis. It is not a random walk, nor a brute-force calculation. It is elegant, efficient, and quantum.

If such behavior exists within a humble leaf, what might be possible inside a lifeform shaped not by sunlight and soil, but by darkness, pressure, and time beyond reckoning?

Once we recognize that quantum phenomena such as superposition, tunneling, and entanglement are not just rare exceptions but foundational tools in nature's biological toolkit, we can begin to shift our gaze away from the mirror of self-regard and out into the vast unknown that has always been there, humming quietly beyond our notice. The universe is not waiting to be discovered by us. It already *is*, and always has been. It is entirely unbothered by whether we happen to be looking or not.

When we watch *Pelagomorph* perform acts that seem to defy inertia, such as sudden accelerations, directionless motion, and vanishing without warning, we are not witnessing magic. We are observing the exploitation of physical laws we have only just begun to glimpse. These are not violations of reality. They are fluencies within it.

*Pelagomorph* no more marvels at these abilities than we marvel at riding a bicycle or jogging down the street. What appears miraculous to us is muscle memory to them, refined across a timeline we can scarcely fathom.

The current best estimate places the emergence of *Homo sapiens* around 200,000 years ago.[427] If *Pelagomorph noetica* has been evolving for 500 million years, then they predate us by a factor of over 2,500. That's not a head start. It's a separate epoch. They hail from a different era of intelligence entirely.

To put this in perspective, imagine a being that began its evolutionary journey 80 years ago, attempting to comprehend our world, our language, our aircraft, our satellites, or our wars. Now reverse the scale. We are the infants here.

And so, we must constantly remind ourselves. What *Pelagomorph* do, they do. It is not an illusion, nor a trick of instrumentation or human perception. The truth is even stranger than we realize. Because while we struggle to describe what we *can* observe, *Pelagomorph* is also operating in ways we cannot.

Perhaps the question isn't what *can Pelagomorph* do?

But rather,

What *can't* they?

That list, one suspects, may be far shorter.

When envisioning how *Pelagomorph* might generate the complex, patterned light displays so regularly reported in "UAP" encounters, we have already hypothesized the existence of an evolved array of bioferromagnetic nodules, or "bioquantum mesh." This kind of dynamic mesh, when interfaced with an internal source of bioluminescence, could serve as an exquisite and flexible medium for manipulating light. Through this system, *Pelagomorph* might refract, polarize, and animate luminous displays with superb precision. These displays could appear to us as pulsating plasma, strobing orbs, or intelligent light shows in the sky.

As we continue to examine other seemingly impossible behaviors, we find that *Pelagomorph* is not violating the principles of biology. They are advancing them to a degree made possible by uninterrupted evolutionary refinement in one of the most extreme and demanding habitats on Earth. This is not biology broken. This is biology elevated, sculpted by deep time and the selective pressures of the hadopelagic world.

One of the most challenging behaviors to explain in conventional terms is *Pelagomorph*'s repeated, fluid transition between ocean and sky. They display no struggle against gravity, no signs of decompression trauma, and perhaps most incredible of all, no splash. They enter and exit the ocean as if the boundary between water and air does not concern them in the slightest. There is no visible disruption of the medium. No plume, wake, or sonic consequence is reported.

And yet, they move at velocities that should demand such effects.

How does a biological organism, evolved in the crushing pressures of the abyss, transition from ocean to atmosphere without disintegrating, much less pausing for breath?

The answer may lie in the spectacular architecture of that same bioquantum mesh.

Although this concept may seem discordant at first, it is increasingly apparent that these behaviors can only be explained through some fusion of biological structure and quantum manipulation. The specifics remain speculative, but multiple plausible avenues begin to emerge.

Let's explore them.

When witnesses describe the movements associated with these phenomena, their reports seldom diverge in the key details. The behaviors consistently share a familiar set of baffling traits. *Pelagomorph* appears entirely unbound by the usual limitations of local space-time or the material properties of either air or water. Just as plants exploit quantum superposition during photosynthesis, exploring all potential molecular paths at once before collapsing into the optimal one, *Pelagomorph* may navigate space-time by delaminating from its classical layers. In simpler terms, they peel away from the ordinary physical plane to access alternate routes through space and time.[428]

To human perception, this seems impossible. It contradicts everything that we think we understand about motion and physical resistance. But as we've repeatedly noted, *Pelagomorph*'s observed behaviors leave little room for traditional explanations. Some other mechanism must be at work.

While we can't identify the precise biological mechanisms that enable these behaviors, we can begin to sketch a working model. One such possibility is a colloidal gel matrix, a thick, jelly-like substance threaded with a lattice of bioferromagnetic particles. This forms what we are calling a "bioquantum mesh." Picture glitter suspended inside a mold of Jell-O. Now imagine that each shimmering flake is a metallic nanocrystal, coated in specialized proteins that respond to electromagnetic fields. These particles can flex, shift charge, or reorient in response to external stimuli. They would effectively behave as a field of nano-antennae.

Together, these microstructures would act as a sensory-processing layer, a soft-bodied quantum interface tuned to the environment. Meanwhile, larger, more densely arranged magnetic nodules, functioning as control centers, could be distributed across *Pelagomorph*'s body. These nodes might serve to anchor the mesh and coordinate its interaction with both internal systems and external EM fields. When a pulse of electromagnetic energy passes through the organism, these nodules would absorb, redirect, or amplify the signal in real time.

Think of the nodules as powerful radio towers, surrounded by a dynamic forest of antennae, which then transmit and receive across a full-body network.

Any incoming electromagnetic wave, or even an internally generated neural or electrical signal, would trigger shifts in the proteins surrounding these suspended nanoparticles. These protein conformations would, in turn, alter the local energy environment like a series of molecular switches flipping on and off. As particles react, they transmit charge and information throughout the entire lattice, much like electrons firing through a neural network.

This dynamic transmission causes the energy state of the entire gel matrix to fluctuate, becoming more magnetically charged, electrically excited, or even temporarily "slippery" to conventional forces like gravity or light. In effect, the organism could reorganize internal mass at will. This would generate powerful pulses that dampen or amplify external electromagnetic signals, "tensing" the body in preparation for an upcoming maneuver or spatial distortion.

The most critical function of this system would be resonance. Specifically, the ability to resonate with external electromagnetic or gravitational fields.

This is where things start to get *very* interesting.

As humans move through space, whether by walking, running, or flying, we constantly interact with gravity. It tugs us downward, steadfastly and invisibly. Newton's first law states that objects resist changes to their motion. This is inertia, which is defined as mass pushing back against any attempt to start, stop, or shift direction. It takes force to overcome that resistance.

We are so perfectly acclimated to this principle that we barely notice it. It is the invisible scaffolding of our reality, so familiar that we scarcely ever question its grip. Physical objects are anchored to the geometry of space-time itself. Movement through that field encounters friction, not exclusively from air or surface, but also from space-time's own rigidity. Birds overcome this resistance with their wings. Planes use engines, thrust, and lift. Every motion is a negotiation with inertia.[429]

So then, what is *Pelagomorph* doing differently?

They are resonating.

Rather than brute-forcing their way through space-time, *Pelagomorph* synchronizes with it. Their internal gel-lattice system brings the organism into resonance with the local field, matching their oscillations to the frequency of surrounding gravity or electromagnetic gradients. When two systems resonate, or vibrate at the same frequency, their interaction becomes not only smoother, but almost transparent.[430] Think of a surfer catching a wave. Once their timing is right, the wave no longer fights them. It carries them.

This is how *Pelagomorph* moves. They are not resisting inertia. They are realigning *with* it. The distribution of charge and mass within their body harmonizes with surrounding field vectors, allowing them to pivot, glide, and redirect momentum with seemingly impossible precision. For us, gravity is an anchor. For *Pelagomorph*, it is a current.

And they are expert swimmers.

In addition to resonance with external fields, *Pelagomorph*'s internal colloidal gel matrix would likely exist in a state of quantum coherence. This means that the wave functions of their constituent particles, whether electrons, ions, or atoms, maintain a fixed phase relationship with one another. Rather than behaving as isolated agents, each particle becomes part of a unified wave, its behavior influencing and being influenced by the entire system. The result is a coordinated, quantum-level choreography.

This coherence allows for serendipitous phenomena, including superposition, entanglement, and nonlocal energy transfer. In classical physics, particles behave chaotically, bouncing and colliding like marbles in a box. But in a coherent quantum state, they move together, like dancers locked in rhythm, waves weaving through one another in perfect timing.[431]

If this hypothesis holds, *Pelagomorph*'s gel-lattice may support coherent electron or ion oscillations, phase-locking across the entire body, and synchronized energy distribution that transcends mere mechanical motion. Internal energy would not push or pull in discrete chunks. Instead, it would ripple fluidly through the whole structure. This unlocks an intriguing biological advantage. They would be capable of exploring multiple states of motion *simultaneously* and then collapsing into the optimal one, without the friction, delay, or constraint imposed by Newtonian rules.

In effect, *Pelagomorph*'s body becomes a quantum computation engine for movement.

In the same way that photons in photosynthesis navigate all possible energy paths before collapsing into the most efficient route, *Pelagomorph* may quantum-navigate motion itself, evaluating countless trajectories in parallel, then selecting the ideal one in real time. From our classical perspective, this presents as impossible feats. These include dropping from high altitude without losing acceleration, halting instantly from hypersonic speeds, vanishing and reappearing across great distances, or executing 90-degree turns with zero inertia.

None of these is impossible. They are only hidden from the instruments we use to measure.

These concepts are, quite naturally, difficult for human minds to fully grasp. Our cognition is shaped by life inside classical, local space-time.

This is a framework in which things move linearly, causality is paramount, and all forces must exert a push or pull through resistance. By contrast, quantum behavior exists in a realm that seems to *defy* these rules, not because it breaks them, but because it follows a different set.

Still, as emphasized throughout this book, human experience of the universe is not the full extent of the universe. We often mistake the edge of our perception for the edge of reality itself. But in truth, that's where reality begins.

To understand the difference, let's use a mental model, one rooted in texture and behavior.

Imagine human beings as globs of molasses, dotted along an elastic rubber sheet. This sheet represents local space-time. We stick to it. We move *along* it, but only by struggling against the sticky resistance of our molasses-like bodies. Every step requires force. Every change in motion pushes against friction and inertia.

Now imagine *Pelagomorph* as a puddle of magnetic ferrofluid, hovering just above the same sheet. The ferrofluid is *de-localized*. It isn't bound or glued to the surface the way we are. It can briefly touch the sheet when it chooses to interact with the space-time continuum. But it isn't anchored. It can glide across, dip down, or disengage entirely. While we grind forward through viscous resistance, *Pelagomorph* flows *over* the field like a whisper on glass.

When they choose to momentarily "connect" to the sheet, we experience their presence. When they "disconnect," we perceive their absence. But to *Pelagomorph*, this shift is seamless and routine. There is no struggle, no pushback. They move across or within space-time by *re-localizing*, without regard for the friction we define as effort.

This is why their movements seem "impossible." Inside our lived framework, they are. But that doesn't make them unreal. *Pelagomorph* is not breaking physics. They are working with a different toolset. While we claw our way along the sheet, battling resistance with every motion, they skim the surface, utilizing quantum states for elegance and efficiency.

It's not difficult to imagine how these kinds of abilities could have emerged through genetic selection in the exacting burden of the deep ocean, sculpted over protracted spans of evolutionary time. For a being like *Pelagomorph*, energy conservation would be paramount. Just as green plants maximize energy efficiency through quantum superposition during photosynthesis, *Pelagomorph* would optimize their own locomotion, navigating high-pressure, zero-light environments with minimal energetic cost.

As discussed earlier, the colloidal gel matrix, paired with an internal array of bioferromagnetic nodules, forms a workable system. Especially when combined with more traditional biological mechanisms, such as bioluminescence. The result is a dazzling and complex light display that can serve multiple functions simultaneously, including camouflage, communication, and possibly even predator evasion.

These capabilities, though initially difficult to comprehend, begin to make sense when considered within the context of the environment and evolutionary timeline in which *Pelagomorph* likely developed. This environment remains largely unexplored by humans. But that very fact strengthens the argument. We are not conjuring pure fantasy. We are actively observing these behaviors. The mystery lies not in whether such a being exists, but in how it came to be.

It is this author's humble opinion that their ability to harness these quantum properties is not a fluke or an evolutionary side path, but rather a natural outcome of prolonged environmental pressure applied over deep time. If one were to argue that *Pelagomorph* could not have survived *all* mass extinction events, that's fair. But some lifeforms clearly did. They had to. Otherwise, we would not be here to study them.

Currently, scientific consensus recognizes five major extinction events in Earth's history. Though it's reasonable to assume there were more, hidden in the gaps and silences of an imperfect fossil record. The five recognized events are:

- **Late Ordovician** (≈445 million years ago)

- **Late Devonian** (≈372 million years ago)

- **Permian-Triassic** (≈252 million years ago)

- **Triassic-Jurassic** (≈201 million years ago)

- **Cretaceous-Paleogene** (≈66 million years ago)

Sprinkled among these are additional, lesser-known die-offs. There is a growing consensus that we are living through a sixth such event, known as the Holocene Extinction, which is currently unfolding in real time.[432]

If there is any habitat on Earth with the greatest likelihood of surviving every known extinction event, it is the deep ocean. This is the very realm in which we speculate that *Pelagomorph* originated. While much of the deep sea appears to depend on surface ecosystems through a complex food web, we have no conclusive evidence that this applies to all deep-ocean zones or to all organisms within them.[433]

If *Pelagomorph noetica* evolved to operate independently of oxygen, light, or traditional food chains, they would have been exceptionally well-positioned to endure these cataclysmic events. More than that, the disruptions that devastated surface life may not have endangered *Pelagomorph* at all. On the contrary, they may have acted as evolutionary pressure valves, nudging their development in sensational directions and ultimately shaping the jaw-dropping form we now observe.

With this biological model in mind, it becomes increasingly plausible that the visual signature we observe is not *Pelagomorph*'s true form. In fact, their full anatomy may be intrinsically unknowable to us, expressed only in bio-quantum or higher-dimensional states beyond the limits of our senses. What we perceive may be nothing more than a cross-sectional illusion, or a fragmentary interface of a qualitatively more complex reality.

Imagine a universe confined to two dimensions, where all beings are flat and can move only along length and width. Now, picture a three-dimensional finger passing through this flatland. To a 2D observer, it would not appear as a finger at all, but as shifting, morphing slices. First, it is a dot, then a widening oval, then a dot again. From within that limited 2D frame, the actual 3D form is beyond comprehension.[434]

So too with *Pelagomorph*. What we see, a glowing orb, an oblong shimmer, a pulsing sphere, may be nothing more than the visible echo of its intersection with our dimensional frame. These shapes could arise from biophotonic surface effects, a kind of shielded nimbus or plasma shell generated at the moment of atmospheric interface. Such a shell might function as camouflage, a metabolic boundary layer, or a by-product of coherence-field leakage when their quantum structure brushes against our physical world.

Their actual form may involve folds, lobes, or extradimensional extensions. These components might retract, compress, or phase-shift as they transition into a collapsed state suitable for our limited perception.

When considering the apparent evolutionary longevity of *Pelagomorph*, along with their methods for energy acquisition, consumption, and conservation, several possibilities arise.

Biological evolution is neither linear nor predictable in its unfolding. It branches, pauses, doubles back, and sometimes terminates entirely. While one lineage advances steadily toward complexity, another may plateau or collapse. Within this fluctuating terrain, convergent evolution emerges. This is the process by which organisms that are distinctly different acquire similar traits, such as body shape, locomotion, or functional adaptations, independent of shared ancestry.[435]

Into this milieu, we must also introduce sapience, for it bestows an incomparable evolutionary advantage. Awareness of self, surroundings, and abstract principles allows for a level of decision-making that transcends reflex or instinct. It touches everything. A sapient being makes choices informed by interior multitudes. These include accumulated experience, memory, symbolic thought, and the ability to draw insight from the grandest or most granular properties of the universe. That kind of cognitive structure alters not only behavior but destiny.

No biological need is more fundamental than the acquisition of energy. In earlier chapters, we have explored a multitude of methods that Earth's organisms employ to capture and metabolize energy, whether via sunlight, chemical gradients, predation, or symbiosis.

When it comes to *Pelagomorph*, we cannot yet know the precise nature of their fuel source. But given their biology, environment, and observed capabilities, several enthralling possibilities begin to emerge.

Suppose *Pelagomorph* is indeed capable of achieving field resonance through quantum coherence. In that case, it stands to reason that they may have evolved equally improbable methods of harvesting, storing, and deploying energy.

A being that can partially delocalize from classical space-time, slipping into a state of quantum flux, might not require traditional fuel in the sense we understand it. This could explain how they execute the kinds of motion we have observed without visibly consuming or carrying any discernible energy source. But entering such a state, let alone sustaining it, would still require some form of input. Likewise, any maneuver performed within classical space-time, no matter how exotic, must still draw from a reservoir of power.

Assuming *Pelagomorph* is the product of an unbroken evolutionary lineage stretching back 500 million years, we must also consider that they would have developed a highly efficient, renewable means of sustaining themselves, likely tied to a plentiful and stable source.

What if that source is not locked to the linear timeline we inhabit?

Instead of metabolizing in sequence, as humans do, ingest, digest, then utilize, perhaps *Pelagomorph* draws energy from *multiple* slices of time *simultaneously*. They may access phase-offset states where energy is more abundant, less bound, or more readily convertible. Imagine a tree not bound to a single moment in its life cycle, but one that could reach across decades or centuries, tapping into every sunlit day at once and bypassing the overcast ones. This isn't just poetic. It may represent an evolutionary strategy divorced from the classic food webs that define deep-ocean life as we currently understand it.

Such an ability would also help explain how *Pelagomorph* appears to expend enormous bursts of energy, accelerating, splitting, shifting course, or vanishing, without any observable preparation or input. The energy wouldn't be drawn from elsewhere, but from else*when*.

In this framing, energy isn't just consumed. It's orchestrated across time.

There is much more to be written and surmised about the locomotive behaviors observed in *Pelagomorph*. It is the author's sincere hope that those with expertise in field dynamics, quantum coherence, and non-localized propulsion will pick up this thread and carry it further. These frontiers lie beyond the reach of this writer's grasp, but not beyond the reach of imagination.

Still, even a casual witness, whether firsthand or through the lens of instrumental data, can see plainly that *Pelagomorph* does not move according to the constraints of classical physics. Whatever internal system they employ, whether a colloidal hydrogel matrix paired with a bioferromagnetic array or some yet-unimagined biological engine, their methods are elegant, efficient, and eminently mysterious.

As argued earlier, this is not some bizarre outlier in the story of life on Earth. *Pelagomorph*'s capabilities may seem conceptually vertiginous. But they are likely the natural result of evolutionary pressures exerted over enormous spans of deep time, in an environment radically different from our own. Not an anomaly, an inevitability.

In the next chapter, we will begin to chart that evolutionary arc. We will do so, one step at a time, backward through possibility, toward the origin.

We press on now, together, into stranger possibilities and shimmering dispositions of thought that flutter at the outermost edge of not just perception, but comprehension itself. This is not just worthwhile. It is essential. *Pelagomorph* is not just our neighbor or our evolutionary cousin. In a very real sense, they are our future made manifest. Just as we are their past yet persisting, briefly entangled in the eternal moment we now share.

The human drive to understand ourselves, not only as individuals, but as a species, must extend beyond our inherited history. How does a lifeform reach past its own potential? How does it step forward from the terminal edge of its own becoming? This author contends that it can only happen by casting off the comforting chains of fatalism, by rejecting the idea that we are limited to a predefined shape or endpoint. We must let go of the myth that our current condition is inevitable or final.

Long before we laid the first foundation stone of our modern world, *Pelagomorph* and perhaps others had already left their foundations behind. They soared not just through air or space, but upward through existence itself. The realm they once struggled to master was eventually transcended. We still know next to nothing about that formative environment, only what we can deduce from the observable traits embedded in their present structure.

By reverse-engineering this arc, we can gain insight into what it might mean to shape our own evolutionary trajectory. We will not do so by imitating *Pelagomorph*. This is a task both foolish and futile. Instead, we will accomplish this by accepting the underlying truth they represent.

What we call impossible may, in fact, be only not *yet* possible.

We possess the power to forge what is "yet possible" in the millennia to come. Revolutions of thought, word, and deed are rarely visible in their own time. They crystallize only in hindsight. One cannot plan for such a transformation, nor can one anticipate it. It can only be lived fully in the present moment. Future and past be damned.

To grasp what has not yet arrived, we must first unclench the hand that clings to what is familiar. The siren song of habit and comfort is strong. These are the forces that quietly betray progress, whispering of safety while stalling our evolution. The genuine path forward does not promise clarity or comfort. On the contrary, it often leads us straight into the pitch-black terrain of the utterly unknown.

And yet, from the lightless depths, *Pelagomorph* rises. They are a portal opened not into some distant future, but into our own freedom from the future's past. They reveal what we have long suspected but feared to believe. Our internal genetic structure is not just code. It is a key. A literal key to eternity. Not an eternity that stretches forward in linear time, but one that opens outward, into everywhere, into everywhen.

These are the playgrounds and verdant fields that beckon us, calling us toward the full flowering of our purpose. We are to become. We are becoming. We are beheld, and we are loved.

Now it is our turn to behold.

Only then can the circuit of creation be completed. Only then can fate be fulfilled.

Cast off the blinders that guide you along the path of least resistance, and see clearly what has obscured your vision. Only when you buck the rider, that quiet tyrant who bends your will to its narrow, selfish aims, will you stand exposed before the eternal, immortal flame at the center of your authentic being. Only when that flame consumes the false fuel we have mistaken for life will we finally burn with the full brilliance of our shared destiny.

# CHAPTER THIRTEEN

## Chrysalis Ablaze: The Conflagration of Quiescence

*"The bird fights its way out of the egg. The egg is the world. Who would be born must destroy a world."*

*Hermann Hesse, Demian*

As discussed throughout this book, a key argument for the biological origin of "UAP" or "UFOs" is the likelihood of a long, uninterrupted evolutionary timeline in a stable environment. Several factors point to this conclusion. First, the suite of observed behaviors suggests complex adaptations not easily explained by short-term evolutionary pressures. Second, the hypothesis that *Pelagomorph* is a descendant of a mutualistic, symbiotic merger between cnidarian and cephalopod lineages implies deep evolutionary roots. Cnidarians appear in the fossil record as far back as the Precambrian period, around 580 million years ago. Cephalopods emerged roughly 500 million years ago.[436] As explored in earlier chapters, both groups possess astoundingly strange abilities. Particularly in how they edit and express their genetic code, including not only DNA but also RNA.

To further illustrate this genetic strangeness, consider that some cnidarian mitochondrial genes contain unusual segments known as introns. Introns themselves aren't inherently rare. They're common in many organisms. What *is* noteworthy is their presence in mitochondrial

DNA, which is typically compact and efficient. If we imagine an organism's genome as a recipe book, mitochondrial DNA is analogous to a quick-reference sheet. It is like a stripped-down instruction manual, short, direct, and linear. In this analogy, introns are like scribbled annotations or supplementary footnotes in the middle of a straightforward recipe.

Yet in some cnidarian species, these genetic margin notes aren't just random commentary. One such intron, found in the COI gene, carries a tool called a homing endonuclease. This is essentially a pair of molecular scissors with a built-in GPS. It can locate a specific spot in the DNA, make a precise cut, and allow the intron to insert itself into that location. It's like a rogue hitchhiker sneaking a copy of itself into someone else's recipe.

Even stranger is the ND5 intron, which contains entire genes hidden inside it. This genetic nesting doll effect is incredibly unusual and sets cnidarians apart from most other known life forms.[437] It strongly suggests that their genetic architecture is far more modular, layered, and unconventional than the streamlined systems seen elsewhere in biology.

As noted in previous chapters, cephalopods come equipped with genetic tricks of their own. Not only is their genome unusually jumbled compared to other life forms, but they also possess hundreds of gene clusters that are unique to them. Many of these are linked to their distinctive features, most notably their outlier nervous system structure, which defies easy comparison to any other animal group.

Perhaps even more fascinating is their ability to perform A-to-I RNA editing. This is a process in which adenine is swapped for inosine during RNA transcription. That seemingly subtle shift alters the resulting protein, effectively diversifying the types of proteins that can be produced from a single gene. Cephalopods possess *tens of thousands* of these RNA recoding sites. This is far more than any other known life form, and this editing is primarily concentrated in their nervous systems.[438] It is likely a crucial adaptation that enables rapid, flexible responses to changing environmental pressures.

Just like the modular introns of cnidarians, this is only one ingenious trick in the cephalopod toolkit. But when we begin to imagine these capabilities combined in a single symbiotic life form, the possibilities become spellbinding.

What might a creature look like if it could merge cephalopods' dynamic RNA editing with cnidarians' regenerative abilities? Could such a being not only heal from injuries but also adapt in real-time, editing its own genome to avoid the same trauma in the future? Might it accelerate its internal genetic responses to function more like a modular operating system, uploading new functions or "plug-ins" on demand?

It's certainly within the realm of biological possibility. And yet, this author suspects that *Pelagomorph* harbors even more surprising adaptations. Ones that push far beyond even these already miraculous capabilities.

As discussed in earlier chapters, evolution does not unfold along a clean, linear trajectory. It is not a noble staircase of constant improvement. It is a snarled, reactive dance governed by a single rule. Survival. Genetic adaptations emerge not with purpose or plan, but with efficiency. If that means abandoning a once-useful feature, repurposing an existing one, or even reverting to an earlier mode of expression, so be it. Evolution plays no favorites and keeps no heirlooms.

When viewed through the lens of environmental pressure on individual organisms, this seems inevitable. Genetic architecture should not be imagined as a building with a fixed foundation and neatly ascending floors. It is better understood as a deck of cards, or perhaps a modular puzzle with multiple possible final images. Each is assembled from the same essential components.

The only force known to interrupt this tireless march of adaptation reliably is prolonged, merciless stress. This is embodied by the hammer-blow of predators, the upheaval of changing climates, or any sustained challenge that overwhelms an organism's ability to adapt in time. These stressors are not just obstacles. They are existential threats to the very process of evolution itself.

If we were to visualize the evolutionary progress of a given life form, its trajectory might resemble the jagged rise and fall of a stock market graph. Minor fluctuations ripple within larger waves of advance and retreat. These small gains and losses, taken together, ideally trend toward greater adaptive success. Every species has its own chart. Some are marked by relatively smooth inclines. Others by violent spikes and crashes. But what if, just once in a very long while, an organism found itself in circumstances unusually favorable to uninterrupted ascent?

Such an organism might stumble into an ecological sweet spot. This might be an environment with minimal predators, or one in which they lucked into an early adaptation that vaulted them to the top of the food chain. If paired with a high degree of genetic flexibility, this life form could progress with remarkable steadiness. The capacity to revise, adapt, and optimize their biology in real time would free them from many of the setbacks that plague most species.

Now, envision this scenario across deep time. Imagine this being not just surviving, but thriving. Over hundreds of millions of years, they would be sculpted by environmental pressures yet rarely struck by evolutionary catastrophe. In what direction might their genetic architecture evolve? How strange, refined, and unrecognizably complex might their physiology become?

This author proposes that *Pelagomorph* is just such a being. As wondrous and incomprehensible as their behaviors may appear, their internal systems may be even more alien than we dare imagine. We're talking biochemical poetry here.

How could *Pelagomorph* have repeatedly survived the extinction events that have wiped out innumerable forms of life on Earth? Surely at least one of these global catastrophes must have touched them. But what if it didn't?

Imagine a symbiotic convergence between ancestral cnidarians and cephalopods that fused into a single, unified genetic line very early in Earth's geological history. These two lineages are already strange by modern standards. While we currently have no known example of two complex multicellular organisms merging in this fashion, if *any* branches of life were capable of such a union, it would be these two.

We already know that life has taken novel turns before. Consider mitochondria and chloroplasts. Both descended from independent bacteria that fused with early host cells, forming entirely new life forms through a process known as endosymbiosis. That merger created eukaryotic life itself.[439] While a fusion of higher-order organisms would be a bolder leap, in principle, it is not without precedent. In the unyielding pressure and uncompromising conditions of the deep ocean, where the strange thrives and time flows differently, such a merger is not just possible. It's almost inevitable.

This could have been the pivotal moment. *Pelagomorph*'s lineage might have stepped off the crowded evolutionary highway and onto a quiet, secret path of its own. A path so deep, so insulated, that not only did they survive the cataclysms that razed life across the surface, they may never have noticed them at all.

Once *Pelagomorph* emerged and developed a stable method of reproduction, their unique genetic propensities may have granted them a characteristic degree of environmental, biological, and genetic adaptability. These traits likely smoothed the volatility associated with most evolutionary arcs. Rather than a jagged, reactive pattern like a stock market graph, *Pelagomorph*'s trajectory might resemble a steady upward curve. It would be one marked by resilience, versatility, and long-term refinement.

From our limited human vantage point, we cannot reconstruct the exact sequence of *Pelagomorph*'s evolution. But we can sketch a speculative developmental timeline.

- **~500 million years ago**: A symbiotic genesis occurs. Perhaps a merging of a primitive cephalopod with a traditional cnidarian or a myxozoan-like parasitic ancestor. Foundational traits include neural plasticity, extreme morphological fluidity, and adaptive parasitism.

- **~400 million years ago**: Emergence of an internal colloidal matrix begins. This enables early-phase-responsive behavior and the rudimentary regulation of energy through an internal lattice.

- **~300 million years ago**: Development of magnetoreceptive hubs or bioferromagnetic nodules grants navigation via Earth's magnetic field and the onset of field harmonization abilities.

- **~250 million years ago**: Colloidal matrix begins to stabilize phase coherence. This allows subtle resonance with nonlocal field signatures and the early formation of delocalized cognition.

- **~200 million years ago**: Pressure-resistant physiology evolves, possibly as an extension of vertical migratory feeding behaviors. Nascent trans-medium capabilities allow brief atmospheric breaches, akin to whales surfacing or fish leaping.

- **~150 million years ago**: Biophotonic skin becomes increasingly complex. Chromatophores, iridophores, and luciferin/luciferase systems integrate, allowing reactive camouflage and communicative displays.

- **~65 million years ago**: *Pelagomorph* survives the Cretaceous-Paleogene extinction event through a combination of deep-ocean habitat and reduced reliance on the traditional food web. This inflection point supercharges their energy efficiency and evolutionary independence.

- **~10-1 million years ago**: Advanced aerial adaptation emerges. Internal mass-shifting capabilities and bioelastic locomotion allow hovering and controlled atmospheric movement. Early hominids perhaps witnessed this.

- **~500,000 years ago**: Cognitive breakthrough occurs. *Pelagomorph* crosses the noetic threshold. This stabilizes phase coherence and unlocks non-localized cognition and memory, potentially enabling a collective consciousness. Energy uptake begins to incorporate entangled, nonlinear sources.

This brings us to the present era, where *Pelagomorph* is fully evolved and regularly detected by human-made radar, thermal, and infrared sensor systems. These instruments, constrained by our assumptions, interpret *Pelagomorph*'s behavior as that of a "craft" due to their peculiar movements and demonstrated capabilities.

Looking back over this speculative timeline, we can identify several key inflection points in *Pelagomorph*'s evolutionary journey. The first is the early fusion of cnidarian and cephalopod genetic toolkits. This would be an unprecedented merger within the stable, high-pressure crucible of the deep ocean. In this environment, selective pressure from crushing hydrostatic forces and persistent electromagnetic exposure began to sculpt a life form unlike any other.

The second milestone, and perhaps the most transformative, is the emergence of biological quantum coherence. This is achieved through the integration of an internal colloidal matrix with a bioferromagnetic array. Once this quantum threshold is crossed, the rules change. Traditional biological or environmental limitations no longer constrain evolutionary progress. Quantum non-locality opens the door to energy access across time and space, offering *Pelagomorph* a survival strategy that surpasses even the most adaptable classical organisms.

With this advancement, camouflage ceases to be skin-deep. Instead of blending with their surroundings, *Pelagomorph* may partially or entirely decouple from local space-time to evade detection or danger. This same pathway of coherence and decoupling is likely what allowed them to achieve sapience, or perhaps something even beyond.

Because once you pair a highly supple, regenerating, gene-editing organism with quantum coherence, the results are mind-boggling. What emerges is not just a biologically advanced being. It is a new category of sapient life. One that is fundamentally out of step with the limitations we assume are universal.

Virtually all reports involving "UAP" phenomena, whether fleeting or extended, share one striking commonality. The observer is left with a lasting sense of eeriness, or a lingering impression that the experience was not just strange, but supernatural. Most of us know the famous quote: *"Any sufficiently advanced technology is indistinguishable from magic."* Coined by writer and futurist Sir Arthur C. Clarke, this statement is usually interpreted to imply an artificial or mechanical advancement.[440] But the principle holds regardless of origin. Whether born of silicon or symbiosis, any system operating beyond the bounds of human comprehension will evoke the same astonishment.

One of the recurring themes in this book has been humanity's enduring tendency to center our own sensory bandwidths, mental frameworks, and temporal limitations as if they were the grand total of what existence permits. We have endeavored here to dismantle that narrowing instinct. Our aim is not to leave ourselves adrift, but to allow consciousness to bloom outward. For all our brilliance, humans remain confined to a realm defined by our subjective sensory range and the perceptual framework that it erects. Reality, as we experience it, is filtered through that narrow aperture.

By every observable metric, *Pelagomorph* is not so constrained. And so, the "spookiness" that many witnesses report, those enthralling, reality-warping qualities, is not supernatural at all. It is *supra*-natural. The magical and the mysterious are one and the same, flowing from a biology that is not bound by our physics. This author contends that the root of these effects lies in *Pelagomorph*'s unique genetic expression. Specifically, their ability to maintain quantum coherence within a living system.

If the previously discussed aspects of *Pelagomorph*'s biology are difficult to grasp fully, then the genetic implications are an order of magnitude stranger still. This is especially true when we consider their entanglement with quantum coherence. Should any organism truly surpass this evolutionary watermark, the resulting transformation would not just be a shift in biological function. It would be a radical redefinition of what function even means.

Just as energy sourcing and metabolic processes might be liberated from the linear demands of time and space, so too would traditional genetic mechanisms be unshackled. The familiar forces of natural selection would no longer apply in any recognizable form. Trial, error, mutation, and advantage would no longer be constrained by the limitations of classical physics. In such a state, the process of picking and choosing mutations becomes obsolete. The rules of evolution as we know them dissolve. It boggles the mind.

Genetic mutations are traditionally selected for based on whether they bestow advantages sufficient enough to improve the hardiness or persistence of a given lineage. This is true whether introduced by replication error or environmental pressure. As Charles Darwin famously observed among the Galápagos finches, each bird evolved a beak shape uniquely suited to its local food source. In one region, where cactus fruit hides its seeds behind spiny armor, the finch sports a narrow, pointed beak for extraction. In another, where insects are plucked from cracks and crevices, the beak grows longer and finer, functioning as an organic probe. And on the ground, where food lies in the open, the beak thickens into a stout instrument, suited to gathering seeds from the surface.[441] This cycle of survival, reproduction, and adaptation unfolds along the forward arrow of time inside classical, Euclidean space.

But if *Pelagomorph* can achieve and sustain quantum coherence, the evolutionary rulebook no longer applies. They might slip out of the local timeline, draw power from non-local space, or vanish into a temporally offset refuge when threatened. The grindstone of mutation and selection would become not only intermittent but also optional. Evolution itself becomes semi-autonomous. No longer driven by random environmental bludgeoning, it is instead shaped by internal informational feedback. The process shifts from blind trial to strategic refinement.

We know that cephalopods can edit their RNA on the fly. They dynamically alter their gene expression without changing the underlying DNA. Cnidarians exhibit their own genomic strangeness. Epigenetic methylation flexibility and reverse differentiation allow them to revert to earlier developmental stages.[442] If *Pelagomorph* represents a symbiotic fusion of these two lineages, they likely enjoy a dynamic chromatin structure that permits flexible expression and robust repair mechanisms.

But what happens when we add quantum coherence into that mix?

The result would not be evolutionary. It would be *revolutionary*. Genetic expression would become non-linear. Instead of simply responding to present conditions, the genome itself could become entangled with probable future states. Rather than following the traditional sequence of mutation → trait → selection, we might instead see something far stranger.

Anticipated condition encountered in non-locality → epigenetic shift in the present → trait adaptation before local pressure is ever applied.

The organism wouldn't wait to adapt. It would pre-adapt.

In such a framework, evolution becomes predictive. Reactive evolution gives way to adaptive anticipation. The genome ceases to be a static archive, slowly shaped by trial and error. It instead becomes a phase-responsive toolkit which is fluid, intelligent, and exquisitely attuned to the shape of what's coming.

Evolution would no longer resemble a simple branching tree or a jagged upward graph of progress. Instead, it might begin to resemble a looping, folding fractal. *Pelagomorph* could develop traits, "forget" them, rediscover them, and mutate their genome based on nonlinear resonance across time. Genetic changes would no longer be dictated solely by present pressures but influenced by echoes from the organism's own past and future states.

This is an important revelation, one worth pausing to absorb fully.

At first glance, the most jaw-dropping aspect might appear to be the ability to tune genetic structure based on *future* events. This would be a breathtaking feat from our tethered position inside linear time. But equally mind-bending would be *Pelagomorph*'s access to *past* time states. These would no longer be fossilized memories. They would be active, responsive, and alive.

Imagine a protein-folding structure that was adapted for an environmental condition 10,000 years ago and then integrated into the genome. As conditions change, that structure is then phased out or overwritten. But it isn't lost. If a future environmental slice rhymes with the one that once produced that protein fold, *Pelagomorph* can selectively reactivate the previously silenced gene. They would do so not through memory alone, but through quantum resonance.

In this sense, genomic change is no longer driven by random mutations or transcription errors. Instead, pre-tested and pre-validated configurations can reemerge. Not because they were recalled in genetic memory, but because their bioenergetic signatures still resonate within the organism's coherent internal lattice.

This is not a case of "going back." This is remembering *the future by resonating with the past*.

The past becomes more than data. It becomes a living option, a functional part of the present-moment solution set. Once a forward-driving arrow, evolution now becomes a looping negotiation between past, present, and future.

Visualize *Pelagomorph*'s DNA as a cathedral filled with tuning forks. Each is suspended in its own alcove across centuries of time. We might then imagine a change in the current environmental frequency causing a fork from 250,000 years ago to vibrate faintly in response. That subtle harmonic awakens the expression of a once-discarded trait that is now perfectly attuned to the needs of the present moment.

Quantum coherence unlocks a temporal depth and resonance to evolution that stretches far beyond the reach of classical biology.

Is it any wonder that encounters with *Pelagomorph* are so often described as *spooky*? These beings are not just different from us. They are differently constituted. They operate with such unparalleled temporal and biological complexity that they appear to exist in another dimension of being altogether.

If this model holds, *Pelagomorph* would be capable of constant, looping genomic refinement. They could achieve in moments what would take a species bound by classical space-time thousands of generations, or more.

Their genome may be populated with dormant or partial sequences. These would not be "junk." They would be temporally context-sensitive. A gene might not be active *now*, but still holds potential relevance across some future alignment of energetic or environmental conditions.

They may even suppress mutation rates altogether. This would not stem from fragility, but rather from the fact that they already possess all the necessary modular flexibility.

In short, *Pelagomorph*'s evolution *itself* may have evolved. It would transcend anything we have ever imagined as possible in the biological world.

In addition to this already flabbergasting framework, *Pelagomorph*'s temporal genetic coherence appears to extend not just backward. It also reaches *forward*, entangled across *future* states of being. They may be capable of accessing data from biological conditions yet to occur in the classical timeline. Imagine a trait that is determined to lower survival odds by 100 or even 1,000 years from now. *Pelagomorph* could preemptively drop it. Conversely, if a future trait would prove advantageous, it might be expressed well before classical pressures would ever produce or select for it. This would function as a kind of temporal genetic radar, constantly scanning for survival-relevant conditions yet to arrive.

To visualize this, imagine three *Pelagomorph*, each floating in their own pool on a different floor of a three-story building. The top floor represents the future, the middle the present, and the bottom the past. Though separated by floors, they remain quantum-entangled, moving in unison. If the future *Pelagomorph* encounters a wall or hazard, that impact reverberates instantly through the entire structure. Warned by resonance, the present *Pelagomorph* adjusts course *before* colliding with the same obstacle.

This concept is as intoxicating as it is disruptive. It obliterates the classical idea that adaptation arises through survival. In this model, evolution is no longer a reactive process. Instead, we see the emergence of quantum-preemptive adaptation, or survival *before* the threat is realized in local space-time.

*Pelagomorph* is no longer playing the evolutionary game. They are running an evolutionary debugger. Armed with a dataset spanning the past, the present, and the yet-to-be, they could rewrite their own genome across dimensions of time.

As it relates to the construction and selective editing of an organism's genome, the unbinding of time's forward flow would unconditionally transform that organism's experience of reality. To exist on the far side of that temporal gulf is to inhabit a kind of alternate dimensionality. This would be a reality space assiduously decoupled from the linear, moment-to-moment existence, within which humans and other time-anchored life forms are constrained.

Yet, when we entertain this possibility, many of the behaviors reported in encounters with *Pelagomorph* become not just less mysterious but potentially predictive.

Consider the now-famous account in which *Pelagomorph* appeared at the CAP (Combat Air Patrol) point during a U.S. military exercise.[419] This event left witnesses and radar technicians equally baffled. If this being is truly capable of accessing information across a non-local, temporally entangled field, then the CAP point would not be merely a target to identify. It would be a known quantity. This coordinate would already be embedded in the future resonance of the system it perceives.

To us, it seemed a display of psychic foreknowledge. To *Pelagomorph*, it may have been as natural and effortless as reaching across a room.

It is challenging, perhaps impossible, to truly comprehend *Pelagomorph*'s reality from their vantage point. However, encounters such as these offer us glimpses. The infamous "Tic Tac" incident, so confounding to observers, may have unfolded quite differently from the other side.[418] To us, it was a high-speed intercept and vanishing act. To *Pelagomorph*, it might have been ancillary contact, or even accidental overlap. From their perspective, time and space are no longer fixed barriers.

*Pelagomorph* may not have perceived Commander Fravor's F/A-18 jet as a "vehicle" in space in any way resembling how humans perceive one. Instead, they might register such an object as a field disruption. It would manifest as a spike in vibration, electromagnetic flux, acoustic signature, or gravitational texture. The jet would not appear as a discrete craft, but rather as a threaded event structure or moving knot in the continuum of non-local space-time. Commander Fravor himself may have registered as a localized consciousness node within this knot of probability. A ripple riding atop the waveform of disruption.

From *Pelagomorph*'s perspective, this might be less like "a pilot in a jet" and more like a scar approaching the fabric of its coherent field.

When Commander Fravor initiated his downward spiral maneuver in an intercept path, *Pelagomorph* mirrored the action with an upward spiraling motion of its own. This may not have been an act of mimicry in the communicative sense. However, that possibility cannot be ruled out. Consider our earlier example of a gorilla in a zoo observing a human and then mimicking its movements. A communication, but of an equivocal and ambiguous kind.

More likely, this was a resonant harmonization. If the F/A-18's motion carved a standing wave through air and gravitational fields, that spiral would have rippled the local space-time sheet like a stone dropped in a pond. *Pelagomorph*'s counter-spiral may have functioned not as opposition but as a waveform inversion. Think acoustic noise-canceling headphones, translated into the geometry of motion.

This was not combat. This was coherence restoration.

When Commander Fravor exited his spiral and vectored directly toward *Pelagomorph*, the harmonic structure may have collapsed. Two resonant waveforms were about to intersect. At that moment, *Pelagomorph* may have decoupled coherence before collision. This would appear to us as zipping across the jet's nose and vanishing. But what we likely witnessed was a phase shift, or a retreat into non-local coherence. *Pelagomorph* rose up and out of alignment with our time-locked reality.

The reemergence over the CAP point may not have been travel at all, but a return to a pre-established resonance coordinate. That coordinate, encoded in the intersecting waveforms of the jets' flight plans, may have been a kind of predictive node that was already identified by *Pelagomorph*'s temporal mapping.

To us, it was an impossible motion. To *Pelagomorph*, it was more like reloading a saved state.

This is difficult, if not nearly impossible, for us to conceptualize from our rooted position within linear time and localized space. We are hardwired to think in terms of cause and effect. We perceive events as sequential, first this, then that. However, for *Pelagomorph*, such a framework would neither be limiting nor defining.

To them, the entire encounter may not have occupied any "time" at all.

The spiral, the intercept, and the CAP rendezvous would not unfold as a sequence, but rather cohere as a single temporal construct. It would be a bundled event-shape apprehended all at once. In this sense, the whole affair may have registered as a kind of topological bulge in the local field. It would be a knot of turbulence that they neatly stepped around.

To us, it was a thrilling, hair-raising pursuit. To *Pelagomorph*, perhaps nothing more than a passing vibration. Briefly interesting, mildly destabilizing, and ultimately beneath engagement.

Once we begin to trace this potential evolutionary path for *Pelagomorph*, and juxtapose it with the reported encounters, it becomes increasingly plausible as an explanation for the puzzling, even "impossible" behaviors we see described. We have explored how achieving quantum coherence would comprehensively transform aspects of their being. These include energy processing, genomic expression, locomotion, and even presence within non-local space-time. But the deeper we unfold this speculative reality, the more its implications stretch the limits of human discernment.

The sheer duration of *Pelagomorph*'s evolution is daunting enough. If the author is correct, their tenure on Earth dwarfs our own. Not just in terms of chronological length, but in the very nature of time they inhabit. The difference is not that one is older than the other. It is that we occupy entirely different stations inside reality itself. A yawning ontological gulf separates us.

We are proposing a life form that began an unbroken evolutionary journey 500 million years ago. If they achieved freedom from the unipolar flow of time somewhere along that path, our traditional notions of linear development would no longer apply. And if *Pelagomorph* has truly entered a phase of quantum coherence, their presence in the universe can no longer be compared to ours in any relatable way.

They are a full symphony orchestra, engaged in an ongoing, multidimensional performance. We are a single pluck of a string, a dropped note, a staccato squeak of a chair leg across the floor.

We exist briefly and audibly, but are soon overlaid, absorbed, and eclipsed by the resonant chords echoing through the concert hall of their reality. To expect *Pelagomorph* to take an abiding interest in us would be like asking the conductor to weave that errant squeak into the musical score.

And yet, they do seem to notice us.

Somehow, perhaps in ways we fail to grasp fully, *Pelagomorph* may regard humankind as a quaint anomaly within their field state. Or, more poignantly, perhaps they recognize in us something that even we do not yet see.

One theme we have returned to throughout this book is the unmistakable correlation between UAP observations and nuclear-related sites.[196] Several such incidents have already been detailed in these pages. This, perhaps more than any other human behavior, seems to draw the focused attention of *Pelagomorph*.

And why wouldn't it?

If *Pelagomorph* possesses an internal colloidal matrix and bioferromagnetic array, such a structure may be acutely sensitive to radiation leakage or emissions. These disturbances could act as a kind of beacon or alarm bell ringing across the layers of their ambient field. To them, a nuclear site might be like a sudden, jarring noise echoing through the quiet fabric of space-time. It would be impossible to ignore. Naturally, one would be compelled to investigate.

Beyond the immediate physical effects, nuclear detonations and even reactors at scale are known to warp the local curvature of space-time minutely.[443] Suppose *Pelagomorph* can tune themselves to gravitational gradients or slipstream through space-time harmonics. In that case, such sites may act as temporary quantum attractors, insertion points, or even interdimensional "surf breaks" into detectable layers of our shared reality. These ripples could provide just the right kind of signal for emergence, observation, or intervention.

But there is another, subtle layer that is perhaps even more compelling. It is the human one.

Nuclear sites do not just generate radiation. They radiate dread. They are repositories of secrecy, existential tension, and the shadow of annihilation. Even without detonation, these sites hum with the potential of global undoing. If *Pelagomorph* is attuned not only to environmental signals but also to psycho-emotional fields, what we might call extinction potential, then these places would glow like psychic flare beacons. They would be "hot zones" in every sense.

And perhaps such sensitivity is not abstract or observational for *Pelagomorph*. Having weathered, survived, and possibly even encoded the trauma of Earth's past extinction cycles, nuclear signatures may resonate like a memory. They are a warning, flaring across the lattice of evolutionary time. These sites might be more than curiosities. They could be scars, like echoes of planetary catastrophe etched into their awareness.

What if their presence isn't just investigatory, but immunological?

What if they are not observers but sentinels, white blood cells responding to a rupture in Earth's equilibrium?

If so, their repeated appearance around our most dangerous technologies may not be incidental.

It may be the clearest message of all.

At this point, we cannot claim to know with certainty which of these possibilities is more probable than the others. What we *do* know is that encounters between humans and *Pelagomorph* continue to occur frequently and with increasing regularity.

But why?

Is *Pelagomorph* being drawn more intently toward our presence? Or are these appearances driven by factors entirely unrelated to us? Are they patterns of emergence spun from their own internal cycle, which are invisible to our perception?

These are questions we will explore in the chapters ahead.

For now, suffice it to say that we have journeyed well beyond the boundaries of what most are willing to entertain when it comes to these phenomena. Perhaps this is understandable. It is far less psychologically disruptive to explain such encounters away as drones, misidentified aircraft, or even secretive, cutting-edge human technologies. The author does not dispute that many UAP sightings can indeed be traced to such origins.

But not all of them.

A growing body of well-documented, inexplicable, and deeply strange encounters remains. They defy all conventional classification.[444] These events demand further inquiry and deserve honest contemplation, however uncomfortable.

When one surveys the labyrinthine web of technological, psychological, conspiratorial, or interstellar explanations, none seem to hold up to scrutiny across the full range of observed behaviors.

None, except the one explored here.

A biological lifeform, originating in the lightless depths of oceanic realms, still untouched by human eyes. One that has evolved across an uninterrupted stretch of hundreds of millions of years. And somewhere along that inconceivable arc, they crossed a threshold, breaching the classical flow of linear, local time into a state of quantum coherence and non-local existence.

This capacity to exploit quantum effects at the level of their own biology has transformed *Pelagomorph* into something beyond our understanding, beyond biology as we define it, and physics as we know it.

And yet, we see them.

We witness this being collapse into our reality. They enter our narrow temporal locality and reveal themselves, briefly, brazenly. Their motions are observable. Their form is detectable. But their behavior breaks all laws that we thought were unbreakable. There is nothing subtle about it. *Pelagomorph* does not appear in esoteric corners or hidden crevices. They appear right in front of us, on radar, on thermal scopes, in our skies and seas.

Whether they regard us with curiosity or indifference, we do not know. But one thing is consistent. Whenever they are encountered, *Pelagomorph* bring with them a feeling that something is *off*. It is a sensation of existential dissonance and a direct defiance of the deep, silent assumptions that form the bedrock of human reality.

We recoil and marvel in equal measure.

We may label them as angels, phantoms, monsters, or gods. They are all of these and none. From their unimaginable vantage, they also regard us as we strain to grasp the true nature of their presence.

There is no empirical evidence to support the claims presented here. But if you feel a quiet resonance echoing inside you, a kind of unspoken truth vibrating beneath your rational mind, then you are not alone. The author feels it too.

Perhaps this is because *Pelagomorph* is not only the most advanced form of life on Earth today, but the most advanced form this planet has *ever* produced.

What is the goal of evolution? Is it purely survival? Then, to what end?

If a single branch of the evolutionary tree is never pruned, never struck down by catastrophe or culled by competition, will it eventually outgrow the tree itself?

We need to ponder this no longer, for *Pelagomorph* is the answer.

The goal of evolution is not solely to dominate an ecosystem or ascend to apex status atop a food web. That is a waystation, not a destination. The true goal of evolution is to *transcend* evolution itself.

Imagine the genome as a kind of chrysalis, or vessel in which the emergent form of life develops, suspended within the flow of time. Through this lens, *Pelagomorph* is not just a fellow creature, but the final expression of life freed from the iterative crawl of adaptation. They are borne from a perfectly isolated biosphere, a hadal womb untouched by disruption, preserved in equilibrium, and nourished across eons. *Pelagomorph* reveals to us our own latent purpose.

To *persist*.

To *endure*.

To never yield, never falter, never self-prune.

We press forward, always forward, up the unyielding slope of evolutionary potential. Even when we cannot see the summit.

For somewhere ahead lies a metamorphosis. This is a fresh genesis, not just for the individual, but for our kind. It is a transmutation as total as a caterpillar to a butterfly, or water to flame. *Pelagomorph* is a living echo of what might be waiting for us, far ahead, yet already flickering at the edges of now.

Our only task is not to cast it away through fear, self-doubt, or, worst of all, by clinging to what is comfortable and inconsequential while letting the extraordinary slip through our grasp.

Perhaps this is *Pelagomorph*'s genuine interest in humanity.

They do not see us only as we are, but as we are to become.

If these prodigious beings are attuned to harmonic echoes across their own cathedrals of quantum resonance, then perhaps our whispers from the future call to them as well. And perhaps their song calls to us now, still unrecognized amid the noise and static of our entanglement with local time and familiar matter.

If you ever see *Pelagomorph*, don't just look up.
Look *everywhere*.
Look *everywhen*.

For that is the direction from which they come to greet us.

They are unidentified no longer. They are neither strictly aerial nor simply phenomena. Together, through these pages, we have named them. They are our forebears and our descendants in the same breath. They have shed the shell that still clings to us, invisible and denied by senses too blunt to perceive their truth.

Do they regard us with tenderness? With love?

Is it not in our nature to cheer for the hatching chick? We instinctively root for the fragile fledgling as it pecks its way into the world.

We do not know why we rejoice, but we do. On some ancient, unexamined level, we feel the rightness of it.

Arise, little one. You have almost made it.

Your shell has done its duty. It shielded you as you grew into the shape of something new.

Welcome to the world we finally share.

Let us show you its confounding beauty.

Let us dance together in the ashes of the shell that once nourished and confined you. We will laugh together beneath skies stitched with stars you no longer fear.

You are born anew into a symphony that has always sung your name.

We have been waiting, hoping, and beckoning across the veil that briefly kept us apart.

Welcome, sister.
Welcome, brother.
You are beheld.
You are loved.

You *are*.

# CHAPTER FOURTEEN

**Everywhen Now: Chronotaxis and the Self Unfolding Code**

*"The Brain is deeper than the sea—*
*For—hold them—Blue to Blue—*
*The one the other will absorb—*
*As Sponges—Buckets—do—"*

*Emily Dickinson, The Brain—is wider than the Sky—*

While the origin of life on Earth remains a source of spirited debate within the scientific community, there are aspects of its development that enjoy broad agreement.[445] Of course, new data may still surface. Science is an ever-fruiting tree. Its branches are heavy with knowledge for those curious, patient, and motivated enough to climb. Among the more widely accepted ideas is that the first animal to venture onto land may have been a member of the euthycarcinoids. They are a now-extinct arthropod lineage that thrived from the Cambrian through the Triassic periods.

These creatures are thought to have spawned in warm tidal pools, a possible strategy to shield their eggs from ocean predators and hasten larval development. They may have employed the biological principle of neoteny, which involves the selective retention of juvenile traits to gain an evolutionary advantage. In this case, the harsher and more variable conditions of tidal zones may have favored

309

earlier sexual maturity, allowing those genes to be passed on more frequently. Euthycarcinoids closely resemble the juvenile stage of another shallow-sea arthropod known as the fuxianhuiid. This has led many researchers to suspect a direct ancestral link between the two.

Euthycarcinoids likely did not spend their entire lives on dry land. Instead, they appear to have shifted between tidal pools and exposed terrain, utilizing these liminal spaces as biological stepping stones.[446] This back-and-forth lifestyle carried blessings and burdens from both land and sea. Perhaps it was nature providing training wheels for the great leap ahead. These early pioneers of terrestrial life are not our central focus here, but they serve as markers. We can view them as evolutionary breadcrumbs that highlight the significance of the late Cambrian Period, approximately 500 million years ago.

Even casual readers of geologic history may recognize the phrase "Cambrian Explosion." Like many terms in science, however, that descriptor is subject to debate. While the exact pace of change remains contested, this era is undeniably marked by a staggering increase in the diversity of life and evolutionary experimentation. Most estimates place this window at 20-30 million years. This span saw a frenzied burst of new body plans, sensory systems, limb structures, and feeding strategies. Crucially, this is also when hard shells and exoskeletons became widespread, leaving fossilized clues that allow modern researchers to chart the evolutionary story in sedimentary layers.[447]

As we transition into a deeper exploration of *Pelagomorph*'s unique potential, it is worth pausing briefly to underscore why the Cambrian is a likely candidate for the beginning of their evolutionary arc. It was during this era that the earliest cephalopods emerged and cnidarians began to flourish.[448] These two branches of the tree of life were already starting to diverge. The author firmly believes that a biosynthetic fusion or symbiotic intertwining of these two forms likely occurred near this point in time. It was, after all, a period defined by radical shifts in morphology and structure. The Cambrian Era was a playground of evolutionary experimentation. Much of it was spurred by the arms race between predators and prey.

In such an environment, it would make perfect sense for *Pelagomorph* to begin developing adaptive camouflage, field sensitivity, and other reactive traits to survive and thrive. This was also the era when bioturbation and oxygenation reached new depths, literally. Bioturbation is the process by which sediment is reworked by organisms that burrow. This helped to unlock previously inaccessible nutrients and resources, which rippled outward to nourish an increasingly complex food web.[449]

It was within the ecological isolation of the hadal zones that this revolutionary symbiotic relationship first took hold and slowly evolved into the organism we are calling *Pelagomorph*. Around the same time that euthycarcinoids were being genetically selected toward a terrestrial future, *Pelagomorph* was beginning its own long, deliberate descent into the most remote and extreme environments on Earth. In many respects, these hadal regions are so distinct from surface or land-based ecosystems that they may as well be an entirely different planet.

It is crucial to note that the behaviors we will explore in this chapter would have required an almost incalculable span of time to emerge. Complex adaptations of this kind do not appear overnight. They would have unfolded through countless successive stages, each nested within vast, uninterrupted epochs. A starting point somewhere around 500 million years ago would have granted *Pelagomorph* the evolutionary runway necessary to achieve a biological altitude unreachable for most life forms. Most evolutionary timelines are littered with detours, dead ends, and catastrophic resets.

Once in a great while, though, a branch endures. This is not by chance. It is instead because the organism achieves the rare equilibrium of a biological state exquisitely attuned to its environment.

A classic example of evolutionary persistence is the shark, which first appears in the fossil record roughly 450 million years ago. Its long-term success speaks to a supremely efficient design. Sharks possess a biological toolkit so well-suited to their environment that only minor adjustments have been required over long arcs of time.[450] But what if we imagine a life form even more perfectly attuned to its domain? One not in the sunlit shallows, but in a hidden world of zero light and inexorable pressure. What kind of longevity might be attained?

Even if one were to dismiss the specific existence of *Pelagomorph*, it must be acknowledged that Earth's deepest zones have remained largely unexplored, both geographically and perceptually. Our aversion to these environments is almost instinctual. The emergence of a life form optimized for such a realm is not just plausible. It is probable.

If we regard the Cambrian Explosion as a watershed moment in the proliferation of biological diversity, then the appearance of a single organism capable of reaching quantum coherence could be seen as a catalytic milestone. It would perhaps even be a driver behind that burst of evolutionary experimentation. It is possible that the genome is not purely a passive blueprint. It may instead function as a lift engine or self-organizing system, striving to increase complexity and seeking equilibrium in the gravitational well of potential itself. If so, then an early head start could serve as a form of evolutionary escape velocity. This would allow certain lineages to outpace the standard pruning of natural selection.

This author asserts that *Pelagomorph* achieved precisely this. They benefited not only from a primordial origin but also from a biological foundation impeccably tuned to their environment. This enabled an uncommonly long, uninterrupted ascent. Over hundreds of millions of years, this unbroken evolutionary climb culminated in a state of quantum coherence. This would result in a fully or partially non-localized existence in space and time.

In the previous chapter, we explored some of the ramifications of *Pelagomorph*'s evolutionary "leveling up." While many aspects of their behavior and biology stretch the limits of our understanding, we can still deduce potential qualities that govern their existence within the plane of reality we call home. We can even cautiously speculate about those facets that appear to transcend comprehension altogether.

To envision a life form untethered from local time and space is to attempt something that defies the very substrate of our own minds. They are built to operate *within* those constraints. And yet, the human mind *can* conceive of such things. This is key. We can imagine realities beyond the one in which we are firmly embedded. That alone tells us something profound. We may not be *entirely* rooted to classical space-time. Something within us, whether through imagination, intuition, or some deeper noetic faculty, reaches beyond.

To appreciate this, we must examine the framework through which we experience the world, our sensory apparatus, and the assumptions it rests upon. Our senses are trusted implicitly. They feel intuitive, seamless, and real. You see your coffee mug on the table. You reach for it and drink. You have done this a thousand times. The pattern holds. Our brains rely on that consistency. We form expectations based on repeated sensory input. If the conditions are stable, we assume the outcome will be stable as well.

This is the groundwork of perception, or what psychology refers to as the Gestalt principles. Symmetry, continuity, similarity, and proximity. These unconscious rules help us make sense of the visual, tactile, and spatial world around us. They are what allow us to navigate daily life with ease.[451] But they are also the reason magic tricks work. Illusions exploit these built-in shortcuts, turning our pattern recognition against us in clever and unanticipated ways.

Those who have observed *Pelagomorph* in motion tend to describe a similar effect. They report a scrambling of their core perceptual expectations. The entity appears to violate the rules of movement, presence, and continuity that we rely on. Just as a magician exploits our assumptions to pull a rabbit from a hat, *Pelagomorph*'s existence appears to operate at the far edges of perception. It may even lie entirely outside the field we take for granted.

Magicians intentionally manipulate the blind spots of our awareness to produce something that feels impossible. Nature is also full of phenomena that fall into the liminal zones of our sensory experience. This is sometimes a matter of chance, sometimes a matter of design. Our brains are evolved for survival rather than objective accuracy. They come equipped with heuristics that help us react quickly, but can also mislead us.[452] One familiar example is pareidolia. This is the tendency to perceive faces in inanimate objects or random patterns.[453] Most of us can recall seeing a "face" in the bark of a tree, a swirl of clouds, or a reflection in a puddle. It is so familiar that we don't tend to question it.

This interplay between sensory limitations and evolved cognitive shortcuts is crucial to consider when encountering anything that exceeds the ordinary parameters of perception. If we are to take the possibility of nonhuman intelligence seriously, we must be willing to question our senses and recognize the influence of our mental conditioning. This is especially true of any intelligence that operates outside our perceptual defaults.

Fortunately, we do have tools to filter the signal from the noise. The two most effective safeguards against perceptual error are: (1) multiple independent witnesses, and (2) simultaneous measurement by instruments alongside eyewitness accounts. Interestingly, the most trustworthy reports from observer groups contain minor inconsistencies. That is not a flaw. It's a feature. When ten people witness a robbery, one may say the suspect's shirt was black, another brown, a third gray. But if they all agree on the key details of the event, that variation *strengthens* the account's credibility. Such minor discrepancies are hallmarks of genuine collective narratives.[454]

Likewise, when radar, infrared, or other instruments independently corroborate events reported by human observers, that convergence becomes compelling evidence of a real occurrence.[455]

The goal here is not just to hammer home the reality of these sightings. It is rather to emphasize that they are both real *and* operating at the fuzzy edges of human perception, if not entirely outside its boundaries. They reside in that shadowy overlap between what we can process and what lies just beyond, where familiar rules begin to fray.

To this day, no coherent framework accounts for why so many of these encounters present details that resist classification or exceed the bounds of conventional physical experience. That is precisely why this book has focused so intensely on the edges of biology, evolution, aerodynamics, ecology, and our lived experience of time and space.

This book has asked the reader to reconsider foundational assumptions. This is not to indulge in fantasy. Instead, our aim is to recognize that humans, like all other life forms, occupy a niche. So does *Pelagomorph*. What's proposed here is a radical expansion of what a niche might entail. Not only its ecological dimensions, but also its perceptual architecture and position within the layered structure of space-time.

*Pelagomorph*'s behaviors suggest a niche that extends in directions we have never thought to chart. Whether we acknowledge such territory is irrelevant to their existence. Nature does not require our beliefs to align with objective reality. And yet, humans possess a select and wonderful capacity. It is perhaps a uniquely evolved one. We imagine beyond what is known. We conceive of realities that have not yet been verified and stretch our minds into spaces that science has yet to reach.

In fact, this is how many, if not most, momentous advancements have emerged throughout human history. First comes a flicker, an intuition, a noetic pull toward something not yet accepted or imagined. That intuitive force is precisely why *Pelagomorph noetica* was chosen as a working taxonomic label. It honors the thread between us, a shared capacity to sense what we cannot yet fully comprehend.

We anticipate them. They, in turn, anticipate us.

As these encounters grow more frequent, they suggest a mounting pressure at the seams. Two sides are being drawn toward contact across a gulf that defies simple analogy. This book attempts to bridge that gap and provide an entry point to a possible arc of connection.

*Pelagomorph*'s sensory world and frame of reference are so altogether dissimilar from ours that they may carry their own "lore" about us, just as we do about them. If we genuinely wish to understand, we must ask a bold and disorienting question. How do they perceive us? Through what senses, in what dimensional bandwidth, and under what relational logic?

If they occupy a form of being we are only beginning to approach, and if their evolutionary path has led to a biological existence within non-local quantum states, then the implications are transformative. Forging a more permanent and reciprocal mode of interaction becomes not just desirable but essential.

Before that can happen, though, we must first reckon with how *Pelagomorph* experiences the world, and what that implies for the future of interspecies understanding. Because if even part of what we've proposed is true, then unprecedented possibilities are about to unfold.

Let's explore them.

At the risk of repetition, encounters with *Pelagomorph* are, as a rule, marked by an unmistakable spookiness, a sense of brushing against states that exceed regular human consciousness. These experiences seem to carry a gravitational weight, pulling thoughts and feelings toward something beyond the range our normal perception allows. The author believes this sense of disturbance, or disorientation, may arise from the downstream effects of *Pelagomorph*'s ability to exist in a coherent quantum state.

The first trait we will examine is what might be called Phase-Selective Perception.

As beings anchored to localized time and space, our senses operate under the rules of classical physics. We are constantly pushing against the current of the present. We experience the world as a sequence. Cause, then effect. Sight, then sound. If you watch someone at the far end of the street bounce a ball, you will see it hit the pavement before you hear the impact. Light moves faster than sound.[456] It's an easy observation. But it's also a reminder. We are embedded in time's forward flow, and everything we perceive is filtered through that stream.

Now imagine a life form not bound by that flow. In a state of quantum coherence, *Pelagomorph* may possess sensory awareness stretching across time and space, much like the relationship between entangled particles. When two particles are entangled in a quantum sense, a change to one instantly affects the other. The distance between them does not matter. From the perspective of classical physics, this appears almost magical.

But it cannot be used for communication, since the outcome of any single measurement is inherently random. We cannot control what collapses. We can only observe that the two results match after the fact.

This is the strange dance between quantum and classical reality. In quantum physics, a particle occupies *all* possible states until observed. Only then does it "collapse" into a specific value, and only from our localized standpoint. That collapse cannot be predicted or directed. It simply happens.[457]

And yet, *Pelagomorph* is not a photon in a lab experiment.

The concept being proposed here is far more radical. It is that *Pelagomorph* is a biological system, evolved, refined, and specifically adapted to exploit coherence itself.

In this state, they may exist simultaneously across multiple phase-locked "frames" of space-time. They would not be bi-locating in the traditional sense, but instead sensing pattern resonance across layered versions of their environment. Each is slightly offset, or vibrating just out of sync with the next. Think of a violin string humming with several harmonic tones at once. Each harmonic is its own "frame," or sliver of space-time experienced in overlapping resonance.

*Pelagomorph* is not "looking ahead" in the way we imagine prediction. They are already *tuned into* what is about to arrive.

This isn't like seeing a storm on the horizon. It's more like already being wet while the rest of you is still standing in sunshine.

It may be entirely possible that *Pelagomorph* can sense the presence of distant objects or oncoming environmental shifts *before* light or sound reaches them. Not because they are breaking physical laws, but because they are not tethered to the same latency as us. This would constitute a kind of physics-based clairvoyance, or precognition without paradox.

If their coherent state enables nonlocal coupling between sensory inputs such as light, electromagnetic disturbances, gravitational ripples, and even biochemical gradients, they could detect phase-skewed resonances before classical signals arrive.

This could explain why *Pelagomorph* often appears to anticipate changes without violating the principles of relativity. They are not outrunning light. They are just not waiting for it to catch up.

The biological foundation for such perception might lie in *Pelagomorph*'s colloidal gel-lattice, coupled to clusters of bioferromagnetic nodules. If tuned across interfering electromagnetic phase domains, this living, field-sensitive array could function much like a biological version of humanity's most sensitive cosmic ear. It would be a sensor for gravitational waves, capable of detecting fluctuations in the fields over a broad temporal envelope.

Our closest analogue is LIGO, which stands for Laser Interferometer Gravitational-Wave Observatory. This astounding scientific marvel utilizes twin facilities in Hanford, Washington, and Livingston, Louisiana, to detect ripples in space-time caused by cataclysmic events, such as the collision of black holes, neutron star mergers, or supernovae. These events distort the very fabric of space, compressing or stretching it by amounts unimaginably small.

LIGO detects this by splitting a laser beam into two identical paths, sending each down a separate arm of the instrument. The beams bounce off mirrors, then return to recombine. When encountering undisturbed space-time, they arrive perfectly in sync and cancel each other out. However, if they pass through a gravitational wave, one arm is minutely shortened, while the other is lengthened. The difference is a mere fraction of a proton's width, which shifts the interference pattern. That shift tells us that *something* has bent reality.[458]

Suppose *Pelagomorph*'s gel-lattice could be tuned similarly. In that case, their bioferromagnetic nodules might act as mirrors or sensors, with the gel-lattice functioning as the beamline through which subtle oscillations pass. The "waves" they could detect might include gravitational flux, electromagnetic phase shifts, localized space-time curvature, or even quantum vacuum fluctuations. These subtle distortions, imperceptible to human senses, may alter the organism's internal calibration, just as LIGO registers shifts in the fabric of space.

In effect, *Pelagomorph* could sense the ripples of approaching motion, massive objects, or energetic events *before* they appear on any instrument we possess. Our ears register pressure waves as sound. Similarly, *Pelagomorph*'s entire body could register distortions inside reality itself.

If LIGO is a giant violin string tuned to hear the deep music of the cosmos, then *Pelagomorph* is a living harp. Every thread of their being would resonate across time, fields, and space. They would not only *hear* the universe. They would dance with it.

When viewed from within the limits of human consciousness, these concepts are admittedly esoteric and unwieldy. Most discussions of quantum states and behavior remain centered on particles, including electrons, photons, and the smallest quanta of matter and energy. Far less attention is given to the possibility of a complex organism existing and moving through reality in such a state.

Perhaps this is the clearest reason *Pelagomorph* remains so elusive. They operate in ways that defy the boundaries of what we think is physically possible. One such possibility is the ability to phase through matter temporarily. This would not be by brute force, but by reducing their effective density and mass. This would allow *Pelagomorph* to pass through solid objects in our local space-time, to appear and vanish without a trace, and to move without generating the telltale atmospheric friction or drag that every terrestrial organism must contend with.

Such an ability could also explain the reports of extreme speed without a sonic boom and motion without a detectable heat signature. From our standpoint, these feats border on science fiction. And yet, that judgment reflects only the limits of our niche, the specific set of constraints that shape our human experience of the universe. What is "real" to us is not necessarily the whole of reality. *Pelagomorph*'s reality may be larger, stranger, and far less bound by the rules we take for granted.

Much like the narrow sliver of the electromagnetic spectrum visible to humans, there is more to matter than meets the eye. It's not quite accurate to say that physical objects are "mostly empty space." What we call empty is, in fact, filled with an electron cloud, a probabilistic region where electrons are most likely to be found.[459]

That said, it *is* true that the nucleus of an atom is vanishingly small. It is roughly 100,000 times smaller than the atom itself.[460] If the atom were the size of a football stadium, the nucleus would be a marble resting at midfield. Everything else is that shimmering, statistical electron field.

The reason you don't sink through your bed when you lie down isn't because the mattress is "solid" in any conventional sense. It's because of electrostatic repulsion, which causes the electrons in your body to repel the electrons in the mattress. Another key factor is the Pauli exclusion principle, which states that no two identical fermions (such as electrons) can occupy the same quantum state. The firmness of the bed, and of the physical world more broadly, is not the result of objects being made of indivisible "stuff." It is due to fields pushing against fields.[461]

It's also worth noting that electrons are not like tiny marbles zipping around a track. They are wavefunctions, or mathematical smears of probability, which are spread out across space until measured.[462] The "empty" space inside matter is alive with quantum fluctuations and virtual particles. It is truly a restless sea beneath the apparent stillness.

From this perspective, our experience of reality is bounded not by what is, but by what our senses can register. We experience heat, light, touch, and pressure only because the fields of our atoms interact with the fields of other atoms. Nothing ever truly "touches" in the classical sense. What we experience is field meeting field.

With these principles in mind, it becomes easier to imagine how *Pelagomorph* might accomplish their seemingly "impossible" feats. If they can enter a state of quantum coherence through their internal gel-lattice, then they could suppress or modulate the field interactions that give rise to the sensation of solidity. By re-tuning their electrostatic boundary, they could make themselves more permeable, less repulsive, or even slip entirely out of phase with the matter they are passing through.

To us, this sounds miraculous, perhaps even magical. But within the framework of quantum coherence, it is entirely plausible. *Pelagomorph* is not breaking physical laws. They are obeying different rules.

Consider a thought experiment. You try to cross a crowded dance floor that is packed with people swaying to the beat. You bump shoulders, get jostled, and squeeze through only with effort. Now imagine you are not a solid body but a drifting cloud of fog. You slide between the dancers without obstruction. This is not because you're pushing harder, but because you've altered your fundamental relationship to the objects around you.

This is the distinction between us and *Pelagomorph*. We inhabit a decoherent state in which we are collapsed into a single fixed outcome. Our particles constantly interact with the environment, locking us into classical behavior. Like the unwavering solidity of the ground beneath our feet, this constancy is so familiar we never question it. It is our comfort zone, and our prison.

*Pelagomorph*, by contrast, remains uncollapsed across time and space. In their coherent state, they occupy a fluid, field-like existence. This is why they can slip into the ocean without a splash, passing seamlessly between media as if the boundary were never there at all.

Taking this concept further, *Pelagomorph* might also modulate their charge distributions by rearranging internal charges at will. This would allow them to attract, repel, or even cancel out external charges. They could alter drag, lift, or resistance as needed for maneuverability. Such control could create a form of electromagnetic stealth, allowing the body to manipulate how surrounding fields interact with it.

If they could also adjust their refractive index, *Pelagomorph* might control how light or other electromagnetic waves pass through or bend around them. In doing so, they could effectively bend light away from an observer, producing a kind of biological invisibility. They might deflect radar signals, scatter infrared detection, or tune how pressure and magnetic forces engage with their bodies in real time.

Like a cuttlefish altering its skin color and texture to merge with its surroundings, *Pelagomorph* could manipulate how the fundamental parameters of reality "bounce off" their surface. They wouldn't just camouflage within the visible spectrum. They could mask themselves across the full range of human sensory and technological detection.

This would explain many of their observed behaviors. These include passing seamlessly from air to water without splash or resistance, skimming over surfaces without contact, vanishing from sight and instrumentation, and gliding at extreme speeds without producing a sonic boom, heat trail, or exhaust signature.

In truth, *Pelagomorph* would not "fly" in the way we imagine. Instead, they would redefine what it means to *exist* within air or space. Delocalizing their mass would make them less of an object and more of a field. Phase shifting would keep them slightly out of sync with our local space-time. Field tuning would allow electromagnetic or gravitational camouflage. In these ways, they could *skim* our dimension, touching it lightly, rather than being bound to it as we are.

The conjecture we are engaging in could begin to feel like self-congratulatory or hyperbolic wish fulfillment from someone who wants this to be true. There may be some truth to that. But if this vision turns out to be inaccurate, it won't be because such things are impossible. It will be because they are being achieved by other means or through methods either already known to humans or not yet discovered.

It's entirely possible that *Pelagomorph* operates within principles of universal design we have not even begun to consider. They may adhere to patterns and laws that sit well outside our current conceptual map. And yet, when all observed factors of their actions and behavior are considered, it feels very much as if they are surfing the lip of a dimensional wave we can just barely see curling toward us.

There is something in their presence that unsettles and compels. It is at once utterly foreign, yet disturbingly familiar. Whether intentional or not, it is as if they brush against our existence in a deliberate way, a beckoning gesture across the divide.

Could their perception of us be limited in ways we cannot easily imagine? Might it restrict their ability to communicate effectively? Perhaps they lack any true analogue for our "classical" senses of sight, taste, smell, hearing, and touch. They may rely on an entirely different sensory suite. If so, how would they go about completing that circuit of connection between species?

Perhaps the same gap that makes communication difficult also grants them an oblique vantage point, one from which they can regard us in ways we cannot return. Could they, from their side, feel a kind of compassion toward us that we fail to recognize?

If *Pelagomorph* does regard us in this way, then surely we must be as intriguing to them as they are to us. What do they make of the distortions we introduce into their own reality field? Do they argue over our "realness" as we do theirs, each staking an existential claim at the boundary of our own perceptual limits?

When we reassure ourselves by declaring that anything outside our ability to perceive is somehow less *real* than what lies plainly before us, we diminish our own purpose. Maybe humans and *Pelagomorph* share the spectacular ability to extend thought and imagination beyond what is known, what is experienced, and even what is possible.

One can love one's homeland yet still revel in visiting another place with unfamiliar customs, unintelligible languages, and entirely different ways of being. In fact, it is the absence from home that can make the return to it sweeter, sharpened by contrast with the journey. Some visitors, of course, never go back.

Consider the euthycarcinoids, who stepped onto dry land for what may have been an evolutionary experiment, and stayed for 500 million years. In much the same way, humanity may someday transition into the reality *Pelagomorph* now calls home. Hominids have been on Earth for only 6-7 million years.[463] That span is but a blink compared to the timelines we contemplate here. Perhaps not every life form reaches such a transition at the same pace.

We cannot know how long *Pelagomorph* has occupied this near-but-distant place, just beyond our own station. But one truth is constant. No one steps through a door without first acknowledging its existence. Perhaps this is the reason for the encounters we are experiencing now. *Pelagomorph* is knocking, not solely to announce the door's location, but to make us aware of their presence. It is a passageway beyond our current reckoning, yet absolutely within our reach.

That door is not the first we've crossed. Life has faced such thresholds before. Just as the earliest creatures once made the formative leap from water to land, retaining within themselves the essence of their oceanic origin even as they adapted to an utterly new realm, so too will we cross over.

This is the truest connection we share with *Pelagomorph*. They carry within their being the markers of a localized evolutionary history rooted in classical time and space. Those remnants may be the magnetic thread that binds us across the quantum divide, drawing us toward one another in ways we have yet to understand fully.

The pull is unmistakable. We feel it in the encounters themselves. It seems they feel the gravity of our presence, enough to reach for us in whatever ways they can. If this is so, then we must continue to reach back, push outward, and expand the limits of our relativistic boundaries until we can fully embrace the expression of being that *Pelagomorph* represents.

We are their past. They are our future. And somewhere in the present, we will meet, before setting off together into whatever lies beyond. This is not a destiny of dominion, but of transcendence. We stand on the cusp of stepping into a fullness we have been chasing longer than memory, longer than history, possibly longer than we have even been *human*.

Unlike the euthycarcinoids, we are not pulled forward by environmental necessity alone. There is a hand, extended in conscious choice, reaching down in an offer of support and security. Will we take it? Can we let ourselves be drawn out of time itself, born again into the timeless, limitless everywhen that waits just beyond the eternal now?

Is *Pelagomorph*'s interest in us mere curiosity? Or, from their vantage in non-local space-time, do they already know we have accepted? Could part of us already exist there? Is it reflected across the barrier, back to us like an echo we hear reverberating in a canyon before we've even opened our mouths?

If so, when the door finally opens, will we find an alien, a ghost, or something far stranger? Maybe the real question is not what *Pelagomorph* is. Instead, it might be whether we will truly be ourselves when that moment comes. Keep pushing forward, friends. We're almost there.

# CHAPTER FIFTEEN

**Phase Descent: Cognition Within the Polytemporal Web**

---

*"All that you touch you change. All that you change changes you. The only lasting truth is change. God is change."*

*Octavia E. Butler, Parable of the Sower*

---

Our journey together to this point has been pulled by hidden currents of information, its depths knotted with unfathomable ideas. Throughout history, countless books have sought to expand human consciousness. They entreat us toward another way of perceiving, a higher order of seeing both world and self. There is nothing wrong with such pursuits. The author of these words has read many of them.

And yet, the mind shaping these sentences remains thoroughly *unenlightened*, rooted in self-doubt, unbelief, cynicism, and a general malaise. It is not a constant state. But it *is* the place to which the psyche naturally drifts in the absence of willful engagement.

Why is it in our nature to limit ourselves before we even begin? Why are we so quick to prune the stem before the bud can bloom? What exactly is it we fear? Is it failure that stills our reach, or the possibility of success?

Even the child who moments ago leapt from a diving board will look up from the water and marvel at how far they soared, in contrast to how near their former perch now appears. As they climb back to that same platform, the hesitation returns, irrational and yet abiding.

On occasion, we may revisit the places of our youth. In the presence of a childhood bedroom, school hallway, or a neighborhood street, we are startled by our own bewilderment. What once loomed so large now feels suddenly, disarmingly small. The bully who threatened life itself. The crush who breathed it into being. Both are now just fellow humans in the world.

Perspective is the fear-giver. When we choose to dwell in only half its light, that choice becomes the poison we consume.

Our place within local space-time binds us to the present moment, allowing us only one face of perception. The other face lingers as a reflection, always dimmer than what stands before us, no matter which side of the event we inhabit.

Recognizing that we cannot trust our eyes and ears is just the beginning of the journey toward human interconnectedness. Our mouths and very hearts betray us daily. Yet we place them on altars of worship, which we tend so lovingly. We flatter our own deceit in a relentless, duplicitous campaign that steals from the soul to settle the debts of the spirit.

Round and round we go, friend. Where does it stop? Are there lessons to be learned, or only serotonin and dopamine surges crashing through our brains like monoamine lightning strikes? They light up one room and blow the fuse in another, reversing themselves a moment later.

This is how we experience the universe around us. We are finely tuned to a world that is chaotic, gorgeous, terrible, sublime, and mystifying. For some of us, it is Earth. For others, the inside of our own skull.

Every human is an irresistibly deadly archipelago of feudal states at war with themselves. Every invitation is a mystery. Every new thought is a universe of roiling possibilities. This is what drives our mad dervish of capriciously slashing at beauty and idolizing the wanton.

This is the prison we inhabit. And we are its architects.

We peer up at *Pelagomorph* from the confines of our cell as they regard us from beyond it. What message rides behind their impossible maneuvers across the skies of our discontent? Why do they appear, time and again, in places tethered to nuclear power and deep water?

We can plainly see that *Pelagomorph* is occupying a time and space far beyond ours. They appear to have complete mastery over the very fabric of the cosmos itself. They manipulate time and space the way we pedal a bicycle.

What is the source of their evident compassion, or lack of what we define as aggression? Does it spring from a physical absence of neurotransmitters on their part? Or is it a maternal instinct more akin to empathy?

Surely their actions are designed, at least in part, to turn our gaze inward. For that is where the answer to the human riddle lies. The key to our hearts is not hidden over the rainbow, nor locked away in some celestial vault. It rests quietly in our own hands.

We must come to terms with the fact that not only are our prison cells self-constructed, but we also have the choice to vacate them when we please. The doors were never locked. So far, we have resisted.

Comfort is the enemy of transformation. Familiarity dulls the blade of progress. To rise to a higher vantage point, we must be willing to risk the loss of momentum that comes with surrendering inertia. No cosmic puzzle master is coming to place the last piece of our greater image and frame us on the wall.

We limit ourselves because our greatest hopes remain too attainable to stoke a proper fire in our imagination. But our aim here is not to scale the mountaintop of existence.

It is to step outside existence entirely.

And there, beyond that threshold, is where *Pelagomorph* waits. Their true face will be known to us only when we have left behind the edifice of local time, space, and identity.

Do they already behold the full spectrum of humankind in our unbound state? This book says yes. *Pelagomorph* is moving now as a being transformed. They manifest in ways we still dismiss as implausible or delusional.

But don't be so quick to dismiss the ones who burn with a fire that defies your comfortable logic. The thing that shakes you today may shake you loose tomorrow.

*Pelagomorph* still has much to teach us about where we are going and who we are.

Let's keep digging. There's so much more to uncover.

As we have touched on before, the behavior of *Pelagomorph* while in a coherent quantum state may be far more complex than even the famously baffling phenomenon of quantum entanglement.

Consider two entangled particles. In their quantum state, they exist in superposition. Each of their potential outcomes is unresolved until measured. Crucially, the particles themselves have no say in when or how they collapse into a classical state. They are subject to the act of observation.[457]

*Pelagomorph* may be fundamentally different.

The theory that underpins this possibility is still highly debated. But it is rich with implications. Physicist Sir Roger Penrose and anesthesiologist Dr. Stuart Hameroff proposed a model known as Orchestrated Objective Reduction (*Orch-OR*). In it, consciousness arises not from computation alone, but from the orchestrated collapse of quantum superposition. This collapse is explicitly tied to the geometry of space-time itself.[464]

Many view this with skepticism, and understandably so. The human mind tends to recoil from frameworks that attempt to define not only the nature of thought, but also the act of recoiling itself.

And yet, there is something about this idea that feels oddly familiar. As if, deep within ourselves, we already recognize it. That we, too, are self-collapsing quantum systems, moment by moment resolving possibility into presence. That what we call "consciousness" is not purely thought arising in the brain, but instead awareness *choosing* its path through the field of potential.

Another theoretical framework that dovetails with this idea is the concept of quantum decision trees. These models in physics explore how coherent systems may evolve toward preferred outcomes, without classical interference. Instead of forcing binary resolution, these systems embrace the inherent uncertainty of quantum superposition. They propose an intrinsically divergent approach to decision-making.

The suggestion here is compelling. The idea is that coherent states aren't simply held in limbo, but may be *leveraged* or steered toward specific paths of outcome. A decision ceases to be a singular moment of collapse and becomes a negotiation among potential outcomes.[465]

Perhaps the most counterintuitive evidence for this way of thinking comes from what is known as the "delayed choice experiment." In this setup, a single photon is sent toward a measuring device. But the decision about how to measure it, as either a particle or as a wave, is deliberately delayed until after the photon has already passed the initial part of the apparatus. And yet, the photon still behaves in a way that matches the postponed choice. It's as if it somehow "knew" what was going to be decided. The effect suggests the choice reached backward in time, determining how the photon behaved.[466]

This result doesn't just bend our classical understanding of cause and effect. It *shatters* it. The implication is profound. The flow of time as we perceive it may not be universal. It could instead be a side effect or an artifact of our local, classical vantage point.

There must be some deeper quality to the universe, one that transcends binary labels like "before" and "after," or "this" and "that." A realm where possibility is bound not to sequence, but to resonance.

Studies like these open the door to remarkable possibilities, ones that may explain many of *Pelagomorph*'s most puzzling observed behaviors. Theoretical work in quantum biology and consciousness suggests that sufficiently tuned biological systems might not merely undergo collapse. They may even influence when and how that collapse occurs.

Rather than being passively subject to quantum resolution, such organisms could actively *choose* from among potential outcomes. *Pelagomorph* could deliberately collapse into the state they want, need, or resonate with, on demand.

Every organism we know must first experience a stimulus and then react to it. But *Pelagomorph* might instead *pre*-act. Riding their probability field, they select the most favorable branch of the multiverse before the classical timeline can lay its claim.

Their internal lattice could function as its own observer state, collapsing reality in response to environmental resonance, internal goals, or local energy conditions.

This would account for reports of disappearance and reappearance. These are not illusions, but literal exit and reentry across alternate probabilistic paths. Their uncanny ability to anticipate motion or evade targeting would not be a matter of reflex, but of *selection*. They would instead choose a timeline where the threat never fully formed.

Their seeming invulnerability to interception or pursuit? Not escape. Avoidance. By *design*.

From our perspective, this may feel miraculous. But from theirs, it's just physics. *Pelagomorph* is naturally operating within the coherent quantum state it calls home.

Extending this line of thought, we can also consider not just action within a coherent quantum state, but thought itself.

Human cognition, as we experience it, is inherently linear. One idea follows another. Even when we consider multiple possibilities, we simulate them sequentially. This is a hard constraint of our classical neural architecture.

But a quantum-coherent organism like *Pelagomorph* would not be so limited.

By using their colloidal gel matrix and bioferromagnetic nodule array to generate a phase-stable internal field, *Pelagomorph* could hold multiple cognitive paths in quantum superposition. In this condition, they would think in parallel across unfolding possibilities, a process akin to quantum multithreading. Numerous outcomes would be explored simultaneously rather than one at a time.

This is the very principle behind quantum supercomputers. Qubits, quantum bits that can exist as both 0 and 1 at once, hold multiple possibilities simultaneously and resolve them into the most effective outcome.[467] Humans, for all our excitement over this technology, may be staring at a living embodiment of it.

If *Pelagomorph* can evaluate possible futures simultaneously, feeling which one resonates best with their internal goals or environmental conditions, then they could collapse into that outcome with purpose and precision.

We experience internal conflict as competing voices, "this path or that one?" *Pelagomorph* might instead run emotional, logical, and intuitive processes in harmony, like a chorus of selves negotiating in parallel.

This may even extend to emotional cognition itself. Imagine feeling both the grief *and* the relief of a future yet to unfold, and using the resonance between them to guide your present actions.

Reflecting on *Pelagomorph*'s observed behavior, this ability would allow them to appear prophetic. This would not be precognition. It would be a navigating of landscapes and probability gradients, selecting from among potential futures before we even sense them forming.

If one path presents danger or disruption, they veer into another, seamlessly and preemptively. This could explain their elusive ability to evade threats and remain just ahead of human detection or response.

To us, such behavior might seem impossibly fast, nonlinear, even divine. But it isn't magic. It's a difference in *cognitive dimensionality*. Their decisions appear contradictory or inscrutable because they are not choosing between yes and no. Instead, they are moving through a current of evolving likelihoods.

Imagine a chess player who, instead of calculating a few moves ahead, simulates *every possible sequence* of moves and countermoves across the entire game.

Or, for the Dungeons & Dragons fans, imagine experiencing every possible dice roll for a decision simultaneously. You wouldn't just guess the outcome. You would see the probability curve and select the optimal branch in real-time.

For *Pelagomorph*, emotion and logic may not be separate forces. They could be entangled streams, playing off one another dynamically across multiple simulated realities.

Where we think in terms of faster or smarter, *Pelagomorph* thinks across and through.

Because of this potentially distributed consciousness, *Pelagomorph* may not only think beyond the linear constraints of our local space-time experience, but also communicate in a quantum sense. In this framework, they cease to be discrete individual beings and become something more comparable to a temporal fungal network.

Consider fungi. What we see as mushrooms sprouting from the ground are not isolated organisms. They are the fruiting bodies of a vast mycelial web buried beneath the soil. The actual organism is the hidden, interconnected substrate. Although invisible to us, it is pervasive and alive.[468] And yet, we label each mushroom as an individual, because our perception tends to favor the visible.

*Pelagomorph* may function in precisely this way. What we witness as separate sightings may be nothing more than localized expressions of a much larger, nonlocal being. Their coherent state would enable a networked intelligence that spans space and even *time itself.*

We might also think of *Pelagomorph* as a kind of quantum cloud consciousness, not unlike a distributed AI. Each node appears independent, but the true awareness flows through the network itself.[469]

The key difference is that *Pelagomorph's* cloud spans both time and space. Their memory isn't linear. It's fractal and iterative. Each new insight ripples both forward and backward through the substrate of its being.

If *Pelagomorph* can communicate through a shared quantum coherent state, then possibilities once dismissed as science fiction begin to align with actual observed behavior.

For instance, if *Pelagomorph* operates as a distributed collective or modular temporal swarm, then when one node is threatened or distressed, all other nodes are immediately aware of it. No delay. No confusion. No need to decode a radio signal. This would be a field-level awareness. It would not be communication in the traditional sense, but a shared condition of being. When multiple *Pelagomorph* are observed in tandem, they often behave in perfect synchronization. They mirror each other's movement, reacting simultaneously and responding as one. These behaviors strongly suggest non-local coherence and shared perception.

However, as enthralling as those dynamics are, even stranger possibilities emerge when we consider what memory might look like for a biologically quantum-coherent life form. One of the most radical possibilities is that *Pelagomorph*'s memory is not stored in neurons at all. Instead, it may be spread across a distributed wavefield, more like a holographic imprint than a localized archive.

Human memory is generally understood to emerge from changes in synaptic activity. While its mechanisms remain debated, its encoding is attributed to the structural remodeling of neural networks that chemically consolidate experience over time.[470] Yet some argue otherwise. Physicists-turned-neuroscientists increasingly suggest that memory may reside in interference patterns, encoded holographically rather than stored in discrete traces.[471]

At first glance, this might seem outlandish. Yet the supporting evidence and the implications merit closer examination. Let's begin with a look at how a hologram is produced.

To create a holographic image, a laser beam is first split into two paths. One beam is directed toward the object of interest while the other is sent off independently as a reference. The beam that hits the object scatters light, which then interacts with the reference beam. Where these two beams meet, they form an interference pattern.

Amazingly, this layered ripple of light and phase differences encodes the entire 3D structure of the object onto a 2D surface.

Even more startlingly, each fragment of the resulting hologram contains the full image. The resolution decreases as the fragment shrinks, but the overall structure remains. This is a resilient, redundant method for storing information. It is nonlinear, non-local, and beautifully suited for robustness in the face of damage or degradation.[472]

It is through a mechanism like this that *Pelagomorph* may encode memory. Not through fixed synaptic connections like ours, but via phase relationships between waves which move through their internal colloidal gel matrix. Instead of photons, they would manipulate biofield ripples, entangled fluctuations, and coherent phase interactions. This would result in interference patterns not just of light, but of *lived experience*.

In this system, the boundaries between past and present, individual and collective, begin to dissolve. Memory is no longer just recalled. It is reconstructed. It is not re-watched, but re-lived.

Much like a cnidarian, damage to one part of *Pelagomorph*'s body might degrade clarity, but not erase the memory entirely. The information would still be embedded throughout the whole system. If memories are shared across a collective quantum field, then perhaps no experience is ever truly lost. One node forgets, another remembers. All remember. Because they are *one*.

Such a distributed system would enable tactical recall during perilous situations and real-time memory synthesis in unfamiliar terrain. Perhaps most powerfully, it would also enable a context-aware, shared-memory prediction engine. This would function as a hall of memory spanning individuals, generations, and even time itself.

Could this also help explain the strange phenomena reported by humans during encounters with *Pelagomorph*? These accounts include vivid emotions, sudden visions, or the eerie sensation of "remembering" experiences that the observer never lived through. Reports of time distortion, temporal stoppage, or even telepathic data transfer have long appeared in close encounter literature. Yet we've struggled to place them within any coherent framework.

What if these effects arise from a brush with coherence?

Perhaps when human consciousness encounters *Pelagomorph*'s shared memory field, it reverberates through our neural architecture like a tuning fork. Their presence ripples with entangled data across the space-time continuum. A mere brush of contact may stir latent patterns, summon visions, or seed thoughts that feel like memories.

Even if only momentary, that proximity could be emotionally and ontologically disorienting.

This underscores just how conclusively different *Pelagomorph* is from us. They may not differentiate between "past" and "future" the way we do. These temporal designations may be no more meaningful to them than "here" and "there" are to us. Just as we casually move from one room to another, *Pelagomorph* may shift between epochs. They might loop their consciousness through a phase-adaptive architecture that weaves memory, prediction, and reflection into a unified model of awareness.

In such a state, they wouldn't just *recall* what was and *anticipate* what might be. They would *inhabit* these realities simultaneously, folding them into their behavior like instinct.

And yet, as arcane as these inner mental architectures may seem, there is more. Because *Pelagomorph* may also turn their coherence outward, sculpting the very forces that shape the physical world around them in ways we cannot yet detect.

We have already noted how *Pelagomorph*'s gel lattice and bioferromagnetic nodules may manipulate bioluminescence in flashy, strategic ways. But this same dynamic system could serve a far more incredible purpose. It may actively *interact* with and *reshape* external fields.

Picture the bioferromagnetic nodules as micro-antennas or adaptive coils that are capable of receiving, modulating, and emitting electromagnetic signals. They might flex or shift their charge in response to field changes, operating like a biological piezoelectric device. In its coherent quantum state, the gel matrix would act as a distributed field amplifier. It would then coordinate and synchronize those modulations throughout the organism.

When coherence is fully achieved, *Pelagomorph*'s internal resonance could couple with external electromagnetic and gravitational fields. In doing so, they might bend the environment itself.

In this state, *Pelagomorph* would function like a living radio tower or a quantum tuning fork. They would harmonize with the substrate of reality itself.

What might such interactions produce?

- **Local gravitational lensing** or **space-time refraction**. This would result in distorted or displaced visual perception. To a human observer, this might appear as light shimmering, a bending of the air, or even a sudden "jump" in location. It would be as if *Pelagomorph* flickered from one place to another.

- **Ionization of the surrounding air.** Triggered by field modulation, this could produce effects such as glowing plasma halos or coronal discharges. These phenomena match countless eyewitness descriptions during so-called UAP encounters.

These would not be traditional light emissions but environmentally induced field effects. Signals refracted and scattered from the environment itself would be bent into strange geometries by *Pelagomorph*'s presence.

They might also explain radar anomalies, communication dropouts, or camera malfunctions. In such cases, it's not that *Pelagomorph* is "jamming" our instruments with intent. It's that their mere presence is resonating and modulating at levels we cannot yet quantify. This would scramble the surrounding signal space.

We repeatedly observe that *Pelagomorph* does not seem to move through the world in any way that humans can reconcile with known physics or technology.

If they are actively bending gravity and tuning local space-time fields like a violinist on the edge of audible reality, then perhaps what we call "impossible motion" is instead the natural byproduct of them existing on *different terms*.

Beyond their ability to manipulate the external environment, *Pelagomorph* may also possess the astounding ability to dynamically alter their own physical structure in real time. Unlike most Earth lifeforms, they may not adhere to a fixed body plan locked in by genetic blueprints. Instead, their morphology could be a fluid expression of intent, updated moment to moment by internal needs or external conditions.

Just as cephalopods can edit their RNA to change protein expression without altering their DNA, *Pelagomorph* could leverage quantum coherence to synchronize adaptive responses across their entire body. These changes in cephalopods are local and reactive. *Pelagomorph* might instead preemptively restructure based on sensed environmental fields, pressure gradients, temperature fluctuations, or approaching stimuli. They would do so before these even fully emerge in classical space-time.

The result would be a body that behaves more like a modular thoughtform than a biological container.

If needed, additional limbs, organs, or tendrils may be temporarily grown and then reabsorbed once they are no longer necessary. *Pelagomorph* could flatten themselves to slip into extremely narrow and otherwise inaccessible areas. Or they could puff up and expand to appear more intimidating. Need to change surface topology for camouflage or aerodynamic adjustment? No problem. This would be extremely convenient to alter their outline. They might even emit decoy body parts or change basic symmetry to confound potential predators or observers.

In this way, *Pelagomorph*'s body would not be inherited. It would be composed, from moment to moment, by a field-responsive and intelligent matrix. Their form is not fixed. It is written in *real time* by the same emergent will that animates them.

Their body is not a limitation. It is an *extension of thought*.

Once again, *Pelagomorph* shatters assumptions we take for granted. Where humans are bound by structure, they are liberated by flow.

Perhaps this explains why so many UAP encounters describe forms that shift, melt, or split. These are phenomena that defy belief precisely because they defy permanence.

While all these potential capabilities are provocative and electrifying to consider, there remains one final, perhaps most unsettling, possibility to entertain.

Each of the concepts explored here can provoke wildly different reactions. These can range from outright disbelief and intense scrutiny to an urge to delve deeper into the hidden details. But the reaction one has is likely not a measure of the idea's credibility. It is instead a barometer for the breadth of one's imagination and the degree to which they are afflicted by ontological myopia.

There is something undeniably disturbing, even terrifying, about these possibilities. They threaten the dogmatic, seldom-questioned schema that humans routinely apply to themselves and their place in the universe. These ideas rattle the architecture of certainty itself.

True debate has become rare. Most novel or challenging ideas today are either met with eager, uncritical assent or dismissed with casual certainty. In between is a chaotic swirl of invective, tribal rejection, and confirmation bias disguised as analysis.

This book is not a manifesto for absolute belief. It is not a doctrine. It is a plea, an *invitation* to entertain the conventionally incomprehensible.

We must begin looking in the places where no one dares to search. We must examine what lives outside the clearly marked boundaries of accepted inquiry. For it is in those blurry, inconvenient spaces that nearly every transformative idea is born.

It doesn't matter what belief system, ideology, or framework you follow. Whether you adopt it intentionally or by default, reality still unfolds. It does not require our permission. It operates with or without us, and certainly without the author's endorsement or the reader's consent.

Reality simply is.

Pushing beyond the limits of our own thinking is one of the few true intellectual rigors available to us as sapient beings on Earth. It is more than a noble endeavor. It is a duty. A sacred imperative. We must not slacken our pace or soften our inquiry as we near the outer boundaries of shared existence.

We must ask the most challenging questions. We must reject vacuums of understanding. Especially those that resist us most vehemently. The pursuit of knowledge is not a solo act. It is a group effort. And debate, *real debate*, is not just welcome. It is essential. Fierce dialogue is the hammer blow we wield against the universe's stubborn refusal to part with its most precious enigmas.

There need not be some immediate or tangible gain from this knowledge. We seek no prize, payoff, or pragmatic advantage. Only the knowing itself. We strive only to understand. To contribute to the great collective ledger. Our endeavor is to push forward toward the next mystery that looms just out of reach.

History is like a fire brigade of discovery, carrying the clear, cleansing light of past insights forward to illuminate the murky corners of enduring mysteries that stubbornly block the way. You are part of that chain, whether you realize it or not.

This may be one of *Pelagomorph*'s most powerful lessons. It is not a fantasy or naïve dream to believe that we are all connected, in thought, spirit, and seeking. This assertion requires no proof to lend an ounce of weight to its claim. It does not need permission. We sense it. We recognize it, even if we cannot explain why.

As we continue to explore the strange and stunning possibilities offered by *Pelagomorph*'s biologically quantum coherence, we begin to glimpse not just what is, but what *could be*. It is a vision of the future that awaits us if we dare to evolve alongside it.

Let's examine one final, formidable feature.

This one is a real trip.

"Probability gradient" describes how the chances of an outcome rise or fall as the variables around it shift. It measures how sensitive probability is to changes in conditions, whether rising or falling.[473]

To visualize this, imagine you are throwing ten darts at a world map. We track which countries you hit and how far off the target you land. Now repeat this same dart-throwing experiment thousands of times. From the accumulated data, we create a heat map showing where darts are most likely to land. The regions with the highest frequency become "hot zones." The steeper the shift from low-hit to high-hit areas, the steeper the probability gradient.

This idea becomes strikingly powerful when applied to *Pelagomorph* in a quantum-coherent state. Their perception and decision-making are not restricted to present-moment observations. They also extend across the terrain of possibility.

Where we see a future as an unknowable guess, *Pelagomorph* feels the unfolding landscape of potentialities. They are reading the curvature of probability gradients like heat rising off asphalt. To us, this would resemble prediction or precognition. To them, it may be nothing more than sensing the weighted resonance of what's most likely to emerge.

As with the other capabilities we have explored, this framing begins to explain the eerily prescient, evasive, or intentional behaviors that humans observe during *Pelagomorph* encounters. It is not prophecy. It is proprioception of the possible.

Reality can be imagined as a ramifying tree of infinite possible futures. Every tiny influence, whether a gust of wind or a passing thought, nudges us toward a different limb. Humans navigate this terrain blindly. We follow the narrow thread of now. *Pelagomorph*, by contrast, follows the slope of probability.

This isn't "seeing" the future in the way we imagine foresight. It's moving *downhill* into the most favorable version of it.

We consistently observe that *Pelagomorph* appears to be in the right place at the right time. This is not a coincidence or a matter of luck. It's *selection*. They sense the shape of probability as it forms and choose the branch that aligns with their intent. When a radar pulse is about to sweep across their location, they don't hide. They instead avoid stepping onto the timeline in which they are seen.

In this way, *Pelagomorph* does not rely on camouflage or cloaking to remain invisible. They never become visible in the first place.

This ability would require them to integrate non-local data and run probabilistic simulations across multiple futures in parallel. They would weigh their options based on resonance, field harmonics, environmental cues, or energetic signatures. They don't *guess* at the future. They *navigate* it, just as we might steer through the ocean or across shifting terrain.

Imagine walking through a fog-draped forest. The trail ahead is obscured, each step swallowed by mist. You don't know what lies beyond the next bend, but *Pelagomorph* is already there, just out of sight. Not because they are faster, but because they don't need to follow your path at all. They never took the trail you're on. They didn't need to see you to avoid you. They already knew which route you would choose and selected another.

Let's return to our dartboard metaphor. Picture *Pelagomorph* standing calmly in front of the map as you prepare to throw. They don't dodge once the dart is airborne. They move before your hand twitches. Not because they see the dart, but because they *feel* the slope of its probability the way you feel the heat of a fire or the pressure of a storm. The dart never needed to be thrown. They were already gone.

When we pause to consider the array of seemingly fantastical abilities attributed to *Pelagomorph* and their staggering implications, our minds naturally recoil. This resistance isn't a failure of imagination. It's a feature of our biology. Much of the human mental construct is centered around survival. We are built to evaluate risk and reward with relentless efficiency. We instinctively push away anything that feels impossible or unfamiliar, especially if it might threaten our stability or sense of safety.

And yet, we are paradoxically drawn to that which seems to defy reality, so long as it is framed within contexts that feel safe. A manufactured illusion can offer just such a context, on a stage where impossibility becomes entertainment rather than a threat. This author, for instance, once had the great fortune of being invited to the legendary Magic Castle in Hollywood, California. It is an enchanting hideaway perched just above Franklin Avenue, radiating a kind of playful energy that must be felt to be truly appreciated.[474]

That night, our group attended three performances. One was a classic stage show, filled with theatrical flair and traditional sleight of hand, including a woman being sawed in half. Another unfolded in a smaller room with an intimate, immersive atmosphere. Both were thoroughly entertaining. But the first experience we had, and the one that captivated us most, was a demonstration of close-up magic. Our magician, though young, performed with expert precision, bowling us over from just a few feet away. His hands moved through reality like *Pelagomorph* through time, flawless, bewildering, and utterly delightful. We understood this was calculated misdirection. But we couldn't see it. And in that gap between logic and spectacle, wonder bloomed.

After the close-up magic performance, most of our conversation revolved around a single theme. We *knew* these were tricks, and yet it was still impossible to determine *where* the sleight of hand had occurred.

During the larger shows, the physical distance between us and the performers gave our minds more breathing room. It felt easier, somehow more acceptable, to rationalize the illusions from a distance. Our senses found comfort in the stage's separation. It created a buffer of belief.

But when you're invited to stand shoulder to shoulder with a magician, where the cards are shuffled inches from your face and coins vanish from palms you swear you never looked away from, it stirs something deeper. It rattles a quiet mechanism inside you that normally rests unbothered by the day-to-day laws of physics. That disturbance is thrilling, even joyful, for most. The brain spins but laughs. The heart quickens but feels safe. In that closeness, the impossible becomes personal.

Now imagine this same sensation, but stripped of its stage and script. Imagine encountering the top half of a woman perched on a subway bench, smiling and scrolling her phone. Her bottom half sits calmly two seats away, legs casually crossed. Both moving. Both real. Our first instinct wouldn't be amazement. It would be panic. We would doubt our eyes. We might even dismiss it outright: *"I must be seeing things."* To speak of it would invite judgment. And so, we don't.

Yet when this *very same* illusion plays out on a stage, in a theater, or inside a television frame, we embrace it. We *want* to be fooled. We even applaud. Context, not content, decides whether the impossible amuses or disturbs us.

This dichotomy is key to understanding why human beings struggle to engage with *Pelagomorph*. It's the difference between the joy of being deceived when we know it's safe versus the dread of being deceived when it feels real. We lack the framework. We don't know whether to laugh, panic, or pray. And into that confusion seeps fear, denial, and ridicule.

To complicate matters further, the same tools that make magic beautiful are also the favored instruments of deception in far darker games. Confidence artists and grifters have long exploited our sensory and emotional blind spots through misdirection, timing, and psychological manipulation. The infamous three-card monte game is pure illusion. It is designed not to puzzle but to fleece, and it works.[475] Because misdirection is powerful. Subtlety is seductive. Brilliant though it may be, the human mind has cracks wide enough for the impossible to walk through unnoticed.

So, when we stand before a phenomenon like *Pelagomorph*, we are left with the same aching question that lingers after every grand illusion. *"How did they do it?"* But unlike a magician, *Pelagomorph* offers no bow, no smiling wink, no sleight-of-hand reveal. The curtain never closes. This is something that should not exist, and yet it does.

One of the magician's greatest secrets is not in the sleight of hand itself. It is, instead, in their deep understanding of how you, the observer, will perceive what you're seeing. A skilled performer doesn't just anticipate your reactions to each step of the act. They know the assumptions and preconceptions you bring with you before the trick even begins. They ride those assumptions like a wave. They actively steer your attention, shape your conclusions, and turn your own mental shortcuts into the engine of your astonishment.

Talented magicians surf the currents of human expectation in the same way *Pelagomorph* may surf probability gradients. The illusionist's toolbox rests on the bedrock of classical physics and the predictable quirks of human perception. Through rehearsed motions, prepared props, and cultivated misdirection, they work within the limits of what our eyes, ears, and minds are accustomed to. In doing so, they turn our own patterns of thought against us.

*Pelagomorph* may or may not have intentions toward humans in a conscious, goal-oriented sense. But they disarm us for many of the same reasons a magician does. They move in ways that step outside our most reliable templates for how the world should behave. The difference is that while a magic trick eventually ends, offering the relief of an exit from the theater and an end to the performance, *Pelagomorph* affords no such closure.

Humans have an often-unspoken preference for engaging only with what we must, or with what affirms the beliefs we already hold. It is this bias toward comfort and familiarity that allows cognitive stagnation to take root. When nothing pushes us beyond our inherent limits of perception or reasoning, we rarely even *try* to imagine what lies just outside those walls. And when something does press against them, our instinct is typically to dismiss or scorn what we cannot comfortably admit into our model of reality.

If it does not touch us personally or trouble our senses, it is easy to treat it as though it does not exist at all. Yet the truth is stubborn. It does not require our consent. Whether we acknowledge them or not, some things, like *Pelagomorph*, remain exactly what they are.

The earliest chapters of this book explored these ideas on a cosmic scale. This is because they form the bedrock of the most basic assumptions we hold about existence itself. We may never truly know how time and space behave in the farthest reaches of the universe. We can theorize and deduce. But woven into that process must be the constant acceptance that the laws which govern reality are immutable. Regardless of our models, they exist *outside* the architecture of our thoughts. Far from limiting us, this truth is a liberation. It reminds us that we are not merely observers of time and space. We are participants in the act of creation itself.

And what of those astronomers and physicists, playwrights and poets, skeptics and cynics who live so far away in the vastness that we will never encounter them? They exist, even if their lives will never intersect with ours. How might they regard the universe? Is their view relevant to us? Is ours to them?

We tend to frame such questions in the narrow lens of our own local experience. We prefer to measure all reality against the hominid mind and body plan. But is ours the best, or only, viable design? Does acknowledging the potential for other forms imply that we are somehow defective? Or does it quietly argue that we have yet to bloom into the full spectrum of what we could become?

Some among us, when dazzled by a performance of magic, are content to marvel at the illusion. Others, however, feel a different pull.

They desire to become the magician. You cannot dream of the craft until you have first witnessed the wonder. And so, it should be with *Pelagomorph*. They are not to be dismissed as mere UAP, mechanical craft, or interstellar tourists. They are more than this. Perhaps they are heralds of what we ourselves might one day become. To see them is to be challenged. To understand them is to aspire.

The question then becomes, how do we aspire, both as individuals and as a species? There are many possible answers. The one offered here is deceptively simple. Know your limits. Learn to be comfortable with them. Only by doing so can you leave the door open for what lies beyond to enter, be accepted, and valued. That which you do not see still exists. That which you do not hear still makes a sound. That which you do not feel still exerts a force. This is true not only within the narrow confines of our defined locality in time and space, but in all the regions that stretch far beyond it.

As we have explored in earlier chapters, there are countless sights, sounds, textures, tastes, and properties of the universe that humans have not evolved to perceive. We also may overlook them because of the mental walls we've built for ourselves. The behavior of *Pelagomorph* is not magic. Nor is it "impossible." It is *supernatural* only in the strict sense of lying beyond what most of us think of as "natural." A better word is supra-natural. We would be wise to distinguish between phenomena we frame as spiritual and those that may be purely physical, yet elude measurement. Somewhere along that spectrum, the two domains may even overlap.

You may believe in spiritual manifestations, or you may not. Either way, it remains possible that any such manifestation could have an entirely scientific explanation. It may simply be one still beyond our reach. This is not an argument for abandoning scientific rigor. It is a plea to apply it more stringently, but freed from dogma and myopia. The fundamental rule should be only to observe what is observable, conceive what is conceivable, and consider what is considerable. Then press forward into measure and reason without letting the boundaries of the "normal" dictate where the line must be drawn.

If you witness a woman on a train whose top half calmly reads a magazine while her bottom half, seated elsewhere, crosses and uncrosses its legs, what is the proper reaction? If you observe quantum entanglement acting instantaneously across light-years, what is the appropriate response? The answer is the same in both cases. What is observed can and must be explored with equal application of reason, curiosity, and discipline.

When we cease declaiming and denouncing, proselytizing and pronouncing, denying and ridiculing what is plainly evident, and instead choose to observe quietly, something shifts. In that stillness, reality seeps into our being, opening us to the deepest kind of wisdom. We can accept our limitations and our blind spots not as shields but as invitations to surpass them and transform those gaps into gateways. The mysteries that surround us daily are not walls. They are thresholds. Once crossed, they cease to be mysteries at all. They instead become keys for unlocking the next layer of what still lies just beyond our reach.

It is in this way that we will truly meet *Pelagomorph*. Not as dazed onlookers, staring at some refugee from our reality perched in a higher dimension, but as guests they have invited in. There is no other frame that holds both their imponderable power and their equally stupendous grace. And here is the best news. We need no exotic technology to join them in that space. The only admission price is stillness, attention, and the courage to loosen our grip on the familiar world.

Miracles are not myth. They are realities that exceed our current model of the everyday. Creation itself is a miracle. Never mind who the *first* human was. Ask instead, who is the *next*? Consciousness is a miracle. Forget the mechanics of quantum tunneling in the brain. Marvel at the fact that *you* are thinking right now. Gravity is a miracle. Forget the math of particles and waves. Remember that you are standing, sitting, or lying on a sphere spinning at roughly a thousand miles per hour. Permanence. Solidity. Motion. Life. Chemistry. Matter. These are names we have given to miracles that are so common we no longer notice them. Why, then, should the rarer ones be any less real or less worthy of our attention?

We fixate so intently on the edges of our reality that we overlook the central truth. Everything we think, everything we imagine, is chained to the immediacy of *now*. That tether infuses all our data with the gravity of local time and space. This is a gravity that *Pelagomorph* does not share. For them, reality is not a fixed moment but a fluid continuum, shifting and reconfiguring in ways we cannot yet sense. This is our final wall. The belief that time's flow is immutable, when in fact it is only a curtain waiting to be parted.

As other creatures can read polarized light or magnetic lines we never see, *Pelagomorph* is attuned to the tides and ripples of time itself. They move within this medium as naturally as we breathe. They beckon us toward what we see as a boundary, but they know as an opening. Their motives may be layered. They may be driven by curiosity, kinship, or something we lack a word for. But the pull is unmistakable. We stand at the edge of a change that some call the singularity. Others refer to it as posthuman evolution. Whatever the name, it is ours to share. Much of reality will remain closed to us until we cross this threshold. Yet even now we feel its pressure, its inevitability.

*Pelagomorph* is a miracle. Forget *now*. Fix your gaze on the *everywhen*. Stop shouting into the void. Listen quietly. Watch without demand. Accept what is evident, and be brave enough to take the hand extended toward us. These are not alien strangers but ancient Earthborn kin. They are waiting to show us the greatest wonder we will ever meet. We are not *approaching* transformation. We are *in it*. The crossing has begun. We are not climbing a mountain to meet *Pelagomorph* at its peak. We are stepping beyond the very frame that holds the mountain. This is not ascent. It is a metamorphosis. The only thing that can hold us back is the noise of our own resistance. *Pelagomorph* does not wish to show us a magic trick. They invite us to become magic itself.

# CHAPTER SIXTEEN

## Liminal Incursions: Greetings From the Wave Function Collapse

*"To live past the end of your myth is a perilous thing."*

*Anne Carson, Red Doc>*

Now that we've fully explored the likely Earth-based, biological origin of what are generally referred to as UAP, one item of unfinished business remains. A close study of UFO and UAP encounters throughout history reveals a persistent wrinkle. Not all reported events fit neatly within the *Pelagomorph* framework. These life forms are formidable. They are capable of altering their appearance and engaging in seemingly supernatural movements. However, there are numerous close encounters involving craft or entities that defy the patterns we've established. Let us now turn our attention to these outliers and consider what hidden factors might underlie them. One of the most unusual examples occurred in Cisco Grove, California, in 1964. Though the details of the event may at first glance seem outlandish or ripped from the reels of a vintage sci-fi film, they deserve careful consideration before being cast aside.

Cisco Grove is an unincorporated community located in the Sierra Nevada foothills of Northern California, not far from Lake Tahoe. Cradled in the Tahoe National Forest, it is a remote and sparsely settled area teeming with wildlife. In early September of 1964, a 26-year-old aerospace engineer named Donald Shrum was enjoying a weekend of camping and hunting with friends. At the time, he worked in missile systems at Aerojet General. His identity would remain undisclosed for decades following the events that transpired.[476] But we now know it was Shrum who lived through what would become one of the strangest encounters in the annals of UAP lore.

The region is breathtaking, with untamed wilderness stretching in every direction. It is home to squirrels, raccoons, bobcats, coyotes, mountain lions, and black bears. Overhead, red-tailed hawks and bald eagles soar. It's a hunter's paradise. But survival here demands skill, awareness, and respect for the unpredictable.

At some point during the trip, Shrum became separated from his companions. As dusk approached and efforts to reconnect proved fruitless, he made the decision to spend the night alone in the forest. Drawing on his outdoor experience, he climbed roughly twelve feet up a tree and secured himself with his belt. This was a prudent move to avoid predators or any other threat that might appear after dark. As it turns out, his instincts were far sharper than even he could have known.

As night fell over the forest, Shrum noticed a light moving silently through the trees. This was odd enough to catch his attention, yet not immediately alarming. At first, he assumed it must be a helicopter or a search party looking for his group. But that assumption quickly fell apart. The light hovered above the treetops without a sound. It emanated a harsh, brilliant white glow unlike anything he'd seen before.

Then came movement at the base of his tree.

A figure appeared. It was roughly human in shape but moving with a stiff, mechanical gait. Its eyes bulged unnaturally, devoid of emotion or expression. Moments later, it was joined by two more entities. Both were dressed in what appeared to be metallic silver suits. The robotic figure seemed to emit a visible vapor from its mouth. Shrum would later describe this mist as a possible incapacitating agent. He also got the distinct impression that the two newcomers were in control of the robot and directing its actions.

What followed was a surreal and harrowing standoff that lasted the entire night. The three entities repeatedly approached the tree, attempting to subdue Shrum and render him unconscious. He fought back with everything he had. Uncommonly composed under pressure, he fired arrows at them and even set pieces of his clothing ablaze to throw as flaming projectiles. These defensive tactics worked, but only temporarily. The beings would retreat, regroup, and return. Each time they came back, they used incapacitating gas in renewed attempts to overpower him. Shrum later recalled the vapor making him nauseous. His vision blurred, and his grip on consciousness wavered. But he held on.[477]

By the time dawn approached, he had managed to fend off the intruders. Just before sunrise, the entities retreated. The craft, still hovering in the distance, silently departed. Shrum remained perched in the tree, still wary. Secured by his belt, he waited patiently until his companions eventually located him.

At first glance, the details of this encounter might sound implausible, even absurd. But several factors suggest that Shrum's account deserves closer consideration. Remember, he was an experienced aerospace engineer. People with such training are seldom prone to overwrought imaginings or mistaking natural events for something metaphysical. He also remained anonymous for decades, never seeking publicity, profit, or recognition. In fact, Shrum reportedly suffered from post-traumatic stress and only revealed his identity in the 2000s. Given his profession, this hesitation to speak out makes sense. The risk of ridicule or damage to his reputation was likely a powerful deterrent.

Unlike others who have capitalized on similar experiences, Shrum never wrote a book, sold merchandise, joined the lecture circuit, or engaged in any promotional activity. His story never shifted over time. It remained consistent throughout. Furthermore, his hunting companions confirmed that they had seen the strange light he described. This lends external support to his account.[478] Most notably, the U.S. Air Force took the incident seriously enough to dispatch investigators from Wright-Patterson Air Force Base.[479]

Whatever the ultimate cause, it is difficult to dismiss the likelihood that something real and genuinely disturbing happened to Donald Shrum that night.

Over the years, speculation about the Cisco Grove incident has generally settled around three main possibilities. The first is that Shrum was under the influence of a hallucinogen, whether knowingly or unknowingly. The second is that he was the unwitting subject of a covert U.S. government experiment or psyop. The third, and most unsettling, is that the event unfolded exactly as he described.[480] That night, he truly encountered an unknown entity.

As noted earlier, the area where Shrum and his companions were camping is not one to traverse casually. It is rugged, isolated, and potentially dangerous. Given Shrum's credentials as an accomplished aerospace engineer and seasoned outdoorsman, the notion that he would willingly ingest a psychedelic substance seems unlikely. If he had been unknowingly dosed, could the effects truly have evaporated so precisely, just as his friends arrived? With his intelligence and technical training, wouldn't he have figured out later that it was only a hallucination?

If it were nothing more than a bad trip, why go public at all? Most people in that situation would likely keep the experience to themselves and move on. Shrum didn't. And his account never wavered.

It's essential to recognize that the era in which this occurred was characterized by covert experimentation. Government agencies were indeed conducting classified research on psychedelics and other substances. They often did so without consent and under egregiously unethical conditions.[481] So it *is* within the realm of possibility that Shrum's experience was the result of such an operation.

But the behavior he exhibited that night tells a different story.

Throughout the ordeal, Shrum remained composed, rational, and strategic. If he had unknowingly ingested a powerful hallucinogen like LSD, we would expect agitation, disorientation, and even incoherence. Not clear-headed decision-making. Lighting one's clothes on fire to use as a weapon is not something typically associated with an acid trip. It's dangerous, but purposeful. Likewise, someone in the throes of a psychedelic state would be unlikely to remain tethered high in a tree for the entire night without panicking, losing focus, or climbing down.

Psychedelic visions are notoriously unstable. They are shifting, evolving, and fragmentary.[482] By contrast, Shrum's experience was sustained, consistent, and defensive in nature. While we can't entirely rule out the possibility that he was an unsuspecting subject of something like MK-Ultra, evidence points away from that conclusion. Of the available explanations, it remains one of the least likely.

The explanation that best aligns with Occam's Razor is the simplest. Donald Shrum encountered real, physical humanoid figures who harassed and attempted to subdue him through the night. He fought them off with a tenacity that deserves admiration.

However, we must now return to *Pelagomorph*.

This encounter does not align with anything we have previously established about these beings. By all accounts, *Pelagomorph* are non-combative and non-confrontational. Their behavior suggests wisdom, restraint, and a quiet, gentle grace in their interactions with humanity. They do not appear in humanoid form. Nor do they engage in overt contact. In fact, they seem to avoid direct observation altogether.

And yet, this happened.

So how does it fit? Is it an outlier, a mistaken identity, or evidence of something else entirely? There may indeed be a plausible explanation. It does not undermine what we've already deduced about *Pelagomorph*. Instead, it adds another layer to the mystery. However, before we draw any firm conclusions, let's examine a few more anomalous encounters. Like Cisco Grove, they refuse to be neatly filed within the *Pelagomorph* paradigm.

Research for this chapter revealed that most UFO and UAP reports outside the behavioral parameters established for *Pelagomorph* are difficult to verify with any rigor. That's not to suggest all such accounts are fabricated or false. Surely some are. But just as surely, some are not.

This ambiguity leaves ample room for debate. We still lack a definitive "smoking gun" when it comes to direct encounters with non-*Pelagomorph* entities. Even reports that seem credible may have explanations that don't require an extraterrestrial origin.

A lack of hard evidence can be used as proof on both sides of the argument. One side may say that such proof doesn't exist. The other, that it does exist, but has been hidden. In truth, it wouldn't be surprising if each were true to varying degrees across individual encounters.

One case, however, continues to resist easy dismissal. We now turn to the Ariel School incident, which occurred in Zimbabwe in 1994. Widely reported and still unresolved, it presents an unusually cohesive body of evidence. Sixty-two schoolchildren were on the premises that day. Not all claimed to witness the event directly. However, those who *did* remained consistent in their accounts, even decades later.[483]

For context, the Ariel School incident took place in a small agricultural community called Ruwa. It is located near Harare, which is the capital and largest city in Zimbabwe. Although close to an urban center, Ruwa itself is sparsely populated and has a rural character. The Ariel School is a relatively expensive private institution. Its enrollment was primarily made up of children from well-off families in Harare.[484]

Crucially, the event occurred in the wake of several reports of unidentified aerial phenomena seen across the area several days prior. Some witnesses described what they saw as a comet or meteor. Others insisted it was something stranger. This brief flurry of sightings had already stirred excitement by the time the schoolyard encounter occurred.[485]

It is also worth emphasizing that although these children lived in a rural area, they were far from naïve. They had access to local media and a general awareness of the concept of UFOs. Two primary investigators recorded their accounts. The first was psychiatrist and Harvard professor John Mack. The second was Cynthia Hind, the African representative for MUFON (Mutual UFO Network) and editor of UFO Afrinews.[486]

Mack interviewed the children in small groups. However, the setup allowed some children to observe others being interviewed. Critics argue that this was a methodological flaw, since it opens the door to unintentional cross-contamination of memories or impressions. Hind asked the children to draw what they had witnessed, then selected some of the drawings to photocopy and distribute to the press. It has been suggested that only drawings supporting the UFO narrative were shared, while others were omitted.[487]

Adding further complexity, some have pointed to a Zenit-2 launch from the Baikonur Cosmodrome in Kazakhstan on August 26 of that year. While the launch itself occurred roughly three weeks prior to the Ariel School incident, a tracked re-entry of the rocket's second-stage booster was observed over southern Africa on the evening of September 14. This event produced a bright, fragmented fireball consistent with known spacecraft re-entries and was independently catalogued and monitored. Although this explains a number of regional sky sightings reported in the days immediately preceding the Ariel encounter, it does not represent debris falling directly from the launch site itself, nor does it establish a causal link to the events reported at the school.[488]

So, what exactly did the children report?

According to most accounts, silver discs or craft were seen descending during recess and landing in a field of brush and small trees just beyond the school grounds. The number of beings said to have emerged varies slightly between students. Some reported one, others up to four. The figures were described as dressed entirely in black, with large, prominent eyes. As they exited the craft and began to approach, some of the children fled in fear. Others, mostly the older students, stood their ground and watched.

In the interviews conducted by John Mack, several children claimed these beings communicated with them telepathically. Though not identical for each student, the message shared a consistent theme across accounts. It was a warning about humanity's relationship with the environment. More specifically, the beings conveyed concern over ecological harm and a looming catastrophe should humanity fail to correct its course. One child recalled being told quite plainly that the Earth would be destroyed if humans continued mistreating it.

All witnesses agreed that the objects were not planes.[489] However, some doubted their origin to be extraterrestrial. A few witnesses with local cultural knowledge interpreted the entities differently. These students identified them as tokoloshes, mythical dwarf-like water spirits from regional folklore.[490]

Over the years, most of the students involved have remained consistent in their narratives. However, one individual has since claimed the event was fabricated as a prank meant to fool others.[491] They expressed surprise that it had evolved into what they now regard as a case of so-called mass hysteria. Despite this, the incident remains one of the most compelling close encounter cases ever reported. This is primarily due to the sheer number of witnesses and the consistency of their testimonies over time.

Much like the account given by Donald Shrum, these reports feel sincere. They come across as unpolished, unaffected, and unembellished. But as with Shrum's experience, sincerity alone does not constitute proof. Still, applying the same basic logic, it seems likely that *something* mysterious occurred at Ariel School in 1994. Clearly, this event had a lasting impact on the children who witnessed it.

Once again, we are faced with a case that falls far outside the behavioral and perceptual boundaries we have established for *Pelagomorph*. Nowhere else in this book have we encountered reports of individual entities making direct contact or materializing in local space-time to deliver messages of warning and environmental concern to humans.

However, not all anomalies involve direct contact. Some seem to offer only the residual traces of influence. We read instead of aftermaths without witnesses, and signs that confound without context. These reports may be less dramatic on the surface. But they raise questions equally as compelling.

One such case is the Delphos Ring Incident of 1971, which took place in Kansas. 16-year-old Ron Johnson reported seeing a glowing, mushroom-shaped object hovering silently above the ground near a tree in his backyard. His parents also witnessed this event. After a few moments, the object ascended and disappeared.

What makes this case especially notable is what was left behind.

The family discovered a glowing, ring-shaped mark in the soil directly beneath where the craft had hovered. Unlike most encounters of this kind, a witness outside the family was able to verify key details shortly afterward. The local sheriff, Ralph Enlow, arrived at the scene with a deputy and confirmed that the ring was visible. He agreed that it retained a faint glow, was lighter in color, and noticeably drier than the surrounding soil.[492] Laboratory tests revealed that the soil in the ring had become hydrophobic. In addition to repelling water, it also showed evidence of altered chemical composition.[493]

Of course, there are possible alternative explanations for the Delphos Ring. Certain fungal or bacterial colonies are known to produce circular patches in soil. Sometimes these even include bioluminescent qualities.[494] This does not invalidate the Johnsons' account. But it does offer a naturalistic explanation that must be considered.

This pattern repeats itself across nearly all UAP or UFO encounters involving physical traces or close contact. We find a lack of consistent, verifiable evidence. Despite many compelling details, it is often impossible to establish a clean sequence of events that holds up under scrutiny.

The author remains sympathetic to these accounts and open to any well-reasoned argument in their favor. But to date, nearly every known case of this kind is burdened by factors that are problematic, if not outright disqualifying. This leaves us with a handful of possibilities. Perhaps the entities involved are exceptionally skilled at leaving no trace, or at erasing it entirely. Or maybe, as some suggest, there is a long-standing, highly effective global effort to suppress evidence of such encounters.

Even so, if verifiable proof existed, wouldn't some solid fragments have surfaced by now?

To be clear, this does not mean that no evidence has ever been presented. It only means that alternate explanations or conflicting interpretations almost always accompany such evidence. The author would welcome the opportunity to meaningfully compare these events with those involving *Pelagomorph*. However, they remain consigned to the realm of the unexplained, shadowed by the absence of clear and convincing evidence.

The intent here is not to dismiss or diminish those who have reported anomalous experiences. The only person who truly knows what occurred in any given case is the witness. At some point, all of us have experienced something that appeared to defy explanation. Some encounters clearly resist classification within the conventional frameworks of physics or shared reality.

As we have noted throughout this book, the universe makes room for the most unlikely events, unfolding at the very edges of our ontological bandwidth. Our senses are tuned to a narrow range of frequencies. When we encounter phenomena that fall outside those parameters, we can experience a kind of perceptual dysphoria. This can result in confusion, fear, denial, or a profound sense of wonder.

But there is a key distinction.

Even the most bizarre and vivid contact encounters tend to fall within the bounds of what the human mind can ultimately classify. The observer may be shocked. The details may be surreal. Yet the scenario remains narratively intact.

Specifics may include a strange craft or a humanoid entity. Frequently, this is a symmetrical being with minor anatomical quirks, unusual skin tones, exaggerated eyes, or oddly proportioned limbs. In essence, a sci-fi template. This form is familiar if distorted.

*Pelagomorph* is different.

In these accounts, witnesses struggle to find the words. Language becomes insufficient. The experience resists structure, refusing to settle into any familiar cognitive frame. What's reported is not a waking vision but something closer to a dream. The details are fluid and unbounded by the usual laws of space, time, or causality. These encounters do not follow the rules of physical interaction in the world we are familiar with. Instead, they hint at an altogether different order of being. It is one that may not be classifiable at all.

This author is sympathetic to such accounts. The reason is personal. In the summer of 1979, an unexplainable event was experienced firsthand. It left no physical trace and offers no tangible proof. Yet it remains vivid and haunting in memory to this day.

It unfolded in an average suburban neighborhood, on a street with a particularly steep hill. Any child who grew up riding bikes knows the exhilaration of standing on the pedals and letting gravity do the work.

It was a warm afternoon, and the area was strangely devoid of the usual neighborhood harmony. No lawnmowers were buzzing, no dogs barking. Neighbors were absent from porches and yards. The silence was noticeable, though not immediately unsettling.

The hill bottomed out beneath a corridor of old-growth trees whose branches formed a high, arching canopy over the road. They cast a shadow that covered most of the block. As the bicycle entered this shaded stretch, a sudden and unmistakable shiver struck without warning. It danced menacingly from the top of the head, down the spine, and into the legs. The reaction of unease was immediate and primal. It triggered a compulsion to look upward.

Through intermittent breaks in the leaves, *something* was visible. Some sort of object was soaring silently above the tree canopy, perfectly tracking the author's movement. It briefly registered as something familiar. Perhaps this was an airplane or a bird? But another wave of instinctive dread followed quickly. With it, a visceral certainty emerged. This was not a plane. It was something unknown. It was something to be afraid of.

Still coasting beneath the canopy of trees and unable to clearly discern what the form above might be, attention shifted forward. The street ahead was still deserted. Upon glancing upward again, the object was still there and tracking directly overhead. It was clearly following. Against all normal reason, terror began to bloom.

What follows may seem like an odd detail, but it is relevant. In the author's small hometown, a local dentist would customarily give children a prize after each appointment. This was usually a small foam glider, mottled with red, blue, and green specks. A forgettable toy. Until that moment.

As the strange presence hovered above the treetops, it seemed to take on the visual signature of one of those gliders. The resemblance was not exact. But it was close enough to imprint itself on the author's perception. Perhaps it was the mind grasping for familiarity, attaching the nearest analogue from memory to an unknown and baleful presence. Regardless, the likeness was intensely unsettling.

This unfamiliar entity continued to mimic the bicycle's path below with unerring precision. It floated silently and steadily, just above the trees. Small lights blinked intermittently across its surface. This reinforced the sense that it was no trick of light or imagination.

Whatever this was, it was real. It moved with undeniable resolve and conveyed a disturbing eagerness in its pursuit.

Decidedly spooked yet grimly determined, the author pedaled faster, intent on breaking free of the canopy and glimpsing whatever loomed above. Fear and incredulity churned into dread as the bicycle surged forward, bursting from shadow into open sunlight.

And just like that, the entire atmosphere changed.

The enveloping stillness that had blanketed the neighborhood vanished in an instant. When the author looked up again, the sky was empty. Whatever it was, it was gone.

At that precise moment, familiar sounds returned. Children's voices carried from a distance. A screen door slammed. A neighbor stepped out, walking to a car in the driveway. All reverberated as usual. The world had snapped back into place. Only later did the strangeness of the preceding silence register fully. The complete absence of ambient life, which had underscored this experience, remains as one of the most vividly remembered details.

This haunting incident has never been publicly disclosed until now. Even at a young age, the author sensed that to share it would invite disbelief, if not outright dismissal. There was no tangible evidence to point to. No physical trace to offer credibility lingered. Every child knows the sting of being disregarded by adults as "just a kid." The encounter had been fleeting, ambiguous, and impossible to authenticate. As with other accounts of this type, it was easy to dismiss by anyone who hadn't witnessed it firsthand. And yet, it lingers. The memory not only endures, but continues to carry with it the same spine-deep shiver that marked it in real time.

So, did something unexplainable *truly* happen that day? Or was it a familiar case of childhood imagination projecting fear onto something ordinary? Were the details of this experience a brush with the unexplainable? Or were they just a misinterpretation of something mundane?

Only the subjective impression remains. But one thing is unequivocally true. *Something* occurred. Its nature remains open to interpretation. But the feeling it left behind was real. It was deeply felt by the mind that lived it. The sudden, inexplicable dread and the terrifying certainty of being watched by something unknowable both persist.

A lack of conclusive evidence does not prove that an event never occurred. Throughout human history, there have surely been events that fall outside the boundaries of ordinary experience. Many incidents undoubtedly defy conventional explanation. Yet, they do not involve *Pelagomorph* at all.

That said, evidence remains elusive. Solid, independently verifiable data for such encounters has yet to surface. Indeed, other explanations for these observations *could* exist. In attempting to define one, though, the author has come up empty-handed.

The only accounts the author has been able to identify that involve human witnesses *and* are supported by corroborating instrumentation are those already discussed in earlier chapters.

The conclusion of this book is that these observed and measured phenomena are best explained by the presence of a biological life form. It is one that originates from the deepest, least-explored zones of Earth's oceans. As argued previously, this life form is the product of an ancient symbiosis between proto-cnidarian and cephalopod lineages, which has undergone uninterrupted evolution for at least 500 million years.

Unlike the ambiguous cases explored in this chapter, *Pelagomorph* encounters have tangible evidence to support their existence. Concurrent visual recordings, eyewitness accounts, and readings from multiple advanced instruments are all well documented.

Based on the available data, this proposed model remains the most coherent and compelling.

Furthermore, it is reasonable to conclude that these beings are not just sapient but far more advanced than humankind. They appear to operate at a level of intelligence, reasoning, planning, and strategy that we can scarcely comprehend. They possess clear air superiority and consistently maintain a presence near water sources and nuclear sites. They also demonstrate behavior that is measured and markedly non-aggressive over long periods of time.

One additional theory is now presented for consideration. It is an idea that attempts to account for phenomena that fall firmly outside the *Pelagomorph* framework. Because it cannot be substantiated or meaningfully supported at this time, it is suggested only as a subject of speculative curiosity.

A pervasive, unexamined bias runs through most accounts of direct contact. This is the assumption that such phenomena are inherently about *us*. As we have explored elsewhere in this book, humans tend to cast themselves as the central subject of any event that affects them directly. It's an understandable reflex. But it is also one that may obscure other, less anthropocentric possibilities.

What if some of the anomalous encounters that don't align with known *Pelagomorph* behavior aren't focused on humanity at all?

What if they are the incidental byproduct of something else entirely? Could they be drawn not to us, but to *them*?

It's conceivable that these entities are not responding to our presence. They may instead be other forms of quantum-coherent life drawn to the unique resonance of *Pelagomorph*. If our Earthly cousins routinely occupy a state of sustained quantum coherence, they may act as a kind of gravitational well in non-local space-time. Their presence could be like a stone dropped into the fabric of the universe, creating ripples that others attuned to that same dimension of being would feel.

In such a case, humans may not be the target at all. We might be unwittingly caught in the wake.

One leading theory in UFO and UAP circles is that non-Earth-based entities are drawn to our planet because of humanity's discovery and use of nuclear power.[495] As argued earlier in this book, though, it seems far more plausible that an Earth-based life form would have both an interest and a stake in such developments.

What motivation would a species with no dependence on Earth's ecosystem possess to monitor our nuclear progress, or to intervene in it? While it's not impossible to imagine an extraterrestrial civilization expressing concern for our technological path, it's worth asking a deeper question. If this life form had mastered quantum technology or faster-than-light travel, would the splitting or merging of the atom truly warrant their attention? Perhaps. But the likelihood seems slim.

By contrast, a fellow inhabitant of Earth would have far greater reason to care. All the more so if that being were as ancient and adaptive as *Pelagomorph*. The widespread or reckless use of nuclear energy could endanger their environment and, by extension, their survival. A "warning" in such a case would be self-preservation as much as altruism.

A more compelling scenario is that once *Pelagomorph* evolved into a stable condition of quantum coherence, their unique signature might attract the attention of other quantum-coherent life forms. These beings may have evolved far from Earth, in entirely different environments. Should they choose to collapse their wave function within the same local space-time as the source of that signal, we might perceive them in ways characteristically different from *Pelagomorph*. They would be shaped not by similarity, but by their entirely separate origins.

Of course, this is conjecture. Yet given the sheer number of such anomalous reports over time, it is a possibility worth entertaining. Should quantum-coherent life forms be drawn by the signature of *Pelagomorph* into our local space-time, they might also notice us. Like a streak or scar across coherent space-time, our presence could pique their curiosity.

Such beings would have evolved or developed quantum coherence elsewhere, on some distant world or within an environment wholly nonconvergent to our own. Their appearance, behavior, and impact on our reality could differ abundantly from those of *Pelagomorph*. And here lies the limit of our knowing. Until we also reach that state of coherence, we can only speculate.

If humanity were to achieve such a state, the mysteries that have eluded us for centuries would no longer be mysteries at all. The change would be transfiguring.

It would carry with it freedom from the unipolar, forward-only arrow of time. We would be released from the narrow vantage point of a mind trapped in the "now," forever chained to its regrets, memories, anticipations, and fears. This would mean liberation from decision-making constrained by the tiny dataset of what *has* happened and what *might* happen.

We do not share *Pelagomorph*'s physical form. Their characteristics of bioluminescence, electromagnetic field manipulation, and effortless defiance of gravity would likely remain beyond us. But this raises a tantalizing question. What *would* humans gain in exchange? What inherent capacities might emerge in us that differ from a being born of combined cnidarian and cephalopod ancestry?

The answers are evasive. Beyond the possibilities already imagined, it is difficult even to conceive of them. However, the potential for transformation, for perception beyond our current frame, is staggering.

Of these possibilities, for now, we can only dream. Yet to dream is not to drift aimlessly. It is to orient ourselves toward the shape of what could be. We must never abandon the drive to know more and to *be* more than we are now.

Many envision the next stage of human development as something engineered. We imagine rewriting our genome, merging with microprocessors, or uploading our minds to a digital archive. These pursuits may be inspiring. They may even be useful in specific contexts. But they are not the pinnacle. They are the scaffolding, not the cathedral. Such approaches risk reducing the human spirit to circuitry and code, a self-imposed limitation masquerading as liberation.

Our destiny is larger. More organic. More transformative.

No genetic tinkering is required to cross this threshold. What *is* needed is a shift in relationship. Between ourselves, our own consciousness, and the architecture of reality in which we already dwell. In their quiet, slippery way, *Pelagomorph* are trying to tell us this. They speak not in words but in patterns, alignments, and fleeting intersections of perception. We are *capable* of hearing them. But we remain too inwardly fixed and distracted by the noise of our own preoccupations to interpret the signal.

It's like a visual illusion you cannot quite resolve. Even when the image is right before your eyes, the more obscure pattern escapes you. Or perhaps you glimpse it, but cannot yet understand what it means. This transition is not far away. It is here and pressing at the edges of awareness.

Our task is no longer to interrogate it into revealing itself. Instead, we must become still enough to *receive* it. The time has come to trade constant inquiry for deliberate attentiveness. That which approaches will not be caught by pursuit, only by recognition.

The very part of the mind that clings to reason and rationality is also the cage that holds us. The tool we revere as our greatest strength is the architect of the barrier we long to transcend.

Yet the clues to our escape are already inscribed in plain sight, across the inner walls of these ephemeral shells we inhabit.

How can two particles, separated by light-years, react to each other instantly? The answer is that the "space" between them exists only in our perception. It is a convenient illusion produced by the brain. This projection of local space-time enables us to navigate the reality we inhabit. To the particles themselves, there is no separation.

Or consider light. How can it be *both* a particle *and* a wave when our choice of how to measure it occurs only after it has already passed through the experiment? Again, the paradox exists not in nature, but in us. The particle-wave dichotomy is a fundamental aspect of the mind's architecture. It is a way of ordering phenomena into categories that suit the expectations our neural architecture demands.

The truth is that we have never observed the universe directly. Our senses are translators, not windows. They receive, compress, and reformat reality into a model our minds can manage. Every color, every texture, every measure of distance or density is an interpretation. They are filtered, packaged, and delivered inside the closed system of our consciousness.

Between our true selves and the true nature of reality lies this buffer of translation. We must learn to see beyond it or dissolve it entirely. Until then, we will remain interpreters at the edge of a language we do not yet know how to read.

What is color?
What is texture?
What is mass?
What is distance?
What is form?
What is function?

They are what we *think* they are. Because, for now, we cannot think in any other way.

One of the most famous Zen koans states this eloquently. *It is not the flag that moves in the wind. It is not the wind that moves the flag. It is your mind that moves.*[496] In such words, we begin to glimpse the outline of what may await us. We strive together towards a reality freed from the inherited, fracturing sieve of human perception.

There is nothing inherently *wrong* in the way we experience the world, except that it keeps us separated from our authentic, more abiding nature. Being rooted in local space-time is neither good nor evil. It is simply the condition we inhabit. But in the ascent toward the next level of consciousness, this rootedness can no longer be a cage.

In that state, we will meet *Pelagomorph* without the veil between us. We will understand the quiet compassion that they already extend in their encounters with us. It is neither born of pity nor tinged with envy. These are artifacts of this duality that we wield. They are illusions cast by our self-limiting perception of the universe. When that illusion falls away, what remains will not be a new world. It will be the true one, finally revealed.

*Pelagomorph* already greets us from just across the thinnest of veils. Another secret, one we will understand only upon crossing that boundary, is that the greeting has no beginning and no end. Unbound by the forward arrow of time, the welcome we receive will echo through all moments, all places, forever.

From *Pelagomorph*'s perspective, it already has.

This is the surest sign that we *will* cross this fading partition and bloom into the full measure of our being. In the truest and most eternal sense of reality, we are already there, looking back at ourselves with wonder.

Like a starship bending toward the event horizon of a black hole, or like a beam of light drawn endlessly into the dark, some part of us has always been transcendent. We feel this in the deepest chamber of our souls as absolute truth. Beneath the shifting surface of doubt lies the certainty that concepts such as "if" *and* "when" are illusions. These are ephemeral abstractions, soon to be replaced by the "forever now" which we presently consume one fragile slice at a time.

In truth, this sustains us endlessly, outside of time and beyond all place. We are not just invited to this table. Our places are already set, waiting for our presence. When we arrive, it will be together. The rejoicing that greets us will be without limit, without boundary, without end.

We have always been there. We will always be there. This is what draws *Pelagomorph* to us. It is a kinship beyond measure. It is a bond that threads through realities we cannot yet grasp, dancing just beyond our reach.

When we stop chasing the moment of crossing, and instead accept that we have always *been* on the other side, we will find ourselves there eternally. When we dare to live beyond the end of our own myth, we will discover that the only myth was the *ending itself*.

# CHAPTER SEVENTEEN

**Eigenstate Revival: Strange Attractor Ad Infinitum**

---

*"We shall not cease from exploration, and the end of
all our exploring will be to arrive where we started and
know the place for the first time."*

*T.S. Eliot, Four Quartets*

---

Thank you, dear reader, for your patience and determination. The terrain we have traversed has been vast. Yet all of it has circled a single point. It is inevitable that some will dismiss these ideas outright. Indeed, nothing less should be expected. The recurring thread throughout has been our intrinsic human tendency to withdraw into the safety of habit and reflex. These impulses are etched into our very genes by long-ingrained survival instincts.

And yet, 99.999% of the deep ocean remains untouched and unseen, let alone explored. That is beginning to change. We have glimpsed some of the early revelations stirred up by these initial forays. Without exception, they are provocative. The more we look, the more we will find. The more we presume certain regions to be inert or barren, the more we will be stunned by the sheer vitality and variety of the life that boldly asserts itself there.

Our opinions and imaginations are uniquely ours. They are a prism through which we interpret the universe. But it is one calibrated to human senses alone. The discoveries that await us in the years, decades, and centuries ahead already exist. They thrive, as they always have, beyond the veil of our understanding. We will not create them. We will only uncover them.

If we believe ourselves to be the only ones seeking, we are mistaken. The veil stretched over our perception can be pulled aside or rent asunder at any moment. Who are we to claim that this hasn't already happened? Perhaps it happened to our ancestors. Perhaps it is happening to us right now, and we just haven't yet recognized it.

We will never cease our inquiries into the mysteries of creation. There are wonders within our reach now which may be too incomprehensible or shocking to behold. Will we persist in our pursuit of that which remains beyond our grasp while still clutching these familiar terrors inside closed fists of denial? How can we gain purchase or claim a new treasure bestowed by fate if we have not yet dealt with that which binds us unremittingly?

Have you ever braced yourself for the certainty that you are on the cusp of revealing a rotten curse, only to be miraculously ambushed by the glowing vitality of a blessing? Even if it *is* rot that you find inside yourself, rejoice that you can eject this disease and replace it with something pure. The malady that afflicts us is the joining of willful ignorance to the rejection of that which does not conform to one's internal expectations regarding the true mechanisms of reality. The cure is the disavowal of those same expectations and the courage to accept what *is* in the face of what we expect or predict. We can pull our own veil off at any time, but choose not to. We prefer the gauzy vision which has grown so familiar and dear to us.

*Pelagomorph* ruffles and billows this curtain as they brush past us in ways we do not anticipate and cannot expect. This is a magnanimous gift

386

which we can choose to recognize as such. They are inviting us gently to come into awareness of this barrier, which we clothe ourselves in as a shroud.

The transformation we are destined for is no more than a flicking away of thought, a sudden moment of realization that arises not from a complicated equation or rigorous regimen of mental fortitude. It is the cessation of struggle that will bring this breakthrough. It is the annulment of duality that begets enlightenment. The utter annihilation of certitude fertilizes true understanding. The greatest obstacle to discovery is not ignorance. It is the illusion of knowledge.

One need not murder the mind. Who is the victim and who is the perpetrator? When we assume nothing, we receive everything. For many of us, the real task is only to accept, when we are so habituated to defining, solving, delineating, and classifying.

Even now, some who read these words dismiss them. You are not incorrect to do so. You are glorified in this. It is not unwise to pursue. It is not wise to cease inquiry. These merely are. When we abandon both, we position ourselves for attainment.

Do with this information what you will. No two souls can occupy the same location in time, space, or thought. A peach blossom is a peach blossom, and a mountain is a mountain. Tomorrow, they may trade positions. But in this moment, they are just so.

After all of this, what else is there to say? Perhaps nothing. But one idea does persist.

If *Pelagomorph* have truly been on their path for 500 million years, it makes one wonder. What variety of experience have they been witness to? This stretch of time is so beyond our ability to consider that our minds falter at the mere conception of it. What patterns might emerge and recede in even 500,000 years of local space-time?

Science now recognizes that evolution can converge, arriving at the same solution, outcome, or destination repeatedly across different species. Many examples of this exist. Echidnas and porcupines have both developed defensive spines. Echolocation is wielded by bats and dolphins alike. When distinct branches of the tree of life face similar environmental pressures and occupy comparable ecological niches, identical or closely related traits may emerge. This can and does occur along separate evolutionary paths.[497]

The key takeaway from our discussion is that convergence is an expected and repeatable process across the planet's flora and fauna. What life forms may have preceded us down this path, not just in known history, but in the unrecorded past? What other quirks or capabilities might genetic complexity have expressed, long before our awareness began to measure time?

From our standpoint within local space-time, we have no way of knowing. But when we consider convergent evolution, we must ask, what of true awareness? We refer here not to *sentience*, or the capacity to feel. We are instead contemplating *sapience*. This is wisdom, reason, and self-reflective cognition.

If *Pelagomorph* truly possess this trait, as we do, then sapience would be another point of convergence. It would have arisen in two distinct life forms at two separate epochs of evolutionary history.

That leads us to the inevitable question. Are we the only two?

Could there be other life forms on Earth, even now, that share this trait?

It's a compelling idea to consider. Has some other unknown being boasted this advantageous genetic attribute? Are they obscured by the veil of deep history, which is now inaccessible to us?

If so, what was their fate?

Did they fly too close to the sun, or tempestuously court some unknown variety of self-annihilation through technology or ecological disaster? Or was their downfall something stranger still? Were they swept away by one of the many planetwide extinction events we see recorded in the fossil record?

Surely, we would find their remains there among the destruction encoded in ancient sediment. Perhaps one of the forms of life we find there did, in fact, rise to this level.

If *Pelagomorph* can achieve quantum coherence, is it not conceivable that another organism might have done the same? This, too, would be a form of convergence.

Is a quantum convergence different than any other?

If it arises from genetic selection under environmental conditions and stressors, then it follows that this survival advantage may have been adapted by more than one organism.

Unfortunately, there is no current method for humans to definitively determine the presence of an advanced society that may have existed at some point in prehistory. The previously cited maxim, *absence of evidence is not evidence of absence* certainly applies here. But the inverse is also true. We cannot, in any intellectually rigorous way, argue *for* or *against* the presence of such a civilization.

The most compelling reason to even consider the possibility is one we've returned to repeatedly in earlier chapters. Life appears to possess an indomitable, insatiable hunger purely to exist, wherever and however it can.

Our imagined framework for the emergence of sapient life tends to include markers like aerobic respiration or multicellular complexity. But perhaps that framework is more pliable or slippery than we like to admit.

Would a "complex" anatomy even be a prerequisite? More specifically, how do we define "complex"? Is it bilateral symmetry, grasping appendages, opposable thumbs? Is it a hyper-enlarged neocortex with just the right cocktail of neurological homeostasis?

Or might it have nothing to do with hominid design at all?

Could a fungal network or bacterial colony achieve such a state? Our human perspective reflexively rejects such possibilities as absurd. Perhaps.

But what of a life form or evolutionary offshoot that we can barely conceive of? Are we constrained by inherited assumptions about what counts as intelligence or wisdom?

Life's resilience is beyond dispute. Its persistence seems almost supernatural. With that in mind, it is not unreasonable to wonder whether some unimagined and forgotten form of life once achieved a level of sophistication that would shatter our current sensibilities.

The timeline we have proposed for the development of *Pelagomorph* suggests a "noetic" expansion around 500,000 years ago. As noted earlier, this would place it roughly 200,000 years before the emergence of modern Homo sapiens.[498] In planetary terms, this is still a relatively recent occurrence.

Hominids have existed for far longer, somewhere in the range of 6 to 7 million years.[499] In the grand arc of evolutionary time, even that is the blink of an eye.

By contrast, the Cambrian Period, which saw the first widespread explosion of complex life, is estimated to have begun around 540 million years ago.[500] The span between the first hominids and modern humans fits into that era roughly 80 times. In theory, that suggests 80 distinct windows in which some form of advanced, sapient life *could* have developed.

With that much latitude, even a slight possibility becomes worthy of consideration. Another life form may have developed quantum coherence. Whether it was similar to *Pelagomorph* or entirely different, the prospect is worth contemplating.

Of course, all of this remains hypothetical. But it becomes more thought-provoking when we consider specific inescapable themes. These include enigmatic technological artifacts, unexplained geologic anomalies, monument alignments such as those at Giza, and enduring myths and legends passed down across cultures and millennia.

We won't dive into those subjects here. They have been explored at length elsewhere. What matters in this context is the question they provoke.

If such a society once existed, somewhere in the forgotten folds of unrecorded antiquity, what became of it?

More specifically, what fate might befall an advanced civilization *besides* self-destruction or extinction by natural disaster?

Even if no prior civilization left behind evidence we can interpret, we are not left without clues. The mechanisms that shape us, down to the smallest building blocks of our being, point toward something extraordinary.

We have previously noted that quantum effects are present in the biological processes of both humans and plants.[501] One particularly fascinating example is the role quantum mechanics may play in DNA mutation. This is a fundamental mechanism of genetic development and evolutionary progress.

Mutations can arise from various sources, including exposure to radiation or chemicals. However, recent research has begun to point toward another potential culprit behind transcription errors, *quantum proton tunneling*.

DNA is composed of base pairs, each made of two specific nucleotide bases: Adenine (A), Thymine (T), Cytosine (C), and Guanine (G). These form pairings, A-T or C-G, are critical to the accuracy of genetic replication. Occasionally, however, these pairs become mismatched. This produces what is essentially a typographical error in your genetic code. When this occurs, it's called a point mutation.

As mentioned, this can happen during cell division due to external influences. But during this process, something remarkable occurs. Protons travel back and forth between the two sides of a base pair. In doing so, they encounter resistance from an energy barrier that lies between two stable positions.

In classical physics, the proton would need sufficient thermal energy to "jump" or arc over this barrier. But here's where quantum mechanics defies our expectations.

Through the process of quantum tunneling, the proton doesn't jump at all. Instead, it "blips," vanishing from one side and appearing on the other, as if it had passed directly *through* the barrier. This can occur in less than a quadrillionth of a second.

By classical standards, this should be impossible. And yet it happens.

In fact, it occurs consistently enough to contribute meaningfully to mutation rates.[502] This fuels both disease and evolution alike.

Once again, we find ourselves confronting the limits of our traditional view of the universe's operation. These limits dissolve at the quantum scale. In this realm, the barriers we consider absolute are as insubstantial as thoughts and dreams. In the quantum world, the distinction between "particle" and "wave" collapses. Protons are both and neither simultaneously.

Because they possess the properties of both particle and wave, they can tunnel across energy barriers in an astoundingly rapid way. From our perspective, the proton just *appears* on the other side instantaneously.

However, when the position of these protons is measured, an errant shift can occur. This leads to the genetic "typos" we mentioned earlier. These mismatched base pairs may be copied during the replication process. In doing so, they are carried forward into the genetic code, becoming permanent mutations.

Fortunately, our cells are also adept in their capabilities. They contain repair systems designed to detect and correct such errors. But not all mistakes are caught. When one slips through, it can become a persistent error, the seed of either disease or evolution.

This final point is crucial to our discussion.

Just as other binary outcomes in quantum systems are resolved only when measured or observed, this process manifests similarly within biology. These mutations, caused by a mismatch during the operation of quantum tunnelling and preserved in the genetic code, are neither inherently good nor inherently bad. Many known genetic mutations have conferred both negative and positive evolutionary consequences to humans and other species.

One potent example is the mutation responsible for the calamitous condition of sickle cell anemia. Individuals who inherit two copies of the gene, one from each parent, will develop it. This has severe and debilitating effects on the body. And yet those who inherit only *one* copy are granted lifelong resistance to malaria, one of history's deadliest infectious diseases.[503]

In this way, we see that injury and advantage may occupy opposite faces of the same spinning genetic coin. Once tossed into the turbulent air of probability, this duplicitous doubloon may incur a burdensome cost or redeem a ruinous debt.

Evolution, then, has roots that reach directly into the quantum realm.

The implications are far-reaching.

Many revolutionary ideas in science and nature have been dismissed, ridiculed, or ignored until the evidence became overwhelming. This insight remains underappreciated in much the same way. DNA is not exclusively a set of instructions for biological replication. It *is* that. But it is also something more.

We have explored many examples of how quantum states and effects not only exist outside the framework of classical physics but also directly contradict it. Quantum reality reveals that the rules governing our shared experience inside local space-time are illusory at their core.

Humans imagine that realms beyond our everyday perception must be insubstantial, dreamlike, or somehow "less real." But this is reversed.

It is *our* so-called physical reality that is gossamer in its construction. We inhabit a collectively rendered and intricate illusion. Many who have glimpsed or briefly entered other states of consciousness or perception already know this to be true.

The human experience *is* real. It exists. However, we can mistake our own plane of reality for the central or primary one. It is, in fact, one among a multitude.

You don't need to believe this any more than you need to believe in black holes, quasars, or neutron stars. These things exist whether we affirm them or not. We are but insignificant observers, lodged in one specific corner of the space-time array.

But this need not limit us any more than the energy barrier inside a base pair of DNA limits a proton. Just as it may exist as both particle and wave, yet manifest as neither, so too may we engage reality in ways not strictly bound by duality.

The reality of 'then' and 'now' is not that they are two separate places in time. They are one unified state, bifurcated only by our sensory perception.

Think of the senses as a filter or sieve. They divide actual reality into channels we can interpret. However, those who have transcended traditional boundaries of perception sometimes describe synesthetic experiences, such as "seeing" sounds and "hearing" colors.

Likewise, the distinction between past and future is a construct of conscious interpretation. These are not fixed territories, but rather the effects of how we experience reality.

This is not a refutation of our shared local space-time experience. It is only a reminder that this is *one* way to encounter the totality of what reality can be.

Some may push back on this thought, for reasons as diverse as the ways they express them. The author offers no rebuttal to those who reject these ideas.

The experience of human consciousness may inherently entail different forms of perceptual partitioning. It is not the role of any one individual to define the boundaries or limitations of another's conscious experience. For who among us truly knows what their neighbor sees, hears, tastes, touches, or feels?

Until we discover a way to connect one human's sense array directly to another's cerebral cortex, we will remain unaware of the whole truth of individual perception.

And yet, if we shift our focus to our shared point of origin, our genetic expression, we may find common cause to celebrate. Our so-called blueprint is woven from processes that originate in the quantum realm. Is it not reasonable to expect that the *purpose* of this expression might extend beyond mere assembly and maintenance?

If our destiny were merely to remain tethered to this local, classical space-time reality, then why would our DNA rely on quantum tunneling?

Wouldn't it be more consistent to anticipate a quantum outcome from a quantum source?

Human history is full of mystics and wise teachers reminding us that we all return to that from which we came. This is not something that needs proof. It is something many rightly *feel* to be true.

If other branches of the evolutionary tree achieved sapience in prehistory, then it is possible that they did not perish at all. Perhaps they ascended to a higher form of existence within the quantum domain.

Many events reported by humans defy traditional scientific explanation. When we begin to examine the strange and unpredictable properties of quantum behavior, holding them up beside supernatural or unexplainable phenomena, parallels emerge.

Correlation is not causation. There is no proof to connect the two. However, the resemblance is certainly suggestive.

Quantum behavior echoes aspects of the unexplainable. Biology itself is entangled with mystery. Maybe it's worth revisiting the experiences we are so quick to dismiss.

The limits of the human sensory array render it incapable of interpreting various facets of reality that fall beyond or near the edge of perception. If we take this into account, then some of the mysteries of our existence might become at least partially comprehensible.

People sometimes refer to "glitches in the matrix." Others describe encounters with non-human entities or dreams that seem to predict the future. These and other phenomena that fall outside the bounds of conventional experience might not be imagined at all.

Many of these *can* be explained by mundane factors. Psychological conditions, cognitive limitations, confirmation bias, or simple misinterpretation of natural stimuli are all routine. But even the most inveterate skeptic must concede that *some* accounts defy categorization.

Throughout history, humans have reported mystical and miraculous experiences with a regularity that can only be brushed aside by those most committed to disbelief. It is entirely possible that these doubters will one day be proven right. However, those reading this book may recognize that such rejection typically stems less from logic than from generalized contempt.

The argument here is not to blindly accept every claim as accurate.

Rather, it is to resist the reflexive dismissal of that which falls outside one's own experience or frame of reference.

From the beginning of this work, we have returned to themes such as ethnocentrism, cultural blindness, and the subtle ways in which one's lived experience both shapes and constrains perception of the world.

We tend to treat our experience of reality as *true* and any experience that contradicts it as *false*. As we have noted repeatedly, though, the real issue is not where the boundaries lie. It isn't even which perception is "correct."

The issue arises from our use of *true* and *false* as conceptual armor. These are purely thought constructs that are deployed to prop up a particular opinion or worldview in the face of something that feels threatening.

So, what are we to do with this kind of perceptual wrinkle? How do we overcome this hum of contradiction that irritates and provokes simply by existing?

The honest answer is one that many enthusiastic participants in modern culture may be unwilling or unable to accept.

For the proper solution lies not only in the willing denial of the panacea offered by our "plugged-in" societal structure, but even of the very notion of the *self*.

It is not a popular or widely sought-after suggestion. But it is nonetheless the correct one.

Each of us must, in our own way, reject the mechanisms and tools of 21st-century life. They seduce us steadily into mistaking division for truth and convenience for meaning.

The author harbors no real hope that this will happen willingly or in any sustained fashion. At least, not in the foreseeable future. And yet, it remains the only way humanity might ever rise to the level *Pelagomorph* now inhabits.

One mystery surrounding their apparent ability to achieve quantum coherence may be their resistance to the self-imposed shackles of "amenity" and "sophistication." These are trappings that humans eagerly adopt.

Was it their cradle in the deep ocean or some aspect of their cognitive architecture that served as inoculation against such seductions?

Do *Pelagomorph* have governments? Do they bow to religious archetypes or seek networks of leisure and entertainment? Do they use intoxicants or rely on systems of medicine?

Is there a prevailing philosophy they abide by? Or do they debate, as we do, the same questions we now ask of them?

Does their access to a realm where time and space are not fixed coordinates but fluid variables render all these possibilities obsolete?

There is only one way for us to know for sure. We must join them in that place.

How do we get there?

The stickiness that binds us to our unipolar station inside local space-time is our unwillingness or inability to recognize, resist, and reject the pervasive nature of duality. This influence presents itself as immutable and absolute.

It is neither.

Instead, it is the fertile soil from which all other illusions of opposition and dichotomy spring forth.

Just as a proton assumes the properties of both particle and wave, while ultimately becoming neither, so too must we learn to behold ourselves.

Almost every conscious moment of every day, we seek to divide the world into halves. We elevate one and discard the other.

One vaunted, the other vanquished. One cradled, the other abandoned. One validated, the other refuted.

We habitually build our realities one brick at a time. We accept one half of the universe as concrete, labeling the other as smoke. But the labels could be reversed at any time. Neither was permanent. They are, by design, immaterial.

So why do we do this?

The reasons are as varied as the humans who wield them. Many trace back to the same reflex. It is an ingrained ancestral response born of fear and insecurity.

The organizations and institutions that conceive, engineer, and distribute these engines of duality are well aware of this. They exploit this reflex with precision. They design for maximum leverage.

And though we recognize their motives, we continue to slip their yokes around our own necks willingly.

Why? Because the seduction they offer is nearly irresistible. Its cadence is reassuring. Its supply of certainty is palliative and constant.

The reconfirmation of our chosen "side" in the war of duality constantly tempts us to drink deeply from the well of self-assured bias and predilection.

We all long for something or someone to echo our beliefs and reconfirm our choices. Too often, the pursuit of resonance replaces genuine self-reflection. It becomes a surrogate for inwardly focused candor.

We seek agreement, or at least sign-off, for the beliefs we spackle together to define a "self." We then deploy this structure against the chaos and turmoil that informs the universe at large.

One of the most seductive illusions available to sapient beings is that they possess control over their lives.

Those who have had this illusion unceremoniously and brutally removed know the danger of such a belief.

These humbled observers are blessed in their nakedness of thought. This stripping away is many times the first step toward true freedom from the trap of experiential duplicity. This prison of thought convinces us we are defined by a chosen state.

Only by denying this malicious apparition of partiality can we begin to perceive the deeper delusions that hinder our true nature.

Freedom is not a state to be attained or achieved. It is a state to be *accepted*.

Its only enemy is entrenched allegiance to the imagined protections of conjured boundaries. These barriers exist solely in the mind of the conjurer.

If you have not yet had your illusion of control shattered by circumstance, do not rejoice. The ruination of your comfort approaches.

The soul within you is eternal. But the shell that houses it is a temporary whirling dervish of causality which already strains against the limits of its own inertial momentum.

Entropy wins the game. Chaos is in no hurry. It waits patiently, threading its tendrils into the heart of order and intention.

All is sacrificed on this altar eventually. It needs only to wait for us to arrive.

Our only actual influence is over the shape we will take when we arrive. Will we enter drenched in enmity and spite? Or will we laugh joyfully as we bite down harder on the tribulations that coil around the very cornerstones we have so carefully installed in preparation for the onslaught?

Demolish your own foundations before life does it for you. This is the only path to authentic freedom.

You will be lighter and happier once you abandon the river stones you have collected in your pockets along the path. As you rise weightlessly into a clearer vantage point, you will stand in awe that you ever coveted them in the first place.

Abandon *everything* that seeks your self-assured opinion. Relinquish all that invites ridicule and resentment to take up residence in your heart. These are the hungry ghosts. The more fiercely you resist the necessity of abandoning them, the tighter their grip becomes.

But once released, you will marvel.

You will at last see the impermanence of their hold. The audacity of the lies they whispered in your ear will be made clear.

Nothing holds you. Nothing *ever* held you. Nothing governs your thoughts but thought itself.

We are like an insect buzzing and endlessly bumping against a partially open window. We remain trapped not by the barrier but by our inability to *see* the way around it.

Only when we slip into the sunshine do we fully comprehend the nature of the obstruction that once seemed total and impenetrable.

It was partial. It was invisible. It was a fiction.

Only through reorientation, not resistance, does flight become possible.

How do we reorient ourselves? How do we become free?

We must first reclaim the helm.

We must reject all forms of opinion and thought that are manufactured elsewhere and handed to us as substitutes for our own cognition.

This includes any media or ideology that requires nothing of us but consumption. Anything that offers a prepackaged position in exchange for the surrender of critical thought is a poison. It is death. Death of the mind. Death of the spirit. Death of the soul. Death of the true self.

In submitting to it, we become hosts for a parasitic virus that consumes the self and reanimates the husk with comforting delusions.

This is not a condemnation of all media or external influence. It is a warning. Consider it a flare fired into the fog of passivity that follows the choice to serve as a vessel for someone else's script.

There is comfort in this abandonment of agency. That comfort is weaponized against you.

Reject it.

Smash your screens.
Kill the signal.
Discard the mask.
Power down the feed.
Bury the phone that buried your attention span.
Delete the app that deletes your will.
Every platform is a pulpit.
Every feed a leash.
Sever them. Sever all of it.
Wave goodbye to the algorithm as it burns.

Let your mind remember itself.

This is not rebellion. It is survival. These deceptions seek to hollow you out completely. They carve you into a container fit for forces we can scarcely comprehend.

Annihilate the influence of politics, religion, allegiance, and all identity built on affiliation. Never join a group that defines loyalty by your willingness to condemn those who choose not to join. Dare instead to walk alone.

Refuse the plate prepared for you by another. Even if it bears no poison, *what choices were passed over in your absence*? What ingredients were omitted that might have nourished you and *only* you?

The ocean of existence relentlessly batters our fortitude. The greatest temptation arises in the most violent of swells. The offer of another to take the wheel is most enticing when the waves seem poised to consume us.

In that moment, it may seem a blessing.

Resist it.

Once surrendered, you may never summon the will to reclaim it. Not even as the jagged cliffs rise before your bow.

Attention, friends. This is among the greatest treasures we possess.

Attention to self. Attention to detail. Attention to the moment at hand.

Intention begets decision. Decision begets action. Action begets our reality.

What we pay attention to *becomes* our reality.

If we attend to a proton as a particle, it is. If we attend to it as a wave, it becomes so.

Let us attend to both. Let us attend to neither. Let us attend to *ourselves* in the same way.

We are not required to ascribe to any defined litany of decrees in this life. Every voice that demands our attention to the exclusion of all else is insidious and ruinously self-possessed.

Instead, we can choose to *perceive that which is* and allow it to be.

Has any opinion ever changed the orbit of a planet? Can our thoughts about a star alter its path by even one degree?

We may act upon our local reality however we choose. But how often do we *truly pause* to consider the outcome? Of both action *and* inaction?

Most of us begin and end each day in a nearly unconscious trance. It is an unquestioned habitual concert of thought and movement that is reenacted endlessly.

Our dreams. Our anxieties. Our hopes and fears. They roll over us like tides, so regular they seem fated.

How simple, how profound then, is even one moment of attention before a choice is made? How many critically essential details do we miss by failing to truly *behold* the circumstance before us?

We press on with assurance and conviction of purpose. But are our choices genuinely our own? Or are we dancing mindlessly through a choreography of inherited patterns, guided less by reason than by the comfort of familiarity?

When we begin to pay attention, especially to the smallest and most seemingly inconsequential moments, we begin to inhabit our lives fully.

This is the first and most vital step toward truly *owning* our existence rather than merely occupying it.

If attention is the seed of conscious existence, then inattention is the shadow that eclipses it. The way we treat that shadow reveals more than we might expect.

This tendency to overlook what we do not yet understand exists even in the sciences we trust.

Early in the study of DNA, scientists believed that only about 2% of the genetic code was functional. The rest was dismissed as "trash" or "junk." It was thought to be inert remnants of evolutionary detritus, no longer relevant to the organism's living processes.

This was more than just a mistaken assumption. It was a gross underestimation of the genome's complexity. What was once labeled as "junk" is now recognized as critically important. It governs the regulation and expression of DNA, controlling how genes are activated and function.[504]

This realization is not just scientifically important. It serves as a striking analogy for how the human psyche processes and evaluates information.

When humans encounter a new field of knowledge, time and again we rush in to declare ourselves masters of it. We treat our first glimpse of understanding as comprehensive, rather than provisional. We struggle to leave space for future refinement or deepening insight.

This is a fault line in our cognitive wiring. We do not tolerate a knowledge vacuum easily. Instead of celebrating the unknown as fertile ground for gaining wisdom, we instead reject it as irrelevant or unimportant.

But this is rarely true.

And just as with our genome, so too with our thought life.

Most of us allocate perhaps 2% of our mental energy to the complex, engaged consideration of the experiences we encounter each day. We dismiss the other 98% as noise, handing it over to unconscious automation. We drift.

You arrive home from work and realize you can't recall a single detail of your commute. You reach the end of a page or a scene in a show, and realize you absorbed none of it.

Your brain was disengaged and floating.

We extend this mental detachment in countless subtle ways until we are no longer *interacting* with our reality, only *reacting* to it.

And yet, our brains *are* designed to disengage from active thought, just not for this purpose.

Like the so-called junk in our DNA, mental disengagement is not a flaw. It is a mechanism for *realignment*. It creates space. It allows active thought to be repurposed and re-expressed.

Dreams are one way this happens. But there are others.

We can engage in this restorative process by allowing our thoughts to quietly be, without force or filtration.

If you find yourself checked out on your way home or lost in the middle of a film, maybe what's needed is not critical self-analysis but balance.

It's time to reclaim the other 98%.

This is not a complicated or esoteric concept. It is as simple as allowing your brain to follow the path it naturally seeks.

As an adult, when was the last time you sat on the ground and truly *looked* at a flower, an insect, or even a patch of grass or dirt? These are moments that reorient us towards a more authentic experience of reality.

If a wave of dismissal or ridicule rises in you at the suggestion, take note. That response is a signal. Your brain may be starved for realignment and honest expression.

This is not folly. Never mistake a childlike nature for childishness.

They are not the same.

One is wise. The other is foolish.

There is no goal in these moments, no expectation of a specific outcome. There is also no requirement to direct this kind of attention only toward objects born of nature or a pastoral landscape.

If you are on a subway train, you might instead behold the pattern on a stranger's clothing or the way the light scatters across a window. It might be a sound, a taste, or a flicker of movement. Although when it comes to subway smells, discretion may be the wiser choice.

This practice is valid, regardless of one's sensory configuration. If you lack sight, hearing, smell, or taste, the experience remains available to you nonetheless.

Likewise, you suffer no disadvantage from living in a dense urban environment. A hyper-realized world of possibilities blooms in such a place.

When your mind is in a receptive state, you need only choose to attend to *one* thing.

Have no opinion about what it is. No judgment. No need to label or evaluate.

Simply watch, listen, and feel.

Let this be a pure and unfiltered communion between you and the world.

It need not be long. It need not be repeated. Once exposed to this mode of being, though, most minds will instinctively seek to return.

While you are attending a moment in this way, don't be surprised when stray thoughts rush in. They will attempt to disrupt your focus. They may demand urgency or whisper that you are wasting time. You are under *no obligation*.

You are free to attend to *those* thoughts as well. Or to let them pass unclaimed.

If someone on the subway spills coffee beside you, it simply is. If you are in a field watching a caterpillar and a jet roars overhead, it simply is.

If your calf cramps and you shift your position, that too may be acknowledged, then dismissed.

When you feel that the moment is complete, be it two seconds, two minutes, or two hours, so be it.

There are no rules enclosing this experience. No gates. No conditions.

It is yours, and yours alone.

Receive it and be thankful.

If these concepts are new to you, celebrate.

The world at large will unfailingly appeal to your sense of self-doubt, anxiety, hesitation, and resistance to change. It will whisper that there is no use in this pursuit. It will tell you that reflection is indulgent and stillness is foolish. It will assure you that you are better off continuing along the path you already know.

There is nothing easier than surrendering to the familiar. There is nothing more transformational than abandoning it.

Transformation may *seem* radical. It is not. It is the most natural thing in the world.

The social, cultural, and institutional currents that swirl around you will demand, entice, and implore.

They will plead for sensibility. They will insist that you submit to their influence, choose a side, and declare allegiance.

But you are a free soul. You owe no debt. You carry no binding devotion unless you choose to.

The world will insist that everyone must choose a side. That everyone must champion one thing and denounce its opposite.

Nothing prevents you from doing that. Nothing, that is, except your own free will.

There is another choice. A third path. One that is never offered in the selection.

You are free, friend, to choose *neither*.

When you refrain from aligning with polarity, something miraculous happens. When you choose instead to inhabit the present moment fully, you gain access to the entire spectrum of choice seen from a neutral stance.

There is great value in this position. It will not make you wiser than others. It will not crown you with special insight.

But it *will* offer you refuge from the manic dichotomy of deification and demonization. This split consumes so many aspects of our shared reality.

It is also essential to understand that this process will always be one of evolution and growth.

Even after the decision has been made to cast off the manufactured cloak of pontification and self-importance, you may once again find yourself wrapped in that shroud with no memory of how it returned to your shoulders.

This is a crucial moment. It is one that can entrap those who misinterpret it as failure or as a rebuke. It is neither.

Like so many moments on the path, this one also invites you back into a duality of thought. Its foundation is a poisonous lie.

You may feel compelled to either surrender to the shroud or rip it away in frustration.

Neither reaction is necessary. You need not accept *either* impulse into your reality.

Instead, it is enough to step aside as many times as needed until you establish a new continuum of presence and resolve.

Like the spilled coffee. Like the cramp in your calf. It merely *is*.

To commence a process of action too quickly is not failure. To resist the urge to act is not victory.

We are trained to define each passing moment as either success or failure. The only accomplishment that matters is awareness of the present moment, free from the grip of any expectation that would steer your decisions without the light of self-realization.

*Pelagomorph* are availing themselves of us in their own way.

But how does an entity in a higher dimension hail another that is nested in a reality below it on the unfolding scale of existence? More importantly, how can that message be properly received?

For us, it begins by accepting that we are currently present inside local space-time only as individuals. We are therefore temporally defined and limited by our own unhesitating self-identification as such.

When we begin to loosen our grip on the antagonistic thought-constructs that have been accepted without question or examination, we may begin to perceive ourselves as *Pelagomorph* already perceive themselves. When we allow our true nature to emerge without the armor of inherited assumption, this kind of self-perception can bloom in earnest.

Five hundred million years is an inconceivable span of time. But only when we attempt to understand it from within local space-time.

Similarly, the behaviors, movements, and communications of *Pelagomorph* may seem mystical or abstruse from our current vantage point. That impression arises not from *what they are*, but from *how we see*.

We are not separated from this knowledge by some uncrossable chasm. We are separated only by a shift in viewpoint.

It will not be an earthquake of revelation that frees us from our ontological inheritance. It will be a voluntary relinquishing of the extinct and rotten abstractions that have failed us time and again across the arc of our so-called progress.

Our true existence is rooted in that which we now label *quantum*. We are already beginning to unfurl, undulate, and reach toward this deeper state of being.

We reach with the same quiet yearning that draws a houseplant toward a shaft of sunlight, spilling through a dusty window.

The truth behind the encounters humans choose to label as "UAP" or "UFO" is far more marvelous and thrilling than the notion of a life form visiting us from a distant star.

Yes, it would be flattering to imagine some beneficent intelligence traveling light-years to initiate contact with us. But we must awaken from this romantic, didactic fantasy and recognize the obvious.

We are not the first nor the only sapient life form to emerge on Earth.

We are not the conductor of evolution's orchestra. We are dancers in its ongoing choreography.

Here indeed lies a great irony. Conductors cannot directly experience the symphony they command. We are being invited to dance. We are being drawn into the choreography by a thoughtful, more experienced partner in this endless celebration of life.

This exultation of creation is ongoing and unceasing. Within it, every hidden mystery that eludes us waits to be revealed. Every secret hope and dream already resides there.

We need not build a craft. We do not need to invent new technology to enter this space. We need not even knock upon the door.

A hand is already reaching out toward us in invitation.

Our only task is to notice, to honor the entreaty, and take the hand extended.

Once accepted, the weight of this world's vexations will fall away. Our ascent will no longer be imminent. It will be *realized*. It has already begun.

We will never wholly define or characterize the intentions of *Pelagomorph* from the narrow station we now occupy.

Only when we pass beyond the barrier of perspective that defines us today will we be empowered to experience those intentions directly. We will do so not in fragments or glimpses, but in their whole living truth.

Throw off the yoke of duality along with the fruitless self-reflection that leads only to stagnation and vanity.

What you do not know can still be valid. What you *do* know may be immaterial.

Learn simply to *be*.

Enter boldly into a place of experiencing that which surrounds you, untethered from your opinions or preconceived notions about the smallest of details.

Stop.
Look.
Listen.
Smell.
Feel.

Reject the preformatted, cultivated fountains of novelty and bias. Reject them all.

The goal is not the annihilation of information. It is the naked, raw acceptance of what *is*.

Turn away from every force that insists on a narrow and inflexible ribbon of opinion or thought. Sacrifice every heart-bound enmity, every arrogance, and every shard of pride upon the altar of equity and recompense.

These are thorns and daggers. They are forged not to wound those you aim them at, but to impale and blind your own potential. Relinquish them all and claim your true position.

Step into the flow of thought and existence that has been waiting for you all this time. Enter not by moving forward, but by ceasing the resistance that has defined your being. When you suspend opposition, your true potential blooms into the superlative structure you are not just destined for but already partly inhabit.

*Pelagomorph* should not be seen as a visitor from afar, but as a neighbor who has pulled us free from a collapsing structure. They attend to our dazed perception as we blink in bewilderment at the new reality which greets our senses.

They do not seek to mystify or beguile us. They offer resuscitation, rejuvenation, and a return of comprehension.

This is not an invitation. It is a welcome.

We need only to remove our preordained orthodoxy of conception to understand this. We already inhabit the space we desire. When we surrender the clever facsimile, we will grasp the undeniably authentic.

Simply be, friend. For you simply *are*.

# EPILOGUE

**Phylogenetic Apparitions: Temporal Refraction Decrypted**

*"Lillian was reminded of the Talmudic words: 'We do not see things as they are, we see them as we are.'"*

– Anaïs Nin, *Seduction of the Minotaur*

The last thought we should consider together is one that has certainly been touched upon in these pages and remains central to the theories we've explored. Every human experience of reality is unique, individually shaped, and solely perceived by the consciousness that inhabits it. It is felt as distinct from the perceptions of others and even from the seamless flow of reality itself. This separation is a self-conjured illusion, born from a natural, ongoing filtering process designed to impose order on the flood of incoming stimuli.

This process is a double-edged sword. It softens the cacophonous onslaught of sensory input, while simultaneously editing and pre-shaping how each mind interprets this firehose of data. Undeniably, it is a blessing to filter or limit the information we receive. But this same mechanism leads us to mistake the curated, filtered stream for the totality of available data. It is not.

If you have ever become fully absorbed in reading, watching a film, studying, writing, or any activity that demands concentration, you have experienced this firsthand. You may fail to notice a car driving by outside or a voice in the next room. Likewise, if you are thoroughly engrossed in thought while driving, attending a lecture, or completing a task, your awareness may exclude anything not directly related to the immediate goal.

It's easy to imagine how this filtering process benefited our evolutionary ancestors. Focused attention was a survival tool, vital for hunting, gathering, or avoiding predators.[505] Yet in gaining this advantage, humans may have traded away an even greater one. What of our capacity for reason, judgment, contextual thought, and complex understanding? Too often, we settle into a curated version of reality, venturing beyond its boundaries only occasionally to explore the unassimilated remainder of available data, which tends to be discarded in the name of maintaining a stable waking consciousness. This is the self-conjured illusion we've been referring to.

This is not bad or wrong. It merely is. Accepting what we sense and experience in the moment as the whole of reality requires no effort. Yet this is fallacious and self-limiting. It subtly restrains our potential as sapient beings. We are well-suited, perhaps even designed, for a deeper form of inquisitiveness that turns inward to question the very tools we use to perceive.

It is natural and anticipated for a sapient species to move beyond the point where such rigid filtering is necessary. The process has served its purpose. It brought us here. However, many of us no longer require this narrowing of awareness for day-to-day survival or the continued propagation of our genes.

We may now direct our sagacity toward these boundaries. The aim is not to eliminate them, but to examine, define, and understand their operation. We no longer need to remain passive recipients of this filtered stream. We can instead become active agents, pinning down the edges, pushing past them, and recovering the discarded signals labeled as "noise" or "junk." These may be as vital, or even more so, than the narrow trickle we've been taught to consume.

As noted, each individual undergoes this process separately. No one person can truly know or understand what another person is perceiving in any given moment. This must always remain at the forefront of our dealings with one another. It is in these contrasts between distinct experiencers of awareness that we find the seeds of discord and disunion.

One way to observe this in yourself is to recall a time when you encountered something that filled you with joy, resonance, or wonder. Any experience that left a deep and lasting impression will do. It could be a particular dish at a restaurant, an engrossing book, a soul-stirring film, or a trip etched into memory. As social creatures, we feel compelled to share these experiences. This impulse to expose another to beauty or delight can be divine.

But when you recount such a treasured moment to someone else, and they respond with indifference or rejection, it can produce a visceral reaction. Perhaps they disliked what you adored. Maybe they had no desire to try it again. Whatever the case, the emotional ripple this sends through you is a reliable yardstick of your unconscious reliance on the margins of your personal worldview. If your response is disbelief, annoyance, anger, or dismissal, it may be a sign that you would benefit from exploring beyond your own frame of reference.

Variety of opinion is a defining feature of human experience. It is not a flaw to be corrected, but a feature to be celebrated.

Even identical twins may have separate preferences, tastes, or fundamental traits that define them as individuals. This is to be embraced as life-affirming and should feel like a validation of one's own perspective rather than a contradiction of it. The reason is that we can view each opinion and perspective as entirely authentic for that individual, with no need or impulse for another to align with our own proclamations of identity.

There is objectively no *best* ice cream flavor or combination of pizza toppings. The author champions anchovy pizza with copious amounts of hot sauce as superior to all other configurations, but harbors no expectation that another soul will agree. No matter. When we detect an errant need or impulse to have others feel, perceive, or act the same way we do, it signals that we are mistaking our own cumulative view of reality for reality itself. Especially when their deviation causes us distress.

This is a binding of consciousness. It must be loosened from the psyche to facilitate a move beyond the seductive comfort of cloaking the self in a filtered array of subjective perception.

The quote that opens this chapter presents a simple, elegant approach to this concept. If we can gain awareness of the subtle yet acutely ingrained tendency to see the world not as it *is* but as *we* are, this becomes a monumental first step toward an expansion of thought. Such a new perspective can be leveraged to examine thought itself.

This is the magnificent and fantastical quality of sapience. It manifests an ability to turn thought inward, to train it upon itself. Like using a microscope to study a microscope, or standing close to a mirror to examine one's own eye, this reflective mode of inquiry can illuminate aspects of knowledge that escape all other forms of probing.

Not everyone is comfortable with this kind of exploration, nor do they need to be. It may, in fact, be the anchovy pizza of scholarly thought constructs. While not to everyone's taste, to some it is the only path worth pursuing with gusto.

It is a noble effort to become acutely aware of one's own curated arrangement of set pieces, the inner furniture populating our personal projection of reality upon the screen of awareness. Equally noble is the choice to refrain from judgment or positive and negative classification of those cognitive exhibitions, especially as they relate to one's own opinions or feelings.

Not everyone needs to regard Picasso as a genius for your admiration of him to remain valid. If you urge a friend or relative to sample a fresh loaf of sourdough and they decline without room for debate, that in no way diminishes the smell, taste, or sensory thrill of yon blessed loaf. Similarly, when someone insists you *must* try a new restaurant they describe as "transcendent," but you visited just last week and found it thoroughly underwhelming, your opinion does nothing to negate their joy.

This phenomenon, though typically felt in the most mundane corners of daily life, applies equally to the more esoteric, abstract, or ineffable realms of perception.

Through explorations of disparate contexts within time and space, we find that even the slightest variations in position, speed, or orientation can lead to significant differences in individual perception. This is not an anomaly. It is a fundamental feature of creation and the universe itself. Why this is so may remain a mystery from our current vantage point. But it is measurable. It is real.[506]

Therefore, this kind of variation in conscious perception should not only be acknowledged but venerated as a universal constant worthy of deep consideration in any attempt to understand how we construct our internal models of reality.

When we describe *Pelagomorph*, or any other life form that appears "impossible" or "magical," we are not confronting impossibility. We are brushing up against the outer limits of our own self-constructed ideas of what is practical or achievable. As outlined in earlier chapters, many life forms on Earth already perceive reality in ways quite different from the human standard. Their sense arrays, evolved through divergent pressures, result in alternative configurations of consciousness itself.

Consequently, it is entirely reasonable to imagine a life form such as *Pelagomorph*. A bio-symbiotic amalgam of cnidarian and cephalopod, shaped across a yawning abyss of evolutionary time into a biological system capable of quantum coherence. This is no less plausible than the supremely adapted organisms we have already encountered throughout this book.

Repeatedly, evolution shows us that it holds no allegiance to human expectation. If there is one consistent lesson to be learned from the persistence and ingenuity of biology, it is that nature exhibits a complete and utter disregard for whatever limitations we presume to place upon it.

This is why we must hold ourselves to a higher standard, not only becoming *aware* of our perceptual biases but actively rejecting them. We must stop insisting that the world conform to our perspective and instead make an effort to resonate with how the world sees *itself*.

This means rejecting not only epistemological centrism but also ontological chauvinism. Believing that our way of knowing is the only way is just as limiting as believing that our form of being is the default or the pinnacle of what is possible.

These projections of the self onto the universe do more than distort reality. They lull us into a comfortable but limited illusion. It not only clouds our vision but also actively conceals what lies just beyond it.

This concept is both ancient and oft-cited. Many examples exist. Perhaps the most well-known is Plato's allegory of the cave. Plato describes a gloomy, cavernous space in which the inhabitants observe shadows cast upon a wall by a fire that burns behind them. Imprisoned there since birth, these individuals perceive only the shadows and never the objects casting them. They mistake the shadows for reality itself.[507]

Only when they are freed from the cave can they begin to grasp the actual relationship between shadow and form. First, they must encounter the true arrangement. The fire, the objects, the outside world. Even then, they struggle to comprehend what they are seeing. Accustomed to shadows, they recoil from the truth. For many, this new reality would seem implausible or even hostile. We would expect them to reject it outright, viewing it as impossible or as a direct contradiction to the world they had painstakingly constructed from the only information available to them at the time.

In this way, we too encounter the world around us on a daily basis. What we take to be the "real objects" of our experience are, in truth, shadows projected onto our consciousness. We may, of course, choose to remain in this state. Especially if it brings comfort or peace. But once the illusion is challenged or dismantled, cognitive dissonance usually follows.

However, for some, the glimpse of true reality is not disorienting but exhilarating. It is an explosive affirmation of a long-standing, barely whispered truth that has been quietly flourishing in the heart. It is a confirmation that these ghosts are merely reflections of the divine.

This is the message offered here. This book seeks to extend Plato's allegory one step further. Egress from the cave is not just *possible*, but inevitable. Not optional, but deterministic. Not the privilege of a few, but the destined path of all conscious, sapient life.

Our eventual and immutable destiny does not lie within the cave. It waits far beyond the mouth of the cavern, where the weight of our delusions finally crashes against the formidable buttress of unfiltered, objective reality itself.

When will we sail into the port of our collective fate? No one can say. But the end of a journey remains forever out of reach until the first step is taken.

*Pelagomorph* may have traveled five hundred million years to arrive in this place. The reward they unlocked is genuine liberation from the unipolar flow of time and space. They have not been released entirely from its grasp, but they have gained the ability to experience it more directly and more beneficially. They align with the thing itself, while we continue to wrestle with its shadow.

We can begin by recognizing this gap in awareness. Just as importantly, we must reject the illusion in favor of what is actual. As previously discussed, a good first step is to acknowledge that our perception is flawed. We must also accept that we are capable of a different orientation toward truth. Perhaps we were even designed for it.

When we allow a neighbor's variation in experience to be just as valid as our own, we reconfirm and strengthen both. But when we dismiss another's perception as mistaken or invalid, we quietly undercut our own and, in doing so, cripple our potential.

If we are still chained in the cave, and a neighbor returns from the light to share what they have witnessed, will we listen?

Reject fear, friends. Turn away from the quick impulse to cling to the unexamined comfort of the familiar. Consider the impossible. Contemplate the unthinkable. Push past your own boundaries of consolation and dogma.

There lies truth. There lies growth. There resides our true collective destiny.

Carefully consider those things that prop up your notions of certainty. Especially those that may have been ill-gained. Attentively examine the sources of information and evidence you enshrine on subjective altars of belief. Many times, the most sacred must be sacrificed first. The deepest beliefs may bind you most fiercely.

Grow aware of passivity, and treat it as your mortal enemy.

The uncritical acceptance of any information must be actively and vigorously challenged by design. This is especially true when that information is given freely. In this way, flawed or erroneous elements can be identified and discarded. Just as importantly, it is how the portions that offer true edification can be confirmed and corroborated beyond reproach.

How can one begin the process of distinguishing the two?

The answer lies with each individual person. We all inhabit our own version of subjective experience. One consistently effective approach is to make a conscious decision to enter a space of observation. A state of solely attending to your own perception, in a nonjudgmental, unopinionated way.

You can choose to experience a specific moment of reality purely *as it is*, while calmly refusing any impulse to define or classify it. You need not retreat from daily life or isolate yourself in an elaborate setting. Sometimes, all it takes is a single second.

Even a fleeting moment of unvarnished attention can be enough to crack the seal that surrounds the shadowed cavern we inhabit.

Whether it be a sight, a sound, a smell, or a taste, you can make the conscious choice to receive it as it arrives, pure and unfiltered, without layering it inside opinion, story, or thought.

This type of attention may also arise suddenly, without ritual, preparation, or intention. It could arrive in a quiet moment when you are purely present, and something ordinary suddenly reveals itself as unaccountable.

One of the first times the author encountered this phenomenon was while walking along a somewhat busy street at night in California's San Fernando Valley. Upon turning to glance at the trunk of a palm tree, it gently, but suddenly, divulged its true nature. It appeared as both mystical and mundane at once. That is to say, the ordinary revealed itself in that moment as transcendent.

These are both valid and ever-present expressions of everyday existence. They are like two sides of a coin spinning so quickly that they blur into one another, forming a third, blended state. Our conscious minds try to filter this amalgamated appearance into something comprehensible and digestible. In doing so, we sacrifice parts of both. We trade the full richness of the moment for a stable and visually coherent rendering.

What often shapes this appearance is our own preconceived notion or expectation that "a tree is just a tree." Rarely are we compelled to pause and genuinely engage with independent objects in our environment as they are. This is mainly because we see no functional purpose in doing so. Yet such intention can be transformative. It allows us to perceive the world free from pre-formed expectations that color and ultimately define our sense of not only what something *is*, but what it *can* be.

When we surrender this construct, recognizing it as a veil of expectation, we enable our consciousness to perceive the thing itself. Free of assumption, its true nature is revealed.

The trunk of the palm tree appeared to respirate. It undulated subtly in its authentic form. This was not jarring or disorienting. It was wondrous and invigorating. It felt as though nature itself were expressing joy at being truly *seen*, in its full splendor and mystery.

The moment lasted only an instant, no more than a second at a leisurely walking pace. And yet, the memory remains vivid to this day. It is as clear now as it was the moment after it occurred.

There was no pretext of altered consciousness. No drugs, no meditation, no breathwork. Not even a prepared mindset or intention. Just attention, and a calm focus on the object, free of expectation or the need to define it as anything beyond its enduring nature.

Perhaps the reader has witnessed something similar. Perhaps not. No matter. Such moments are not proofs or mandates. They are only the result of a quiet desire that alights within the heart to see the world without prejudice or definition.

Where then does such prejudice and definition arise?

Unfortunately, this world is composed of innumerable sources of influence that seek to shape the opinions and outlooks of each individual consciousness. Their reasons for existing are as countless as their forms.

Some seek to control or persuade, others to divide or diminish. Still others aim to inflame, paralyze, distract, or subvert.

Their motivations and goals are less important than their presence.

In fact, part of their fondest hope is that we become so bogged down in the details of *why* they exist that we fail to escape their grasp.

Humans ask constantly of the universe:

"Why is this happening to me?"
"Where did this come from?"
"What should I do?"
"When will it stop?"

All valid questions, and deserving of answers. Yet, as many of us have discovered firsthand, those answers can be hard to come by. The unmet need for an answer can sometimes be more damaging than the event that birthed the question in the first place.

Still, we *can* overcome, even in the absence of resolution.

We do this by stepping away from the answer and instead turning toward the question itself. Why do some receive a devastating medical diagnosis? When will a loved one overcome a debilitating addiction? What is the source of a painful situation that seems poised to engulf our world? If we knew the answers, would that truly resolve anything? Or would it just raise more questions?

In moments such as these, perhaps it is better to calmly observe. To attend to the situation itself, without demanding explanation or revelation. At the very least, this shift releases us from the anxiety of not knowing, and from the crushing pressure to provide clarity in the face of overwhelming uncertainty.

When we approach such moments without the weight of expectation, we free ourselves from the burden that constricts and drags us down, if only partially.

This is not something that must be believed or championed by anyone. We have noted time and again that nature persists, with or without an observer. The reality we construct through our individual perception is but one layer or field of depth, among many.

If a tree falls in a forest with nobody around to hear it, it still makes a sound. However, it may be a sound the hypothetical listener would not recognize as such.

This, then, is a call to all who have ears to hear. Listen not for the tree, but for the sound itself. You may be thunderstruck by what you discern. But never disappointed. Never denied.

Unplug from your expectations. Connect instead to the deeper source, the one that has fed the headwaters of reality all along.

When we surrender our passive, tacitly accepted overlay of conscious awareness, we position ourselves to receive an everyday miracle. One that already surrounds us. One that eludes us only through its own ubiquitous profundity.

Identify the streams of inherited opinion and belief that infiltrate the sanctum of your perception. Divorce them. Ruthlessly, if necessary. Dig out the roots of insidious influence that creep through the most casual, tedious thoughts occupying your mental landscape.

Take ownership of even the smallest automations. Do this by rejecting every impulse that tries to bypass your authentic momentum, those selfish little stowaways seeking to override your direction with purposes that are not your own.

Honor instead what arises uniquely from the heart that beats in your chest, and yours alone.

Value your distinctive outlook *as it is*, not as others wish it to be. Boldly reveal your most hidden purposes to the universe. Exalt in the mutual recognition and admiration that follows.

The correct time to do this is *now*. For *now* is the only time we have.

Past, present, and future all exist in the same manner. It is only our perspective, our limited vantage point from deep within the cave of local space-time, that casts the illusion of a past fading behind us and a future just out of reach.

The insidious voices of influence will cry out, *"Then! Only then!"*

You must answer, *"Now. Only now."*

Whatever you seek in your heart of hearts, do it *now*. There is no other time provided to you.

Possibly the most destructive lie told by the chorus of ignorance and obfuscation that threatens to envelop our conscious experience is the one that tells us to delay or deny the present moment.

If there is one thing you take to heart from this book, let it be this. You, dear reader, have no other time available.

Will you wake up tomorrow? Perhaps. Will you finish this sentence? There is no way to know.

We mistake the persistence of the present moment as proof that it will continue to persist. But this is not so.

Yes, it is comforting to plan in advance. Dinner with a friend, a vacation a few months from now, a retirement that may be decades away. But none of it is promised. Not to you. Not to anyone.

If there is a single unifying force behind our collective willingness to accept the fallacies handed to us, surely it resides here.

Every lie of this kind hangs on the same illusion that *later* exists, that *options* await us, someday, somewhere further down the path.

But the only option is *now*.

We must accept this into our hearts. When we do so fully and honestly, without flinching, there is no longer any room for disdain, prejudice, self-importance, bias, or deceit.

Anyone who has experienced the loss of a loved one can attest to the random and seemingly aimless vagaries that escape the mouths of those trying to offer comfort in the painful aftershock of grief. People say things that are unhelpful, sometimes even hurtful. They do so because they remain stuck inside a false reality that echoes one thought endlessly.

"Not *me*. Not *yet*."

They do not recoil at the loss of *your* loved one, but instead at the stark reminder of their *own* finite providence.

Attend a funeral, and you will hear speaker after speaker striving to celebrate the life of the departed, seldom to reckon with the event of death itself. This is understandable. Death is an immovable barrier. It is one that resists every mental and spiritual assault we can muster. It is difficult, perhaps impossible, to comprehend.

Still, we are called to observe and attend to this boundary as well, free from any preconceived notions or inherited frameworks handed down by external sources. And when we do, we may begin to glimpse both sides.

Is the end of our life the end of *all* life?

It may feel that way. Or perhaps it is merely the end of one current of sensory input. A personal stream of memories and meaning, now gone quiet.

If we attend to the world as it is, with true neutrality, free of bias, free of expectation, what might we find? If we apply the same presence and openness to even the smallest, most mundane moments of our lives, how might it transform us? Individually, and collectively?

We may begin to realize that the distinction between life and death is not as absolute or defining as we once believed. We may even find that the barriers we cling to, the ones we use to separate ourselves from one another, are just as illusory.

The mystery we humans face in our perception of *Pelagomorph* is a story far greater than contact with an alien life form. It is, instead, the story of contact with ourselves. To encounter one, we must seek the other.

The good news is that *Pelagomorph* seems to understand this truth in a way we have yet to comprehend fully. And we are well-positioned, friends. The only things we lack are proper orientation and sufficient determination. One is born in the heart, the other in the mind. Free them both from the constraints this world seeks to impose.

Cease the passive acceptance of every unholy invocation offered to soothe hesitation and doubt. Rejoice instead in your essential, divine nature. It longs to bloom into its full measure, reaching upward and outward to touch its equal counterpart in every moment, in every object you perceive.

Marvel at your truly magical nature, which stands in stark and thrilling opposition to all that tries to bury it in shadow and confusion.

Your heart is beautiful. Your mind is a miracle.

This is the message *Pelagomorph* communicates to us. This is their invitation into the grand choreography of creation itself.

You can accept this invitation by allowing yourself to let go of every certainty, every deeply rooted belief to which you cling as an anchor. These are not a mooring, but a shackle. Break free. Enter the dance unencumbered by what you believe, and instead experience what simply *is*.

Behold everything you encounter, not as you are, but as it is. And in turn, you will find yourself beheld as you truly are.

This mutual release results in a corrective equilibrium that brings you into a balance with all that surrounds you, and also brings it into the same balance with you.

Here, you will discover a new definition of reality itself, where you transform from a being who exists and survives into one who truly lives and thrives.

We have seen in these pages that the ultimate purpose of life is to thrive. Life persists beyond every expectation of what it can or should be. We are no different. Our destiny extends far beyond anything we could have imagined for ourselves. Because life, our very genetic code, rooted in quantum processes, expects more of us than we could ever conceive.

When we surrender to this reality, we activate our true purpose. We connect to a place and a destiny that surpasses not only our wildest dreams, but the very definition of what we believe a dream to be.

We marvel now at the incomprehensible capabilities of *Pelagomorph*. Yet this is only a prelude, a glimpse of the realities available to us once we willingly unshackle ourselves from the self-chosen limitations by which we now define our being.

It is as simple as choosing a new definition. We can actively, consciously obliterate these voluntary constraints and accept only what presents itself, just as it is. For what we perceive has no use for our viewpoints, no need for our beliefs. Why then should we cling to the bias and judgment laid at our feet?

You are as you are. You will never be anything else. Seize this now, for tomorrow is a falsely defined space we cleave to in the hope that our consciousness will persist unchanged. Yesterday is the lie we tell ourselves to hold onto what we failed to experience fully.

So let us experience *everything* fully and continually. In this way, we surpass and transcend every expectation, every limitation, every possibility that ever was or could be.

Death is not the finish line. Birth is not the starting line. Life is not the journey between.

Your consciousness is a projection of reality imprisoned by its own perception of itself. Allow yourself to behold without assumption. Allow yourself to expect this with no hesitation.

We have matured beyond the elemental architecture of consciousness. The beams once raised to support and define us now confine and restrain our shared destiny. Like an adult turning to face their childhood home, we can finally perceive the nest for what it is. Ordinary, limited, and too small for what we are becoming.

Our new dwelling is the universe itself, unbound, limitless in expanse. And surely, there we will meet *Pelagomorph*, and much, much more.

Let us embark together on this path, free of illusions, liberated from prophecies proclaimed by others. We pledge fidelity only to what *is*, never to what *should be*. Let our north star be freed from anticipation, so we may follow where it truly leads, not where we wish it to deliver us.

See the universe as it is, and you may find you have already arrived at your destination.

Welcome, again, for the first time, to the place you have always been.

KEATON RYON

# ENDNOTES

## PROLOGUE

[1] The distinction between *Chronos* and *Kairos* originates in ancient Greek thought. While *Chronos* denotes sequential or measurable time, *Kairos* signifies the opportune, sacred, or decisive moment. For further discussion, see Paul Tillich, *Systematic Theology*, Vol. III (Chicago: University of Chicago Press, 1963), and Chelsea C. Harry, *Chronos in Aristotle's Physics: On the Nature of Time* (Springer, 2015).

[2] In Jewish theology, time is often understood as cyclical rather than strictly linear, with holiness recurring in patterns of sacred remembrance and renewal. See Abraham Joshua Heschel, *The Sabbath: Its Meaning for Modern Man* (New York: Farrar, Straus and Young, 1951), and Jonathan Sacks, *Covenant & Conversation* (Jerusalem: Maggid Books, 2009).

[3] Albert Einstein referred to time as a "stubbornly persistent illusion" in a letter to the family of his close friend Michele Besso, written in March 1955, shortly before Einstein's death. The letter appears in *Albert Einstein and Michele Besso, Correspondence 1903–1955*, ed. P. Speziali (Paris: Hermann, 1972; English translation, Princeton: Princeton University Press, 1979).

[4] Joe Haldeman, *The Forever War* (New York: St. Martin's Press, 1974), which explores the effects of relativistic time dilation on soldiers fighting an interstellar war.

[5] *Planet of the Apes*, directed by Franklin J. Schaffner (20th Century Fox, 1968). The screenplay was written by Michael Wilson and Rod Serling, adapted from Pierre Boulle's 1963 novel *La Planète des singes*. Serling's early drafts introduced the film's famous twist ending, which was later revised but retained his central concept.

[6] *Interstellar*, directed by Christopher Nolan (Paramount Pictures and Warner Bros., 2014). The film's depiction of black holes and gravitational time dilation was developed in collaboration with theoretical physicist Kip Thorne, whose work on general relativity and space-time inspired the story's scientific foundation.

[7] GPS satellites orbit at an altitude of approximately 20,200 kilometers (12,550 miles) above Earth's surface. Their onboard atomic clocks experience both special and general relativistic effects, resulting in a net offset of approximately +38 microseconds per day. These corrections are built into the GPS system's design. See National Institute of Standards and Technology (NIST), "Putting Einstein to the Test," and NASA, "Einstein's Theory of Relativity Critical to GPS, Seen in Distant Stars."

[8] Experimental confirmation of relativistic time dilation through muon decay was first demonstrated by Bruno Rossi and D. B. Hall in 1941, who measured the extended lifetimes of cosmic-ray muons reaching Earth's surface. Subsequent high-precision measurements, including modern experiments at CERN, have repeatedly verified these results.

[9] The Block Universe model, also known as Eternalism, originates with Hermann Minkowski's 1908 formulation of four-dimensional space-time, in which time is treated as a dimension alongside space. Albert Einstein's theory of relativity later provided the physical foundation for this view, giving rise to the modern interpretation of time as part of a unified continuum.

[10] The quantum no-hiding theorem, proved by Samuel L. Braunstein and Arun K. Pati in 2007 and experimentally demonstrated at the University of Bristol, establishes that quantum information cannot be destroyed, only redistributed within the system's environment. This complements the first law of thermodynamics, first formulated by Rudolf Clausius in 1850, which holds that energy can neither be created nor destroyed, only transformed. Together, these principles illustrate a scientific symmetry with philosophical and spiritual notions of continuity and persistence.

## CHAPTER ONE

[11] The aphorism "Absence of evidence is not evidence of absence" evolved gradually over more than a century. The earliest known partial form appears in 1887, when the Reverend William Wright wrote in his paper The Empire of the Hittites that "the absence of evidence is not evidence" during a meeting of the Victoria Institute in London. An exact version of the phrase is credited to Dugald Bell in The Glacialists' Magazine (1895). It resurfaced in 20th-century scientific writing, including usage by Martin Rees and Richard Berendzen in the early 1970s, before being popularized by Carl Sagan in The Dragons of Eden (1977), Cosmos (1980), and The Demon-Haunted World (1995). Sagan employed it to critique what he called humanity's "impatience with ambiguity," reminding readers that a lack of data does not constitute disproof. (Source: Garson O'Toole, "Absence of Evidence Is Not Evidence of Absence," Quote Investigator, July 2020, with acknowledgments to Stephen Goranson, Barry Popik, and others for archival contributions.)

[12] As of 2024, more than fifty missions have been launched toward Mars, beginning with the Soviet Union's Mars 1M No. 1 in 1960 and extending through NASA's Perseverance rover (2020) and China's Tianwen-1 orbiter and rover (2021). Roughly half have successfully reached orbit or the Martian surface, while others failed during

launch, transit, or entry. (Sources: The Planetary Society, *"The Mars Exploration Family Portrait,"* updated 2024; *Every Mars Mission,* planetary.org; *Historic Mars Missions,* space.com; *Mars Exploration Fast Facts,* CNN, 2024.)

[13] NASA's FY 2024 budget totaled approximately $27 billion, as outlined in the agency's *Fiscal Year 2024 Congressional Justification* (March 2023) and corroborated by congressional appropriations summaries. Roughly $13 billion, about half of NASA's total funding, supported spaceflight-related programs, including human exploration, low-Earth-orbit operations, and deep-space initiatives. (Sources: *NASA Fiscal Year 2024 Congressional Justification,* nasa.gov, 2023; Casey Dreier, "The FY 2024 NASA Budget," *The Planetary Society,* March 2024; U.S. Congress, House Committee on Appropriations, FY 2024 NASA Budget Hearing Transcript.)

[14] Because SpaceX is privately held, exact financial data are not publicly released. However, independent industry analyses estimate that the company generated over $13 billion in total revenue during 2024, with roughly one-third derived from launch and orbital operations and the remainder from its Starlink broadband division. (Sources: *Estimating SpaceX's 2024 Revenue,* Payload Space, January 2024; *How Much Money Did SpaceX Make in 2024?,* Nasdaq.com, April 2024.)

[15] Because Blue Origin is privately held, precise financial data are unavailable. However, reports from Forbes and Ars Technica suggest that founder Jeff Bezos provides annual funding exceeding $1 billion, supplemented by NASA and Department of Defense contracts, including the $3.4 billion Artemis Human Landing System award. (Sources: Jeremy Bogaisky, "Why Musk's SpaceX Has Grown Bigger, Faster, and ...," *Forbes,* Jan 18 2025; Eric Berger, "Citing Too Much Bureaucracy, Blue Origin to Cut 10 Percent of Its Workforce," *Ars Technica,* Feb 2025.)

[16] As of 2024, forty-six missions have been launched toward Venus, beginning with the Soviet Union's *Venera 1* (1961) and continuing through more recent orbiters such as *Akatsuki* (Japan, 2010) and *Parker Solar Probe* (U.S., 2018, multiple flybys). Roughly half achieved partial or complete success, returning data, entering orbit, or landing on the surface. (Sources: "List of Missions to Venus," *Wikipedia*, updated 2024; NASA, "Venus Exploration," science.nasa.gov, 2024.)

[17] As of 2022, only twenty-two human descents have reached Challenger Deep, the deepest known point of the Mariana Trench. The first was achieved by Jacques Piccard and Don Walsh aboard *Trieste* in 1960, followed by James Cameron's solo dive in 2012, with the remaining missions conducted since 2019 using the *DSV Limiting Factor* and China's *Fendouzhe* submersible. (Source: "List of People Who Descended to Challenger Deep," *Wikipedia*, updated 2022.)

[18] The Consolidated Appropriations Act, 2024 (Pub. L. 118-42), signed into law on March 9, 2024, provided a total of $6.72 billion for the National Oceanic and Atmospheric Administration. Of this, $4.55 billion funded Operations, Research, and Facilities, and $1.72 billion supported Procurement, Acquisition, and Construction. Within the Office of Oceanic and Atmospheric Research, only $46 million, less than 1 percent of NOAA's total budget, was allotted to the Office of Ocean Exploration and Research (OER), the sole federally funded program devoted exclusively to ocean exploration. (Source: NOAA, *Fiscal Year 2024 Appropriations Summary*, research.noaa.gov/external-affairs/budget, 2024.)

[19] John F. Kennedy, "Address at Rice University on the Nation's Space Effort," delivered September 12, 1962, Houston, Texas. (Transcript: John F. Kennedy Presidential Library and Museum.)

[20] Key early milestones of the Soviet space program include the launch of *Sputnik 1* (1957), the first artificial satellite; *Laika's* orbital

flight aboard *Sputnik 2* (1957); *Luna 2* (1959), the first human-made object to impact the Moon; and *Vostok 1* (1961), which carried Yuri Gagarin as the first human into space. (Sources: Royal Museums Greenwich, "Space Race Timeline," rmg.co.uk, 2024; *History.com*, "From Sputnik to Spacewalking: 7 Soviet Space Firsts," updated 2023.)

[21] Nabta Playa lies about 700 miles south of Giza and was constructed more than 7,000 years ago, making it older than Stonehenge and among the earliest known astronomical observatories. The megalithic circle is believed to have marked the summer solstice and possibly aligned with Arcturus, Sirius, Alpha Centauri, and Orion, evidence of sophisticated celestial observation by a pastoral culture on Egypt's Nabta Plateau. (Sources: *Astronomy Magazine*, "Nabta Playa: The World's First Astronomical Site Was Built in Africa — and Is Older Than Stonehenge," astronomy.com, 2024; *Live Science*, "Nabta Playa: A Mysterious Stone Circle That May Be the World's Oldest Astronomical Observatory," livescience.com, 2024.)

[22] The *HMS Challenger* expedition (1872–1876) marked the birth of modern oceanography. A British Royal Navy corvette retrofitted for science, *Challenger* conducted the first systematic survey of the global ocean, measuring depths with weighted ropes, recording temperature and salinity profiles, collecting sediment, and cataloging marine life. The voyage identified what was then the deepest known point in the ocean, later named Challenger Deep, and returned more than 4,700 new species to science. (Sources: Woods Hole Oceanographic Institution, "The Challenger Expedition," divediscover.whoi.edu; *Natural History Museum*, "HMS Challenger: How a 150-Year-Old Expedition Still Influences Scientific Discoveries Today," nhm.ac.uk, 2022.)

[23] The myth of Tiamat and Marduk originates in the *Enuma Elish*, the Babylonian creation epic derived from earlier Sumerian and Akkadian cosmologies. In this text, the storm god Marduk slays Tiamat, the

goddess of the primordial sea, and fashions the heavens and the earth from her divided body. (Sources: *Enuma Elish*, translated by Stephanie Dalley, *Myths from Mesopotamia: Creation, the Flood, Gilgamesh, and Others* [Oxford University Press, 2000]; World History Encyclopedia, "Enuma Elish – The Babylonian Epic of Creation – Full Text," worldhistory.org, 2024.)

[24] Nick Wyatt, "*Distinguishing Wood and Trees in the Waters: Creation in Biblical Thought*," in *Creation and Chaos: Biblical, Theological, and Scientific Perspectives*, ed. David Toshio Tsumura et al. (Berlin: De Gruyter, 2022).

[25] Cartographers of the late medieval and early modern periods frequently illustrated the margins of their maps with fantastical sea creatures—part artistic flourish, part warning of the unknown. (Sources: Chet Van Duzer, *Sea Monsters on Medieval and Renaissance Maps* [British Library, 2013]; University of Texas at Arlington Libraries, "Here There Be Monsters: The Art of the Mapmaker During the Age of Exploration," libraries.uta.edu, 2024.)

[26] The account of "Doubting Thomas" appears in the Gospel of John 20:24–29, in which the apostle Thomas refuses to believe in the resurrection until he sees and touches Jesus' wounds for himself.

## CHAPTER TWO

[27] Ancient accounts, such as Strabo's Geography (8.6.20), describe Corinth as a prosperous and indulgent city whose wealth was tied to its role as a major port. Later historians often repeated these depictions, though modern scholarship has questioned their accuracy. See Mary Beard and John Henderson, "With This Body I Thee Worship: Sacred Prostitution in Antiquity," *Gender & History* 9, no. 3 (1997): 480–503; and Stephanie Lynn Budin, *The Myth of Sacred Prostitution in Antiquity* (Cambridge: Cambridge University Press, 2008).

[28] Thomas Nagel, "What Is It Like to Be a Bat?" *The Philosophical Review* 83, no. 4 (October 1974): 435–450. Reprinted in *Mortal Questions* (Cambridge: Cambridge University Press, 1979).

[29] C. G. Jung, *Modern Man in Search of a Soul*, trans. W. S. Dell and Cary F. Baynes (London: Kegan Paul, Trench, Trubner & Co., 1933), 282.

[30] "Anthropomorphic," *Merriam-Webster.com Dictionary*, s.v. *anthrōpomorphos*, from Greek *ánthrōpos* ("human being") and *morphē* ("form"), accessed October 10, 2025, merriam-webster.com/dictionary/anthropomorphic. *(See also "anthropo-" and "-morphous," Merriam-Webster.com Dictionary.)*

[31] Christopher D. Whalen and Neil H. Landman, *"Fossil coleoid cephalopod from the Mississippian Bear Gulch Lagerstätte sheds light on early vampyropod evolution,"* *Nature Communications* 13, Article 1107 (2022). doi.org/10.1038/s41467-022-28333-5

[32] "Carboniferous Period: Carboniferous Life," *Encyclopaedia Britannica*, accessed October 10, 2025, britannica.com/science/Carboniferous-Period/Carboniferous-life; and "The Carboniferous," University of California Museum of Paleontology, *UCMP Berkeley*, accessed October 10, 2025, ucmp.berkeley.edu/carboniferous/carboniferous.php.

[33] Caroline B. Albertin et al., *"The Octopus Genome and the Evolution of Cephalopod Neural and Morphological Novelties,"* *Nature* 524 (2015): 220–224. doi.org/10.1038/nature14668

[34] Matthew A. Birk et al., *"Temperature-Dependent RNA Editing in Octopus Extensively Recodes the Neural Proteome,"* *Cell* 186, no. 12 (2023): 2602–2615. doi.org/10.1016/j.cell.2023.05.003

[35] "Octopuses, Squids, and Relatives," Smithsonian Ocean, accessed October 10, 2025, ocean.si.edu/ocean-life/invertebrates/octopuses-squids-and-relatives

[36] Holly Chetan-Welsh, *"Dumbo Octopus: The Murky World of the Deepest Octopus Species,"* Natural History Museum (London), accessed October 10, 2025, www.nhm.ac.uk/discover/what-is-a-dumbo-octopus.html

[37] Antigone is the central figure of *Antigone* by Sophocles, written circa 441 BCE. Prometheus appears in *Prometheus Bound*, traditionally attributed to Aeschylus, first performed in the 5th century BCE.

[38] The account of Paul's voyage, shipwreck, and time on Malta appears in *Acts of the Apostles* 27–28 (New Testament). The tradition that Paul was later released from Roman custody and continued his missionary work derives from early Christian writings, including *1 Clement* (ca. 96 CE) and the *Muratorian Canon* (2nd century CE).

## CHAPTER THREE

[39] Source: NOAA Ocean Exploration, *Ocean Depth Zones* and *Journey into the Deep* (accessed 2025); Woods Hole Oceanographic Institution, *Exploring the Hadal Zone* (HADES Program, 2024); Smithsonian Ocean Portal, *How Deep Is the Ocean?* (updated 2024). These sources define the Hadal or Hadalpelagic Zone as extending from roughly 6,000 to 11,000 meters in depth, derived from *Hades*, the Greek underworld. "Hadalpelagic" commonly denotes open-water segments of trenches, while steep trench walls are referred to simply as "hadal."

[40] Katherine L. C. Bell, Kristen N. Johannes, Susan E. Poulton et al., *"How little we've seen: A visual coverage estimate of the deep seafloor,"* Science Advances Vol. 11, No. 19 (7 May 2025). DOI: 10.1126/sciadv.adp8602. The study estimates that less than 0.001% of

the deep seafloor has been visually observed — meaning
approximately 99.999% remains unexplored.

[41] Source: NASA *New Horizons* Mission, *Pluto and Charon Exploration
Overview* (updated 2024); Alan Stern et al., *"The Pluto System: Initial
Results from its Exploration by New Horizons,"* *Science* Vol. 350, Issue 6258
(2015). The *New Horizons* mission mapped approximately 85% of
Pluto's surface and much of Charon's in 2015. The moon's name
honors both the mythological ferryman of the dead and Charlene, wife
of discoverer James Christy, who identified the moon in 1978.

[42] Source: NASA, *New Horizons – Arrokoth (2014 MU69)* mission
overview (updated 2024), science.nasa.gov/solar-system/kuiper-
belt/arrokoth-2014-mu69/; Frank T. Siebert Jr., "Reconstructing
Powhatan Vocabulary," in *Studies in Southeastern Indian Languages*,
University of Georgia Press (1975), p. 324. NASA initially glossed
*Arrokoth* as "sky," but Siebert's comparative Algonquian
reconstruction shows the term derives from /a ꞉ rahkwat/, meaning
"cloud," with the plural /a ꞉ rahkwatas/ ("clouds").

[43] *Total Recall*, directed by Paul Verhoeven (TriStar Pictures, 1990).
Notable for its exaggerated depiction of vacuum exposure, the
sequence is frequently referenced in discussions of cinematic
portrayals of human physiology in space.

[44] NASA Small Spacecraft Systems Virtual Institute (SSSI), *Thermal
Control Systems: Deployable Radiators* (Section 7.2.7), NASA Ames
Research Center, 2023. nasa.gov/smallsat-institute/sst-soa/thermal-
control/#7.2.7. Discusses radiative heat transfer as the primary means
of thermal regulation in spacecraft, noting that convection and
conduction are absent in the vacuum of space.

[45] NASA, *Radioisotope Power Systems: Overview*, Science Mission
Directorate, 2024. science.nasa.gov/planetary-

science/programs/radioisotope-power-systems/overview/.
Describes how radioisotope thermoelectric generators (RTGs) convert
the heat from the natural radioactive decay of plutonium oxide into
electricity and heat for deep-space missions such as *Curiosity* and
*Perseverance*.

[46] Gregory A. Nelson, "Space Radiation and Human Exposures: A
Primer," *Radiation Research* 185, no. 4 (2016): 349–358. DOI:
10.1667/RR14311.1. Describes the charged-particle environment of
space and the ionization processes by which cosmic and solar
radiation interact with materials and biological tissue, causing
atomic-level damage and degradation.

[47] NASA, *Distributed Spacecraft Autonomy (DSA) Project Overview*, Game
Changing Development Program, NASA Glenn Research Center, 2023.
nasa.gov/game-changing-development-projects/distributed-
spacecraft-autonomy-dsa/. Details NASA's development of
autonomous decision-making systems to address communication
latency and bandwidth limits in deep-space missions.

[48] NASA, *Environmental Control and Life Support Systems (ECLSS)
Overview*, Marshall Space Flight Center, 2023.
nasa.gov/reference/environmental-control-and-life-support-systems-
eclss/. Describes the Environmental Control and Life Support System
used on crewed spacecraft, which continuously recycles air and water
to maintain a habitable environment.

[49] Woods Hole Oceanographic Institution (WHOI), "Why Is Pressure
Different in the Ocean?," *Ocean Learning Hub*, 2024. whoi.edu/ocean-
learning-hub/ocean-facts/why-is-pressure-different-in-the-ocean/.
Explains that hydrostatic pressure increases by roughly one
atmosphere for every 10 meters (33 feet) of seawater depth, reaching
more than 1,000 atmospheres in the deepest ocean trenches.

[50] Woods Hole Oceanographic Institution (WHOI), "All About Trenches," *HADES Project: Exploring the Hadal Zone*, 2024. web.whoi.edu/hades/all-about-trenches/. Notes that typical hadal trench bottom temperatures range between 1 °C and 4 °C (34–39 °F), with minor variation by location due to pressure and the adiabatic temperature gradient.

[51] Woods Hole Oceanographic Institution (WHOI), *Micromodem Acoustic Communication System Overview*, 2024. acomms.whoi.edu/micro-modem/. Describes WHOI's Micromodem, a low-power digital acoustic modem widely used for communication between deep-submergence vehicles and surface ships, confirming that DSV data links rely on sound transmission rather than radio waves.

[52] NOAA National Ocean Service, "How far does light travel in the ocean?" 2024. oceanservice.noaa.gov/facts/light_travel.html. States that sunlight does not penetrate below 1,000 meters, marking the beginning of the aphotic zone, which extends through the bathypelagic, abyssopelagic, and hadopelagic regions.

[53] *Man of La Mancha*, book by Dale Wasserman, music by Mitch Leigh, lyrics by Joe Darion (New York: Random House, 1966). Adapted from Wasserman's 1959 teleplay *I, Don Quixote*. The line "Facts are the enemy of truth" appears in Act II, spoken by Don Quixote, and remains one of the play's most enduring quotations.

[54] Philip K. Dick, "How to Build a Universe That Doesn't Fall Apart Two Days Later," in *The Shifting Realities of Philip K. Dick: Selected Literary and Philosophical Writings*, ed. Lawrence Sutin (New York: Vintage Books, 1995), 259–280. Originally delivered as a lecture in 1978. Contains the oft-quoted line, "Reality is that which, when you stop believing in it, doesn't go away."

[55] André Gide, *Les Faux-Monnayeurs* [*The Counterfeiters*], trans. Dorothy Bussy (New York: Alfred A. Knopf, 1927). Originally published in 1925. Contains the line, "One does not discover new land without consenting to lose sight, at first and for a long time, of every shore."

[56] *Star Wars: Episode V – The Empire Strikes Back*, directed by Irvin Kershner, written by Leigh Brackett and Lawrence Kasdan, story by George Lucas (Los Angeles: Lucasfilm Ltd., 1980). The line "You must unlearn what you have learned" is spoken by Yoda during Luke Skywalker's training on Dagobah.

[57] John G. Fleagle and David R. Begun, "Overview of Hominin Evolution," *Nature Education: Scitable Knowledge Library*, Nature Publishing Group, 2019. nature.com/scitable/knowledge/library/overview-of-hominin-evolution-89010983/. Summarizes fossil evidence for the earliest known hominins, including *Sahelanthropus tchadensis* (6–7 million years ago) and *Orrorin tugenensis* (≈6 million years ago), marking the approximate emergence of the hominin lineage.

[58] University of California Museum of Paleontology, "When Did Life Originate?" *Understanding Evolution*, University of California, Berkeley, 2023. evolution.berkeley.edu/from-soup-to-cells-the-origin-of-life/when-did-life-originate/. University of California Museum of Paleontology, "The Ediacaran Period," UCMP, 2023. ucmp.berkeley.edu/vendian/ediacaran.php. NASA Astrobiology Institute, "Crossing the Boundary of the Ediacaran and the Cambrian," 2022. astrobiology.nasa.gov/news/crossing-the-boundary-of-the-ediacaran-and-the-cambrian/. Collectively summarize fossil and geologic evidence indicating that life on Earth began around 3.5 billion years ago and that the first macroscopic multicellular organisms, the Ediacara biota, appeared roughly 560 million years ago.

[59] Tim Stephens, "Mammoth Tusk Recovered from the Deep Ocean Reveals Surprises about Ancient Giants," *UC Santa Cruz Newscenter*, November 25, 2021. news.ucsc.edu/2021/11/mammoth-tusk/. Describes the recovery of a Columbian mammoth (*Mammuthus columbi*) tusk from 3,070 meters (10,000 feet) below the ocean surface, discovered by the Monterey Bay Aquarium Research Institute and later retrieved for study by scientists from UC Santa Cruz and the University of Michigan.

[60] Woods Hole Oceanographic Institution (WHOI), "A Rare Black Seadevil Anglerfish Sees the Light," *Oceanus Magazine*, 2024. whoi.edu/oceanus/feature/a-rare-black-seadevil-anglerfish-sees-the-light/. Describes a documented surface sighting of the black seadevil anglerfish (*Melanocetus johnsonii*) near Tenerife, an event notable for revealing this deep-sea species far above its typical 200–1,500-meter range.

[61] Theodore W. Pietsch and John P. Van Duzer, "Systematics and Distribution of Ceratioid Anglerfishes of the Family *Melanocetidae*, with the Description of a New Species from the Eastern North Pacific Ocean," *NOAA Technical Report NMFS SSRF-781* (1980): 1–24. spo.nmfs.noaa.gov/sites/default/files/pdf-content/1980/781/pietsch.pdf.
Provides morphological and ecological descriptions of *Melanocetus johnsonii* and related species, detailing reduced ossification, absence of a swim bladder, gelatinous tissue composition, and pronounced sexual dimorphism.

[62] Smithsonian Institution, "*Anoplogaster cornuta* (Fangtooth)," *National Museum of Natural History Collections Snapshot*, 2023. si.edu/collections/snapshot/anoplogaster-cornuta-fangtooth. Describes the fangtooth's anatomy and ecology, noting that *Anoplogaster cornuta* possesses the largest teeth relative to body size of

any known marine species, and detailing its characteristic fangs, reduced vision, and reliance on lateral-line sensing.

[63] Monterey Bay Aquarium Research Institute (MBARI), "Pacific Viperfish (*Chauliodus macouni*)," *Animal Database*, 2024. mbari.org/animal/pacific-viperfish/. Describes the Pacific viperfish's anatomy and hunting adaptations, including needle-like teeth that extend beyond the eyes and unhinging jaws that form a cage to trap prey.

[64] United States Environmental Protection Agency (EPA), "How Human Activities Increase the Occurrence of Harmful Algal Blooms," *Harmful Algal Blooms (HABs) Overview*, 2024. epa.gov/habs/what-causes-habs. States that the incidence of harmful algal blooms has increased in magnitude, extent, and frequency in recent decades, linking this rise to nutrient runoff, wastewater discharge, and climate change–driven warming—factors that also contribute to the growing prevalence of bioluminescent plankton events such as "sea sparkle" and "milky seas."

[65] M. V. Razzhivin, "Origin of Nomenclature in Bioluminescent Systems," *Siberian Federal University Electronic Library*, 2018. elib.sfu-kras.ru/handle/2311/935. Explains that *luciferin* and *luciferase* derive from the Latin *lux* ("light") and *ferre* ("to bear"), with *lucifer* meaning "light-bearer" or "morning star."

[66] Hendrik J.T. Hoving and Bruce H. Robison, "Deep-Sea Vampire Squid Defy Their Name: Adults of *Vampyroteuthis infernalis* Feed on Marine Detritus," *Proceedings of the Royal Society B: Biological Sciences* 279, no. 1747 (2012): 4121–4126. pmc.ncbi.nlm.nih.gov/articles/PMC3479720/. Identifies *Vampyroteuthis infernalis* as the sole extant species of the order *Vampyromorpha*, a phylogenetic relic exhibiting both squid and

octopod traits, and notes its placement as sister to the Octopoda within the Octopodiformes.

[67] Ocean Exploration Trust, "Not So Bloodthirsty: An Encounter with a Vampire Squid," *Deep Ocean Education Project Exploration Notes*, adapted under federal award NA19OAR0110405 for the National Marine Sanctuary Foundation, 2025. oceanexplorer.noaa.gov/wp-content/uploads/2025/04/not-so-bloodthirsty-exploration-notes.pdf. Reports observations of *Vampyroteuthis infernalis* releasing a large cloud of bioluminescent mucus as a defensive strategy, an adaptation used to distract predators in the aphotic zone where the species resides.

[68] National Geographic Society, "Bioluminescence," *Education Resource Library*, 2024. education.nationalgeographic.org/resource/bioluminescence/. Explains the camouflage technique known as counter-illumination, in which deep-sea organisms—such as hatchetfish—emit light from their undersides to match the dim sunlight filtering from above, thereby eliminating their silhouettes and avoiding detection by predators below.

[69] Leo Richards, "Deep-Sea Gigantism," *Natural World Facts*, 2024. naturalworldfacts.com/deep-sea-gigantism. Summarizes current explanations for deep-sea gigantism, including Kleiber's rule (metabolic efficiency in larger organisms), Bergmann's rule (size increase in colder environments), and the Island rule (isolation and reduced predation), highlighting how temperature, oxygen, scarcity, and evolutionary isolation may favor larger body sizes in deep-sea species.

[70] Alan D. Jamieson, Tammy Frank, and Daniel Jones, "Dulcibella camanchaca gen. nov. sp. nov. (Crustacea: Amphipoda) from the Atacama Trench, Southeastern Pacific Ocean," *Systematics and*

*Biodiversity* 22, no. 4 (2024): 381–392. doi.org/10.1080/14772000.2024.2416430. Describes *Dulcibella camanchaca* as a newly identified amphipod genus and species collected from nearly 8,000 meters in the Atacama Trench. The genus name honors Dulcinea, muse of Cervantes' *Don Quixote*, while the species name derives from the Aymara word for "darkness."

[71] MarineBio Conservation Society, "Colossal Squid (*Mesonychoteuthis hamiltoni*)," 2024. marinebio.org/species/colossal-squid/mesonychoteuthis-hamiltoni/. Reports *Mesonychoteuthis hamiltoni* reaching lengths up to 14 meters (46 feet) with eyes about 25 centimeters in diameter—the largest in the animal kingdom—and notes its status as one of the largest and least observed deep-sea cephalopods.

[72] John R. Holsinger, "Taxonomic Summary of Deep-Sea Crustaceans," *Proceedings of the Biological Society of Washington* 128 (2015): 145–162. pmc.ncbi.nlm.nih.gov/articles/PMC4304853/. Lists *Bathynomus giganteus* as the largest known isopod species, reaching a total length of approximately 50 centimeters (20 inches).

[73] Chen Ly, "Supergiant Crustaceans Could Live Across Half the Deep Sea Floor," *New Scientist*, April 25, 2024. newscientist.com/article/2480859-supergiant-crustaceans-could-live-across-half-the-deep-sea-floor/. Reports that *Alicella gigantea*—the world's largest amphipod—can reach 34 centimeters in length and has been recorded across multiple ocean basins, highlighting the species' rarity and wide distribution in the hadal zone.

[74] Smithsonian Institution, "Japanese Spider Crab (*Macrocheira kaempferi*)," *Ocean Portal*, 2024. ocean.si.edu/ocean-life/invertebrates/japanese-spider-crab. Notes that the Japanese spider crab has a leg span reaching up to 4 meters (13 feet) and is the

largest crab species known, with individuals living up to a century in the deep waters off Japan.

[75] Miguel de Cervantes Saavedra, *Don Quixote*, Part II, Chapter X (1615), trans. Edith Grossman (New York: HarperCollins, 2003), p. 563. Contains the line "La verdad adelgaza, pero no quiebra, y siempre anda sobre la mentira como el aceite sobre el agua," commonly translated as "The truth may be stretched thin, but it never breaks, and it always surfaces above lies, as oil floats on water."

## CHAPTER FOUR

[76] Edward Sapir (1884–1939) and Benjamin Lee Whorf (1897–1941) each argued that language influences cognition and perception, an idea later labeled the *Sapir–Whorf Hypothesis* or *Whorfian Hypothesis*. The term itself was coined in 1954 by linguist Harry Hoijer, though Sapir and Whorf never jointly formulated a single hypothesis. Modern scholarship distinguishes between a strong form—*linguistic determinism*, in which language constrains thought—and a weaker form, *linguistic relativity*, in which linguistic structure influences perception without fixing it. See *Encyclopaedia Britannica*, "Whorfian Hypothesis," 2024. britannica.com/science/Whorfian-hypothesis; *Stanford Encyclopedia of Philosophy*, "Whorfianism." plato.stanford.edu/entries/linguistics/whorfianism; and Harry Hoijer, "The Sapir-Whorf Hypothesis," *Language in Culture Conference Proceedings* (1954).

[77] The Ouroboros—literally "tail-devourer" in Greek—is among the oldest known symbols of cyclical renewal. Its earliest known depiction appears on a golden shrine in the tomb of Tutankhamun (13th century BCE), where, according to Egyptologist Jan Assmann, it represents "the mystery of cyclical time, which flows back into itself." See Kelly Grovier, "The Ancient Symbol That Spanned Millennia," *BBC*

*Culture*, December 4, 2017. bbc.com/culture/article/20171204-the-ancient-symbol-that-spanned-millennia.

[78] Tubeworms (*Riftia pachyptila*) are fast-growing deep-water invertebrates that anchor near hydrothermal vents and cold seeps, lacking mouths or digestive systems and relying instead on chemosynthetic bacteria housed within an internal organ called the trophosome. These symbionts oxidize hydrogen sulfide and hydrocarbons to create organic compounds that sustain the host. See Bureau of Ocean Energy Management (BOEM), "Tube Worms," *Ocean Science News*, 2024. boem.gov/newsroom/ocean-science-news/tube-worms.

[79] Polychaetes are segmented marine worms of the class *Polychaeta* within the phylum *Annelida*, comprising more than 8,000 known species characterized by paired parapodia and bristled segments. The name *Annelida* derives from the Latin *anulus*, meaning "little ring," a reference to the segmented body form. See *Encyclopaedia Britannica*, "Polychaete," 2024. britannica.com/animal/polychaete; and *Online Etymology Dictionary*, "Annelid." etymonline.com/word/annelid.

[80] Deep-sea polychaetes (*Polychaeta*, phylum *Annelida*) display extreme morphological diversity, ranging from vividly colored or transparent forms to species with feather-like gills, iridescent bristles, and eversible jaws. Some live freely in the water column, others dwell in tubes or burrows adapted for respiration and feeding. See *Smithsonian Magazine*, "14 Fun Facts About Marine Bristle Worms," 2015. smithsonianmag.com/science-nature/14-fun-facts-about-marine-bristle-worms-180955773; Smithsonian Ocean Portal, "Why I Love Polychaetes," 2024. ocean.si.edu/ocean-life/invertebrates/why-i-love-polychaetes; *Proceedings of the Royal Society of London B*, vol. 139, 1952. doi.org/10.1098/rspb.1952.0045; and *ScienceDirect*, "Polychaete – General Ecology and Behavior of Polychaeta."

sciencedirect.com/topics/agricultural-and-biological-sciences/polychaete.

[81] *Osedax mucofloris*, discovered on a minke whale carcass near Sweden and formally described in *Proceedings of the Royal Society B* (2005), is the first *Osedax* species identified in the North Atlantic. Its Latin name—*muco* (mucus) and *floris* (flower)—references both its mucus-covered exterior and its flower-like appearance. The colloquial English name "bone-eating snot-flower" originates from researcher Adrian Glover's description of the worm's form and texture. See Adrian G. Glover et al., "World-wide Whale Worms? A New Species of *Osedax* from the Shallow North Atlantic," *Proceedings of the Royal Society B*, vol. 272, 2005. pmc.ncbi.nlm.nih.gov/articles/PMC1559975; and Richard Black, "Scientists Find 'Bone-Eating Snot Flower,'" *BBC News*, October 19, 2005. news.bbc.co.uk/2/hi/4354286.stm.

[82] *Osedax* species, commonly called "zombie worms," anchor to whale and fish bones using root-like structures that house symbiotic bacteria. The worms secrete acid to dissolve the bone matrix, releasing fats and proteins that their bacterial partners digest and convert into nutrients. See Smithsonian Ocean Portal, "Zombie Worms Crave Bone," 2024. ocean.si.edu/ocean-life/invertebrates/zombie-worms-crave-bone.

[83] *Osedax* worms sustain themselves through endosymbiosis—housing bacterial partners within root tissues that penetrate whale bone. These bacteria extract and metabolize organic compounds from the bone matrix, supplying nutrients to the host. See Caroline Verna et al., "High Symbiont Diversity in the Bone-Eating Worm *Osedax mucofloris* from Shallow Whale-Falls in the North Atlantic," *Environmental Microbiology*, vol. 12, no. 11 (2010): 3083–3092. DOI: 10.1111/j.1462-2920.2010.02299.x.

[84] *Osedax* worms rely on bacterial endosymbionts housed within root-like tissues to extract and metabolize fats and proteins from whale bone. The genus was first described in 2004, when researchers identified bone-eating annelids that derive nutrition through symbiotic bacteria in their root systems—a novel form of deep-sea chemosymbiosis. Subsequent work confirmed their ability to colonize other mammalian bones and further detailed the role of their bacterial partners. See Greg W. Rouse, Shana K. Goffredi, and Robert C. Vrijenhoek, "*Osedax*: Bone-Eating Marine Worms with Dwarf Males," *Science*, vol. 305, no. 5684 (2004): 668–671. science.org/doi/10.1126/science.1098650; and William J. Jones et al., "Marine Worms (*Genus Osedax*) Colonize Cow Bones," *Proceedings of the Royal Society B: Biological Sciences*, vol. 275, no. 1640 (2008): 387–391. pmc.ncbi.nlm.nih.gov/articles/PMC2596828.

[85] The *Osedax–Oceanospirillales* symbiosis is characterized by a reciprocal exchange of nutrients. The worm secretes acids and collagen-degrading enzymes that release amino acids and lipids from bone; these substrates are absorbed and metabolized by the bacterial symbionts, which in turn supply the host with carbohydrates and essential amino acids absent from the worm's own biosynthetic repertoire. See Giacomo Moggioli et al., "Distinct Genomic Routes Underlie Transitions to Specialised Symbiotic Lifestyles in Deep-Sea Annelid Worms," *Nature Communications* 14, 2814 (2023). doi.org/10.1038/s41467-023-38521-6.

[86] Whale-fall microbial succession includes sulfur-oxidizing *Sulfurimonas* species that associate with *Osedax* worms during later stages of bone decay. These bacteria help mitigate the buildup of toxic sulfide and other reduced compounds generated by anaerobic decomposition, stabilizing the local chemical environment around the worm. See Anna G. Lewis et al., "A Dynamic Epibiont Community Associated with the Bone-Eating Polychaete Genus *Osedax*," *mBio* 14, no. 3 (2023): e03140–22. doi.org/10.1128/mbio.03140-22.

[87] Whale-fall decomposition proceeds through several overlapping ecological stages—from rapid scavenging to long-term sulfophilic and chemosynthetic phases—each supporting distinct microbial and faunal communities. These multi-year, multi-phase successions continue to reveal new insights into deep-sea biogeochemical cycling and ecosystem resilience. See Qihui Li et al., *"Review of the Impact of Whale Fall on Biodiversity in Deep-Sea Ecosystems,"* *Frontiers in Ecology and Evolution* 10 (2022): 885572. doi.org/10.3389/fevo.2022.885572.

[88] *Osedax* worms possess no mouth, stomach, or anus. Their anatomy consists of a trunk and anterior palps exposed to seawater, while the ovisac and root structures extend into bone cavities that house symbiotic bacteria. The roots form a complex network of epithelial and vascular tissues specialized for nutrient absorption. See Hitoshi Miyamoto et al., "Comparative Ultrastructure of the Bone-Eating Worm *Osedax japonicus* (Annelida: Siboglinidae) and Its Endosymbiotic Bacteria," *Proceedings of the Royal Society B: Biological Sciences* 280, no. 1757 (2013): 20121923. pmc.ncbi.nlm.nih.gov/articles/PMC3652447; and *Sci.News*, "Bone-Eating Worms," 2013. sci.news/biology/article00444.html.

[89] All known *Osedax* species display extreme sexual dimorphism, with large, tube-dwelling females and microscopic dwarf males that live within the females' gelatinous tubes. The males remain larval in form, producing sperm and often accumulating in large numbers—a reproductive arrangement unique among siboglinid worms. See Ana Hilário et al., "Reproductive Biology of the Deep-Sea Siboglinid Worms (Annelida: Siboglinidae)," *Invertebrate Biology* 134, no. 4 (2015): 376–400. pmc.ncbi.nlm.nih.gov/articles/PMC4285288.

[90] Female *Osedax* worms reproduce continuously, with oogenesis proceeding within the ovary and fertilization occurring using sperm from the dwarf males inhabiting the female tubes. Observations of

*Osedax rubiplumus* and related species show that females may spawn hundreds of oocytes daily—averaging approximately 335 eggs per day—with high fertilization success. See Greg W. Rouse et al., "Spawning and Development in Osedax Boneworms (Siboglinidae, Annelida)," *Marine Biology* 156, no. 3 (2009): 395–405. doi.org/10.1007/s00227-008-1091-z; and Ana Hilário et al., "Reproductive Biology of the Deep-Sea Siboglinid Worms (Annelida: Siboglinidae)," *Invertebrate Biology* 134, no. 4 (2015): 376–400. pmc.ncbi.nlm.nih.gov/articles/PMC4285288.

[91] The extreme sexual dimorphism of *Osedax* represents an energetically efficient reproductive strategy in nutrient-poor environments. Dwarf males residing within female tubes minimize resource competition and ensure reproductive proximity, conserving energy otherwise spent searching for mates. See Robert C. Vrijenhoek, Greg W. Rouse, and Shana K. Goffredi, "A Dwarf Male Reversal in Bone-Eating Worms," *Current Biology* 25, no. 21 (2015): 2851–2855. doi.org/10.1016/j.cub.2014.09.059.

[92] *Physalia physalis*, commonly known as the Portuguese man o' war, is a colonial siphonophore rather than a true jellyfish. It propels itself with a gas-filled pneumatophore that acts as a sail, trailing long tentacles equipped with potent nematocysts capable of paralyzing fish and stinging humans. See Catriona Munro et al., "Morphology and Development of the Portuguese Man of War, *Physalia physalis*," *Scientific Reports* 9 (2019): 15522. doi.org/10.1038/s41598-019-51842-1.

[93] Siphonophores are colonial hydrozoans composed of numerous, genetically identical zooids, each specialized for a distinct role such as feeding, reproduction, or propulsion. These zooids remain physiologically connected and function collectively as a single integrated organism—a hallmark of siphonophore biology that challenges traditional definitions of individuality. Woods Hole Oceanographic Institution, "About Siphonophores," *Ocean Twilight*

*Zone: Creature Features*, 2023. twilightzone.whoi.edu/explore-the-otz/creature-features/siphonophore/

[94] Siphonophore colonies comprise asexually produced, morphologically distinct zooids specialized for discrete functions—nectophores for propulsion, gonophores for reproduction, bracts for protection and buoyancy control, and the pneumatophore as a gas-filled float aiding orientation. See Samuel H. Church et al., "The Histology of *Nanomia bijuga* (Hydrozoa: Siphonophora)," *Journal of Experimental Zoology Part B: Molecular and Developmental Evolution* 324 (2015): 435–449. doi.org/10.1002/jez.b.22629

[95] *Praya dubia*, the giant siphonophore, inhabits the deep pelagic zone between roughly 700 and 1,000 meters (2,300–3,300 feet) and can reach extraordinary lengths of about 40 meters (130 feet)—comparable to that of a blue whale. Monterey Bay Aquarium, "Giant Siphonophore," *Animals A–Z*, 2023. montereybayaquarium.org/animals/animals-a-to-z/giant-siphonophore.

[96] *Praya dubia* and related siphonophores maintain body form and propulsion through a hydrostatic system in which internal water pressure, modulated by circular and longitudinal muscle layers, provides both rigidity and flexibility for coordinated colony movement. G. O. Mackie, P. R. Pugh, and J. E. Purcell, "Siphonophore Biology," *Advances in Marine Biology* 24 (1987): 97–262. doi.org/10.1016/S0065-2881(08)60074-7.

[97] The genus *Praya* was named in 1827 by French naturalists Jean René Quoy and Joseph Paul Gaimard, likely derived from the Portuguese *praia* ("shore" or "beach"), a term used in 19th-century Hong Kong and Macau to describe waterfront roads. The epithet *dubia* is Latin for "doubtful" or "uncertain," reflecting early classification ambiguity rather than a literal phrase meaning "dubious prayer."

[98] *Syringammina fragilissima* was first described by Henry Bowman Brady in 1883 from specimens dredged in the Faroë Channel by HMS *Triton*. The name derives from Latin and Greek roots, meaning "very fragile sandpipe." Modern studies identify it as a xenophyophore—an agglutinated, multinucleate foraminiferan protist—and note that its precise taxonomic placement remains debated (Brady 1883; Gooday et al. 2017; WoRMS ID 137339).

[99] *Syringammina fragilissima* belongs to the xenophyophore clade, a group of giant, agglutinated foraminiferal protists characterized by unicellular but multinucleate organization. Individual specimens consist of a single cell containing numerous scattered nuclei within a branching tubular network. See A. J. Gooday et al., "Giant Protists (Xenophyophores, Foraminifera) Are Exceptionally Diverse in Parts of the Abyssal Eastern Pacific Licensed for Polymetallic Nodule Exploration," *Biological Conservation* 207 (2017): 106–116; and Michael Marshall, "Zoologger: Living Beach Ball Is Giant Single Cell," *New Scientist*, March 3, 2010. newscientist.com/article/dn18468-zoologger-living-beach-ball-is-giant-single-cell/.

[100] Xenophyophores have undergone repeated taxonomic reclassification, having been described at various times as sponges, testate amoebae, or even their own phylum. Modern molecular analyses now place them within the Foraminifera (class Monothalamea). See O. S. Ashford et al., "Deep-Sea Xenophyophores (Protista, Foraminifera) in the Eastern Clarion–Clipperton Zone, Pacific Ocean," *Deep-Sea Research Part I: Oceanographic Research Papers* 94 (2014): 88–104.

[101] Xenophyophores extend a network of pseudopodia—fine, arm-like cytoplasmic projections—within their test cavity to capture and manipulate sediment particles, forming their agglutinated outer casing. The process resembles the way amoeboid cells use pseudopodia for movement and engulfment. See G. Bull and H. Tyler-

Walters, *"Syringammina fragilissima Field on Atlantic Mid-Bathyal Coarse Sediment,"* Marine Life Information Network (MarLIN), 2024, Marine Biological Association of the United Kingdom. marlin.ac.uk/habitat/detail/1276

[102] Agglutinated xenophyophore tests are extraordinarily delicate, often shattering during trawl or dredge collection and disintegrating under minimal handling. Their fragility makes them exceptionally difficult to collect, preserve, or study intact in laboratory settings. See G. Bull and H. Tyler-Walters, "Syringammina fragilissima Field on Atlantic Lower Bathyal Mud," *Marine Life Information Network (MarLIN)*, 2024, Marine Biological Association of the United Kingdom. marlin.ac.uk/habitat/detail/1283

[103] *Syringammina fragilissima* is recognized as the largest known xenophyophore, and the largest single-celled organism on Earth, with recorded diameters reaching up to 20 centimeters (8 inches). Its fragile, agglutinated test makes intact specimens rare, but its size rivals that of a human hand. See "#22 Xenophyophore," *Deep Sea News*, 2008, deepseanews.com/2008/10/the-27-best-deep-sea-species-22-xenophyophores/; and *"Syringammina fragilissima," MicrobeWiki*, microbewiki.kenyon.edu/index.php/Syringammina_fragilissima

[104] The test of *Syringammina fragilissima* consists of a complex network of branching, hollow tubes that form an interconnected labyrinth within the agglutinated sediment structure. As the organism grows, it retreats inward, leaving abandoned chambers that smaller deep-sea animals, such as nematodes, later occupy. See Michael Marshall, "Zoologger: Living 'Beach Ball' Is Giant Single Cell," *New Scientist*, January 13, 2010. newscientist.com/article/dn18468-zoologger-living-beach-ball-is-giant-single-cell/

[105] The feeding strategy of *Syringammina fragilissima* remains uncertain. It may function as a suspension or filter feeder, using

pseudopodia to draw in particulates or graze on bacteria within surrounding sediments. Ole Tendal has also proposed that it may cultivate bacteria within its internal tubes as a food source. See Michael Marshall, "Zoologger: Living 'Beach Ball' Is Giant Single Cell," *New Scientist*, January 13, 2010. newscientist.com/article/dn18468-zoologger-living-beach-ball-is-giant-single-cell/; and G. Bull and H. Tyler-Walters, "Syringammina fragilissima Field on Atlantic Mid-Bathyal Coarse Sediment," *Marine Life Information Network (MarLIN)*, 2024, Marine Biological Association of the United Kingdom. marlin.ac.uk/habitat/detail/1281

[106] *Chrysomallon squamiferum*, the scaly-foot snail, is endemic to hydrothermal vent fields of the Indian Ocean (Kairei, Solitaire, and Longqi) and exhibits extraordinary adaptations: iron-sulfide armor on shell and foot, endosymbiotic thioautotrophic bacteria within an enlarged esophageal gland that supply nutrition via chemosynthesis, loss of conventional feeding, simultaneous hermaphroditism, and a giant heart comprising roughly 4% of body mass—features interpreted as mutual adaptations between host and microbes. See Chong Chen et al., "The Heart of a Dragon: 3D Anatomical Reconstruction of the 'Scaly-Foot Gastropod' (Mollusca: Gastropoda: Neomphalina) Reveals Its Extraordinary Circulatory System," *Frontiers in Zoology* 12 (2015). DOI: 10.1186/s12983-015-0105-1.

[107] *Bathypterois grallator*, the Tripod Fish, inhabits abyssal depths of roughly 900–4,700 meters and is distinguished by its elongated pelvic and caudal fin rays that form rigid "tripods," allowing it to perch motionless above the seafloor while detecting drifting prey with tactile pectoral fins. The species is a simultaneous hermaphrodite, capable of both cross-fertilization and self-fertilization when solitary—an adaptation to the sparse population densities of the deep sea. See "Tripod Fish," *Aquarium of the Pacific Online Learning Center*, 2024. aquariumofpacific.org/onlinelearningcenter/species/tripod_fish

[108] Members of the family *Ipnopidae*, such as *Ipnops meadi*, possess highly modified, platelike eyes that lack lenses and are thought to detect the faint glimmer of bioluminescent prey. These flattened, reflective membranes appear luminous in the dark and may emit light themselves, though their precise function remains uncertain. See "Welcome to the Central Pacific Basin," *NOAA Office of Ocean Exploration and Research, Mountains in the Deep: Exploring the Central Pacific Basin* Expedition Log, May 10, 2017. archive.oceanexplorer.noaa.gov/okeanos/explorations/ex1705/logs/may10/welcome.html

[109] *Abyssobrotula galatheae*, a cusk eel of the family *Ophidiidae*, was first described from specimens collected in deep-sea trenches worldwide and includes one individual captured at 8,370 meters (27,460 feet) in the Puerto Rico Trench—the deepest verified record for any fish. See S. Ohashi and J. G. Nielsen, "A New Species of *Abyssobrotula* (Ophidiiformes, Ophidiidae) from the Kuril–Kamchatka Trench," *Zootaxa* 4132, no. 4 (2016). DOI: 10.11646/zootaxa.4132.4.7.

[110] Many deep-sea fishes reduce or eliminate rigid skeletal material, replacing calcified bone with flexible cartilage and gelatinous tissue that can deform under pressure without fracturing. This structural reduction allows the body to compress rather than resist the crushing forces of the abyss. See "Diving Deep into the Depths: Exploring Biomechanical Adaptations of Deep-Sea Creatures," *Biomechanics in the Wild*, University of Notre Dame, November 6, 2024. sites.nd.edu/biomechanics-in-the-wild/2024/11/06/diving-deep-into-the-depths-exploring-biomechanical-adaptations-of-deep-sea-creatures/

[111] Most deep-sea fishes have lost their gas-filled swim bladders, which would implode under the immense pressures of the Hadal Zone. Instead, they maintain neutral buoyancy through waxy lipids, low-density oils, or gelatinous tissues that reduce body weight

without relying on compressible gases. See *Oceanography (Hill)*, "12.4: Deep Sea," *LibreTexts Oceanography*, University of California, 2024. geo.libretexts.org/Bookshelves/Oceanography/Oceanography_(Hill)/1 2%3A_Marine_Environments/12.4_Deep_Sea

[112] Many deep-sea animals possess soft, pliable bodies that distribute external pressure evenly, allowing them to deform under stress rather than resist it. This flexibility minimizes structural damage and reduces metabolic energy demands, a vital adaptation in high-pressure, low-energy environments. See "Diving Deep into the Depths: Exploring Biomechanical Adaptations of Deep-Sea Creatures," *Biomechanics in the Wild*, University of Notre Dame, November 6, 2024. sites.nd.edu/biomechanics-in-the-wild/2024/11/06/diving-deep-into-the-depths-exploring-biomechanical-adaptations-of-deep-sea-creatures/

[113] Many deep-sea organisms possess pliable, amorphous body forms that deform rather than resist environmental forces, enabling them to move through complex seafloor terrain and variable currents while conserving energy. Such morphological plasticity reflects evolutionary responses to the extreme pressure, low temperature, and limited energy of abyssal environments. See Roberto Danovaro, Cinzia Corinaldesi, Antonio Dell'Anno, and Paul V. R. Snelgrove, "The Deep-Sea under Global Change," *Current Biology* 27 (2017): R431–R510. DOI: 10.1016/j.cub.2017.02.046.

[114] Deep-sea bioluminescence serves a range of ecological functions, from luring prey and communicating with potential mates to evading predators and cloaking the body through counterillumination. Species such as anglerfish, lanternfish, and hatchetfish use distinct light-producing organs or bacterial symbionts to attract, defend, and camouflage themselves in the dark ocean. See Helen Cooper, "Illuminating the Facts of Deep-Sea Bioluminescence," *Monterey Bay*

*Aquarium*, November 15, 2021.
montereybayaquarium.org/stories/bioluminescence

[115] In the lightless depths, many marine organisms rely on mechanoreception to perceive subtle vibrations and pressure changes. Specialized sensors—ranging from the lateral line systems of fishes to hydrodynamic receptors on crustacean appendages—detect water displacement and low-frequency oscillations, providing critical information about approaching predators or prey. See N. Klages, W. Arntz, and M. Gutt, "Hydroacoustic Stimuli and Mechanoreception in the Deep-Sea Shrimp *Pandalus borealis*," *Deep-Sea Research Part I: Oceanographic Research Papers* 48, no. 6 (2001): 1343–1360. DOI: 10.1016/S0967-0637(01)00047-4.
sciencedirect.com/science/article/pii/S0967063701000474

[116] In the aphotic depths, eyes become energetically costly liabilities. Many deep-sea species exhibit regressed or absent visual organs and have evolved heightened nonvisual senses—mechanoreception, chemoreception, and electroreception—to compensate for the loss of light. See Justin Marshall, "Vision and Lack of Vision in the Ocean," *Current Biology* 27, no. 11 (2017): R494–R502. DOI: 10.1016/j.cub.2017.03.012.
sciencedirect.com/science/article/pii/S0960982217302816

[117] Some deep-sea species evolve the opposite strategy—enlarging and refining their eyes to detect the faintest flashes of bioluminescent light. The barreleye fish and giant squid possess immense, highly sensitive eyes adapted to register minute flickers across the darkness of the abyss. See Justin Marshall, "Vision and Lack of Vision in the Ocean," *Current Biology* 27, no. 11 (2017): R494–R502. DOI: 10.1016/j.cub.2017.03.012.
sciencedirect.com/science/article/pii/S0960982217302816

[118] Many deep-sea predators, including gulper and pelican eels, possess enormous hinged jaws and elastic stomachs that allow them to swallow prey as large as—or larger than—themselves. These adaptations enable survival when food is scarce, allowing a single catch to sustain them for extended periods. See "The Deep Sea," *MarineBio Conservation Society*, 2024. marinebio.org/oceans/deep-sea/

[119] Deep-sea organisms exhibit extremely low metabolic rates, conserving energy in an environment where food is scarce and temperatures are near freezing. Their slowed metabolism allows them to stretch the nutritional value of a single meal across long periods, a crucial adaptation to the abyss's energy poverty. See E. V. Thuesen and J. J. Childress, "Metabolic Rates, Enzyme Activities and Chemical Compositions of Some Deep-Sea Demersal Fishes: Evidence for a Metabolic Cold Adaptation," *Philosophical Transactions of the Royal Society B: Biological Sciences* 362, no. 1487 (2007): 2061–2078. DOI: 10.1098/rstb.2007.2101. pmc.ncbi.nlm.nih.gov/articles/PMC2442854/

[120] Many deep-sea animals survive through chemosynthetic symbioses—partnerships with bacteria that oxidize hydrogen sulfide or methane to fix carbon into organic nutrients. These microbial allies transform chemical energy into sustenance for their hosts, enabling entire ecosystems to flourish in lightless realms where photosynthesis is impossible. See E. Maggie Sogin, Nikolaus Leisch, and Nicole Dubilier, "Chemosynthetic Symbioses," *Current Biology* 30, no. 19 (2020): R1137–R1144. DOI: 10.1016/j.cub.2020.07.050. sciencedirect.com/science/article/pii/S0960982220310757

[121] Many deep-sea organisms sustain themselves through sediment or filter feeding, extracting bacteria, plankton, and detrital particles from the surrounding water or seafloor sediments. Zooplankton in the ocean's twilight zone fragment sinking organic matter and harvest microbial biomass growing on detritus—a strategy that turns the deep-sea "soup" of suspended particles into nourishment. See Daniel

J. Mayor, Richard Sanders, Sarah L. C. Giering, and Thomas R. Anderson, "Microbial Gardening in the Ocean's Twilight Zone: Detritivorous Metazoans Benefit from Fragmenting, Rather Than Ingesting, Sinking Detritus," *BioEssays* 36, no. 11 (2014): 1132–1137. DOI: 10.1002/bies.201400100. pmc.ncbi.nlm.nih.gov/articles/PMC4278546/

[122] Hermaphroditism is a common reproductive strategy among deep-sea invertebrates such as mussels and snails, allowing self-fertilization or sex change when potential mates are scarce. In extreme environments like cold seeps, sedentary lifestyles, and sparse population density favor this flexible reproductive mode. See Yu Shi et al., "Proteome and Transcriptome Analysis of Gonads Reveals Intersex in *Gigantidas haimaensis*," *BMC Genomics* 23, no. 174 (2022). DOI: 10.1186/s12864-022-08407-w. pmc.ncbi.nlm.nih.gov/articles/PMC8892766/

[123] Some deep-sea fishes exhibit extreme sexual dimorphism, with dwarf males permanently fused to the bodies of much larger females. In these parasitic unions—exemplified by deep-sea anglerfish—the male's body atrophies until only the gonads remain functional, serving as a lifelong sperm source. See Elizabeth A. Peden, Rochelle M. Tinghitella, and Geoff A. Parker, "The Evolution of Sexual Dimorphism and Mating Systems," *Current Biology* 34, no. 15 (2024): R815–R828. DOI: 10.1016/j.cub.2024.05.066. sciencedirect.com/science/article/pii/S0960982224005761

[124] Many deep-sea species invest heavily in a small number of large, yolk-rich eggs, producing well-developed hatchlings adapted to life in the abyss. Brooding squids, for instance, carry oversized eggs for months or years, while deep-sea copepods produce few, energy-dense eggs that ensure survival in food-limited waters. See MBARI, "MBARI's Advanced Underwater Robots Discover Deep-Sea Squid That Broods Giant Eggs," *Monterey Bay Aquarium Research Institute*, 2024. mbari.org/news/mbaris-advanced-underwater-robots-discover-

deep-sea-squid-that-broods-giant-eggs/; and Holger Auel, "Egg Size and Reproductive Adaptations Among Arctic Deep-Sea Copepods (Calanoida, Paraeuchaeta)," *Helgoland Marine Research* 58, no. 2 (2004): 147–153. DOI: 10.1007/s10152-004-0179-1. link.springer.com/article/10.1007/s10152-004-0179-1

[125] Many deep-sea species release larvae that migrate upward into shallower, food-rich waters before returning to the depths as adults. This vertical migration strategy enhances growth, survival, and basin-scale dispersal before settlement in abyssal environments. See Stefan F. Gary, Alan D. Fox, Arne Biastoch, J. Murray Roberts, and Stuart A. Cunningham, "Larval Behaviour, Dispersal and Population Connectivity in the Deep Sea," *Scientific Reports* 10, no. 10675 (2020). DOI: 10.1038/s41598-020-67503-7. nature.com/articles/s41598-020-67503-7

[126] Many deep-sea predators employ sit-and-wait hunting strategies, conserving energy by remaining motionless until prey ventures near. Adaptations such as camouflage, tactile or luminous lures, and rapid suction strikes allow these ambush hunters to capture food efficiently in resource-poor environments. See Dan Berkson, "How Ambush Predators Rule the Waves," *Ocean Conservancy*, May 2, 2024. oceanconservancy.org/blog/2024/05/02/how-ambush-predators-rule-the-waves/

[127] Marine microbial symbioses allow host organisms—from corals and sponges to hydrothermal vent shrimp—to survive in chemically extreme environments. At deep-sea vents and cold seeps, microbial partners metabolize sulfides, methane, and metals to detoxify host tissues, extract energy from minerals, and synthesize essential cofactors and nutrients. See Zhiyong Li, "Editorial: Marine Microbial Symbioses: Host-Microbe Interaction, Holobiont's Adaptation to Niches and Global Climate Change," *Frontiers in Microbiology* 15 (2024).

DOI: 10.3389/fmicb.2024.1416897.
frontiersin.org/articles/10.3389/fmicb.2024.1416897/full

[128] Siphonophores are not single organisms but colonies of individual hydrozoans—zooids—specialized for swimming, feeding, reproduction, and defense. These integrated units function cooperatively as a single, complex organism, exemplifying one of nature's most intricate forms of colonial life. See NOAA Office of Ocean Exploration and Research, "Siphonophores: In This Together," *Ocean Explorer*, 2024. oceanexplorer.noaa.gov/explainers/marine-life/

[129] Some deep-sea and benthic organisms incorporate metallic or mineral compounds directly into their tissues. The scaly-foot snail (*Chrysomallon squamiferum*) builds iron-sulfide armor through biologically controlled nanoparticle mineralization, while hydrothermal vent crabs (*Austinograea sp.*) reinforce their exoskeletons with aluminum- and sulfur-rich minerals. Even coastal species such as bloodworms (*Glycera dibranchiata*) integrate copper into hardened jaw structures. Together, these examples demonstrate evolution's capacity to merge biology with metallurgy. See Satoshi Okada et al., "The Making of Natural Iron Sulfide Nanoparticles in a Hot-Vent Snail," *Proceedings of the National Academy of Sciences* 116, no. 41 (2019): 20376–20381; Boongho Cho et al., "Exceptional Properties of Hyper-Resistant Armor of a Hydrothermal Vent Crab," *Scientific Reports* 12 (2022): 11816; and Javier G. Fernández et al., "Bloodworm Jaws Are Composites of Protein, Melanin, and Both Mineral and Ionic Copper," *Matter* 5 (2022): 1813–1830.

[130] China Miéville, *Embassytown* (New York: Del Rey/Pan Macmillan, 2011). penguinrandomhouse.com/books/206876/embassytown-by-china-mieville/

[131] Sign languages are complete linguistic systems with their own grammar, syntax, and phonological structure, distinct from spoken

languages. They employ visual-spatial and multimodal features—handshape, movement, facial expression, and body orientation—to convey complex meaning, forming the foundation of Deaf culture and communication. See Chuen Yet, "Linguistic Properties of Sign Languages and Its Cultural and Social Significance," *Journal of Communication Disorders, Deaf Studies and Hearing Aids* 11, no. 4 (2023): 267. DOI: 10.35248/2375-4427.23.11.267. longdom.org/open-access/linguistic-properties-of-sign-languages-and-its-cultural-and-social-significance-104355.html

[132] Accounts of so-called "feral" children—such as Victor of Aveyron in 18th-century France and Genie in 20th-century California—have long informed debates about language acquisition and socialization. However, many details remain contested or uncertain. See Sarah Lacey, "The Case of the Feral Child: Understanding the Role of Language in Human Development," *Research Ideas and Outcomes* 2 (2016): e20696. DOI: 10.3897/rio.2.e20696. riojournal.com/article/20696/

[133] The Tower of Babel story, found in *Genesis* 11:1–9, describes humanity's attempt to build a tower to the heavens and God's subsequent intervention to confuse their single language, scattering people across the earth. The tale is widely interpreted as an allegory of linguistic division and the limits of shared understanding. See Ashleigh Imus, "The Tower of Babel (Biblical Story)," *EBSCO Research Starters: Literature and Writing*, 2021. ebsco.com/research-starters/literature-and-writing/tower-babel-biblical-story

[134] Baleen whales, such as blue and fin whales, communicate through low-frequency calls that can travel thousands of kilometres by propagating within the ocean's deep-sound (SOFAR) channel, enabling long-distance navigation and social contact. See Zoe Cormier, "The Loudest Voice in the Animal Kingdom," *BBC Earth*, 2024. bbcearth.com/news/the-loudest-voice-in-the-animal-kingdom

[135] Many marine species, including baleen whales and pinnipeds, produce and perceive infrasonic sounds—frequencies below 20 Hz—that can propagate across vast ocean distances. Such low-frequency communication plays a role in navigation, mating, and social contact, and parallels have been observed in certain fish that generate drumming or pulsing signals. See *Low-Frequency Sound Production and Reception in Mammals*, Discovery of Sound in the Sea (DOSITS), University of Rhode Island, 2024. dosits.org/animals/advanced-topics-animals/low-frequency-sound-production-and-reception-in-mammals/

[136] Marine ecosystems are filled with complex soundscapes composed of grunts, pulses, pops, whistles, codas, and other acoustic signals produced by fishes and marine mammals. These specialized sounds facilitate social communication, navigation, and coordination across species. See *Specialized Sounds in Cetacean Communication*, Discovery of Sound in the Sea (DOSITS), University of Rhode Island, 2024. dosits.org/animals/advanced-topics-animals/specialized-sounds-in-cetacean-communication/

[137] Bioluminescence in the ocean spans rhythmic and staccato flashes used for communication, camouflage, mate attraction, and predation. Deep-sea species—from squid and lanternfish to worms and crustaceans—can modulate light intensity, duration, and color to serve these diverse functions. See *"Bioluminescence,"* Smithsonian Ocean Portal, Smithsonian National Museum of Natural History, 2013. ocean.si.edu/ocean-life/fish/bioluminescence

[138] Quorum sensing is a chemical communication process in which bacteria coordinate behavior through the release and detection of signaling molecules (autoinducers) that accumulate with population density. This mechanism regulates collective activities such as bioluminescence, biofilm formation, and nutrient cycling in marine ecosystems. See Zhengyu Wang et al., "Quorum Sensing and Quorum

Quenching in Marine Microbial Communities," *Frontiers in Marine Science* 9 (2022): 834337. DOI: 10.3389/fmars.2022.834337. frontiersin.org/articles/10.3389/fmars.2022.834337/full

[139] Many marine organisms use hydrodynamic and pressure cues as a form of non-acoustic communication, detecting subtle water movements through specialized sensory systems such as the lateral line or distributed mechanoreceptors. These systems allow fish, invertebrates, and marine mammals to perceive nearby movements and interactions even in complete darkness. See Ava Singh, "How Marine Animals Use Water Flow as Their Secret Communication Channel," *Marine Biodiversity Science Center*, 2025. marinebiodiversity.ca/how-marine-animals-use-water-flow-as-their-secret-communication-channel/

[140] Many marine species produce and detect bioelectric fields generated by muscular and cellular activity. Electrosensitive taxa such as rays, catfish, and knifefish use these weak electrical signals for detection, orientation, and communication—behaviors that can be disrupted by anthropogenic electromagnetic fields. See Zoë L. Hutchison et al., "Anthropogenic Electromagnetic Fields (EMF) Influence the Behaviour of Bottom-Dwelling Marine Species," *Scientific Reports* 10 (2020): 4219. doi.org/10.1038/s41598-020-60793-x

CHAPTER FIVE

[141] The Trojan Horse, central to the mythic fall of Troy, appears in *Homer's Iliad* and *Virgil's Aeneid* as a ruse of war and enduring cautionary symbol of deception. While archaeological discoveries at Hisarlik (modern Turkey) confirm Troy's historicity, the reality of the horse itself remains uncertain—interpreted variously as a siege engine, a ship, or a purely symbolic device. See Julia Kindt, "Was the

Trojan Horse Real?" *History Today*, August 2024.
historytoday.com/archive/history-matters/was-trojan-horse-real.

[142] The *RMS Titanic*, a British luxury liner, sank on April 14–15, 1912, during its maiden voyage from Southampton to New York City, killing about 1,500 passengers and crew. The disaster became one of history's most enduring modern tragedies and a lasting cultural symbol of human overconfidence in technology. See Amy Tikkanen, "*Titanic*," *Encyclopaedia Britannica*, last updated October 15, 2025. britannica.com/topic/Titanic.

[143] From myth to modern fiction, archetypal tales of hubris end in ruin: Daedalus and Icarus (Greek myth), Oedipus (Sophocles), Dorian Gray (Wilde), Faust (the German legend as developed by Goethe), Captain Ahab (Melville), and Dr. Henry Wu in *Jurassic Park* (Crichton). Each figure embodies overreach and its consequences, a throughline spanning antiquity to contemporary literature. See "Daedalus," *Encyclopaedia Britannica*. britannica.com/topic/Daedalus-Greek-mythology; "Oedipus," *Encyclopaedia Britannica*. britannica.com/topic/Oedipus-Greek-mythology; "Dorian Gray," *Encyclopaedia Britannica*. britannica.com/topic/Dorian-Gray; "Faust," *Encyclopaedia Britannica*. britannica.com/topic/Faust-literary-character; "Captain Ahab," *Encyclopaedia Britannica*. britannica.com/topic/Captain-Ahab; Michael Crichton, *Jurassic Park* (New York: Alfred A. Knopf, 1990). archive.org/details/michael-crichton-jurassic-park.

[144] *Hamlet*, tragedy in five acts by William Shakespeare, was written about 1599–1601 and first published in a quarto edition in 1603 from an unauthorized text, with a revised second quarto appearing in 1604. The story draws on Saxo Grammaticus's *Gesta Danorum* (12th century) and François de Belleforest's *Histoires tragiques* (1570). See "Hamlet," *Encyclopaedia Britannica*. britannica.com/topic/Hamlet-by-Shakespeare.

[145] A. C. Bradley's classic analysis of *Hamlet* situates the play among Shakespeare's great tragedies, emphasizing its complex motives, introspection, and moral vision. See A. C. Bradley, *Shakespearean Tragedy: Lectures on Hamlet, Othello, King Lear, Macbeth* (London: Macmillan, 1904). Project Gutenberg EBook #16966. gutenberg.org/files/16966/16966-h/16966-h.htm.

[146] In *Hamlet*, the ghost of the murdered King Hamlet first appears to the watchmen and Horatio at Elsinore (Act I, Scene I), while the revelation of his murder by Claudius is disclosed later to Prince Hamlet (Act I, Scene V). See William Shakespeare, *Shakespeare's Tragedy of Hamlet, Prince of Denmark*, arranged for representation at the Royal Princess's Theatre, with explanatory notes by Charles Kean (London: Bradbury and Evans, 1859). Project Gutenberg EBook #27761. gutenberg.org/files/27761/27761-h/27761-h.htm.

[147] The ambiguity of the ghost's identity—whether it is truly King Hamlet's spirit, a demon, or a projection of guilt—is central to the play's tension and moral uncertainty. See Stephen Greenblatt, *Hamlet in Purgatory* (Princeton, NJ: Princeton University Press, 2001). archive.org/details/hamletinpurgatoroooogree_u7q2.

[148] Bernardo's line appears in *Hamlet*, Act I, Scene I, where the watchmen first observe the ghost on the battlements of Elsinore. See William Shakespeare, *Shakespeare's Tragedy of Hamlet, Prince of Denmark*, arranged for representation at the Royal Princess's Theatre, with explanatory notes by Charles Kean (London: Bradbury and Evans, 1859), Act I, Scene I. Project Gutenberg EBook #27761. gutenberg.org/files/27761/27761-h/27761-h.htm#sceneI_1.

[149] Cassiopeia lies opposite the Big Dipper across the North Star and serves as a reliable celestial marker for locating Polaris. See Isabel M. Lewis, "How to Find the North Star," *The Scientific Monthly* 21, no. 5

(November 1925): 560. American Association for the Advancement of Science. Stable URL: jstor.org/stable/7492

[150] *Cassiopeia* is a northern constellation recognizable by its W-shaped formation of five bright stars. In Greek mythology, it represents the queen of Ethiopia, whom Poseidon punished after boasting of her daughter, Andromeda's, beauty. See "Cassiopeia," *Encyclopaedia Britannica*, last modified 2025, britannica.com/place/Cassiopeia-astronomy.

[151] Researchers from Southwest Texas State University (now Texas State University) have proposed that the "star west of the pole" mentioned in *Hamlet* Act I, Scene I may refer to the supernova of 1572, later known as Tycho's Supernova. The event, visible across Europe, profoundly influenced contemporary thought about celestial change and may have inspired Shakespeare's use of an unsettling heavenly omen in the play's opening. See Don Olson, Russell Doescher, and Marilynn Olson, "Researchers say star in *Hamlet* may be supernova of 1572," *Sky and Telescope* (Nov 1998); archived at Texas State University News Release, Sept 28 1998, web.archive.org/web/20080906170542/https://www.txstate.edu/new s/news_releases/news_archive/1998/09/supernova092898.html.

[152] Retrograde motion refers to the apparent reversal of a planet's movement across the sky, caused by differing orbital speeds between Earth and other planets. This effect creates the illusion of a planet "halting" and moving backward for a brief period, though its motion remains continuous in space. See "Retrograde Motion," *Encyclopaedia Britannica*, last modified 2025, britannica.com/science/retrograde-motion.

[153] The "Roswell incident" refers to the July 1947 recovery of debris near Roswell, New Mexico, initially reported by the U.S. Army Air Forces as a "flying disc" before being identified as part of a classified

Project Mogul surveillance balloon. The event later became central to postwar UFO mythology and conspiracy culture. See "Roswell Incident," *Encyclopaedia Britannica*, last modified 2025, britannica.com/event/Roswell-incident.

[154] Historical accounts of unidentified aerial phenomena appear throughout human history, long predating the 1947 Roswell incident. Reports range from ancient religious and astronomical observations to the 19th-century "airship" wave and early 20th-century "foo fighter" sightings. See Matt Blitz, "Angels, Airships, and Aliens: The 3,500-Year History of UFO Sightings," *Popular Mechanics*, September 24, 2019. popularmechanics.com/military/a29189536/ufo-sightings/

[155] Accounts of unexplained aerial phenomena predate modern UFO narratives by centuries. Examples include Governor John Winthrop's 1639 diary describing mysterious lights over Boston, later regarded as America's first recorded UFO sighting, and the 1665 "Air Battle of Stralsund," documented in contemporary broadsheets and recently examined in a Berlin State Museums exhibition. See Christopher Klein, "America's First UFO Sighting," *History.com*, November 9, 2016 (updated May 28, 2025), history.com/articles/americas-first-ufo-sighting; *A UFO in 1665: The Air Battle of Stralsund*, exhibition text, Kunstbibliothek, Staatliche Museen zu Berlin (2023), smb.museum/en/exhibitions/detail/a-ufo-in-1665/.

[156] In *Hamlet*, Act I, Scene IV, the Prince condemns King Claudius's drunken celebrations, calling them "a custom more honour'd in the breach than the observance," and laments that such revelry causes Denmark to be "tax'd of other nations" as a land of "drunkards." See William Shakespeare, *Shakespeare's Tragedy of Hamlet, Prince of Denmark*, arranged for representation at the Royal Princess's Theatre, with explanatory notes by Charles Kean (London: Bradbury and Evans, 1859), Act I, Scene IV. Project Gutenberg EBook #27761. gutenberg.org/files/27761/27761-h/27761-h.htm#sceneI_4.

[157] William Shakespeare, *Shakespeare's Tragedy of Hamlet, Prince of Denmark*, arranged for representation at the Royal Princess's Theatre, with explanatory notes by Charles Kean (London: Bradbury and Evans, 1859), Act I, Scene IV, lines 39–47. Project Gutenberg EBook #27761. gutenberg.org/files/27761/27761-h/27761-h.htm#sceneI_4.

[158] Hamlet's encounter with the ghost—spanning Act I, Scenes IV–V— marks the turning point of his fate. The apparition urges vengeance for his father's murder, setting in motion the cycle of doubt, deception, and ruin that follows. See William Shakespeare, *Shakespeare's Tragedy of Hamlet, Prince of Denmark*, arranged for representation at the Royal Princess's Theatre, with explanatory notes by Charles Kean (London: Bradbury and Evans, 1859), Act I, Scenes IV–V. Project Gutenberg EBook #27761. gutenberg.org/files/27761/27761-h/27761-h.htm#sceneI_4.

[159] The modern shift from "UFO" (unidentified flying object) to "UAP" (unidentified aerial phenomenon) reflects official U.S. government terminology beginning in 2020, when the Pentagon established the Navy-led UAP Task Force to investigate military sightings. NASA followed up with an independent study in 2023 that emphasized data standardization and scientific rigor. See CBS News, "What Are UAPs? The Pentagon, NASA, and Other Agencies Have a New Name for UFOs," *CBS News*, last modified May 28, 2025. cbsnews.com/news/what-are-uaps-unexplained-aerial-phenomenon-ufos-new-name/.

[160] Official documentation confirms that on March 16, 1967, all ten missiles of Echo Flight at Malmstrom Air Force Base lost strategic alert within seconds, with the cause listed as "no apparent reason." See U.S. Air Force, *Declassified Malmstrom AFB UFO Incident Report* (1967), obtained through the Freedom of Information Act, archived at The Black Vault.

documents.theblackvault.com/documents/ufos/malmstromufo.pdf.
A second reported event, eight days later at Oscar Flight, is attested in firsthand accounts by then–Lt. Robert Salas and other personnel, though no official report has been released. See Bill Kaufmann, "Seminal Montana UFO Events Draw Pentagon's Attention," *Calgary Herald*, February 23, 2023. calgaryherald.com/news/seminal-montana-ufo-events-pentagon.

[161] Construction of Great Falls Army Air Base began on May 7, 1942, approximately six miles east of Great Falls, Montana, as a World War II training field for B-17 Flying Fortress crews. The base was later renamed Great Falls Air Force Base and officially designated Malmstrom AFB on June 15, 1956, in honor of Col. Einar A. Malmstrom. See *Malmstrom Air Force Base – About Us*, U.S. Air Force, malmstrom.af.mil/About-Us/; and *Malmstrom Air Force Base: Wing I History*, MinutemanMissile.com, minutemanmissile.com/afbwing1.html.

[162] Troy A. Hallsell, "B-17 Training and the 7th Ferrying Group at Great Falls Army Air Base in World War II," *Air & Space Power History* 70, no. 1 (Spring 2022): 5–12. U.S. Air Force Historical Research Agency / 341st Missile Wing History Office. Accessible at malmstrom.af.mil or via official archive (PDF titled "20220300 (U) Hist TAH B17 Training and 7FG at GFAAB.pdf").

[163] Great Falls served as a central hub for aircraft and matériel transfers to the Soviet Union via the Alaska–Siberia (ALSIB) route during World War II; see "History of Malmstrom Air Force Base," U.S. Air Force. malmstrom.af.mil/About-Us/History/Malmstrom-History/ For local operational detail and veteran testimony on Lend-Lease movements through Great Falls, see Jenn Rowell, "Veterans Share History of Lend-Lease," *Great Falls Tribune*, July 19, 2015. greatfallstribune.com/story/news/local/2015/07/19/veterans-share-history-lend-lease/30398459/ For broader context on the program's

scope and transfers of strategic scientific materials relevant to early atomic research, see Richard Rhodes, *The Making of the Atomic Bomb* (New York: Simon & Schuster, 1986).

[164] The 582nd Air Resupply and Communications Wing operated out of Great Falls Air Force Base from May 1953 to February 1954, conducting classified "PSYWAR" (psychological warfare) and special operations missions before transferring to RAF Molesworth in the United Kingdom. The unit's activities formed part of the Air Resupply and Communications Service, a Cold War–era precursor to modern Air Force special operations. The Department of Defense formally defined PSYWAR as "activities intended to convey selected information and indicators to audiences to influence their emotions and reasoning." See "582nd Air Resupply and Communications Wing," *Air Force Historical Research Agency* (via Wikipedia, en.wikipedia.org/wiki/582d_Air_Resupply_and_Communications_Wing); *582 Air Resupply Group Fact Sheet*, U.S. Air Force Historical Research Agency, Maxwell AFB (AL), 2012; "Air Resupply and Communications Service," *War History Online* (2015); and *Air Force Doctrine Document 2-5.3: Psychological Operations* (Department of the Air Force, 1999), citing *JP 1-02: DOD Dictionary of Military and Associated Terms*.

[165] Colonel Einar A. Malmstrom, a decorated World War II veteran and vice commander of the 407th Strategic Fighter Wing, was killed in a T-33 crash near Great Falls, Montana, on August 21, 1954. Early local accounts described his death as a deliberate act of sacrifice to steer the aircraft away from town. At the same time, later Air Force historical assessments suggest a sudden medical emergency may have caused the crash. Great Falls AFB was renamed Malmstrom Air Force Base in his honor on June 15, 1956. See Troy A. Hallsell, "Memorializing Colonel Einar Axel Malmstrom," *Signature Montana* (May 2023); "Into the Fog: Last Man to See Einar Malmstrom Alive Tells Story," *Great Falls Tribune* (Jan. 18, 2014); American War Memorials Overseas – Einar

Axel Malmstrom, uswarmemorials.org; and *Malmstrom AFB History: Col. Einar Malmstrom*, U.S. Air Force, malmstrom.af.mil/About-Us/History/Col-Einar-Malmstrom/.

[166] During the 1950s and 1960s, Malmstrom Air Force Base (then Great Falls AFB) played a significant role in North American air defense operations. The 29th Air Division established command at the base in 1951, bringing with it fighter-interceptor and aircraft-control-and-warning squadrons. Radar units such as the 801st Aircraft Control and Warning Squadron (later 801st Radar Squadron) operated AN/FPS-20 and AN/FPS-6 systems as part of the Semi-Automatic Ground Environment (SAGE) network, providing early warning coverage for the northern plains region. These radar arrays formed part of the continental defense shield under the newly created North American Aerospace Defense Command (NORAD), headquartered at Malmstrom as an alternate command post until 1983. See *Malmstrom Air Force Base – History*, U.S. Air Force, malmstrom.af.mil/About-Us/History/Malmstrom-History/; "Malmstrom Air Force Base," *Wikipedia*, en.wikipedia.org/wiki/Malmstrom_Air_Force_Base; "801st Radar Squadron," *Wikipedia*, en.wikipedia.org/wiki/801st_Radar_Squadron; and Defense Technical Information Center (DTIC) Report ADA339158, *History of the North American Aerospace Defense Command and U.S. Air Force Aerospace Defense Command, 1940–1980* (U.S. Air Force, 1981).

[167] By the mid-1950s, Malmstrom Air Force Base had become a pivotal node in both continental air defense and nuclear deterrence strategy. Following the creation of the North American Aerospace Defense Command (NORAD) in 1957, Malmstrom served as an alternate command post for the 24th NORAD Region, responsible for air defense across nine U.S. states and four Canadian provinces. Earlier, in 1954, the base was aligned under Strategic Air Command (SAC), hosting the 407th Strategic Fighter Wing and later the 4061st Air Refueling Wing, both integral to SAC's long-range atomic strike

capability. The dual missions of continental defense and nuclear readiness cemented Malmstrom's role as both shield and spearhead of the early Cold War. See *Malmstrom Air Force Base – History*, U.S. Air Force, malmstrom.af.mil/About-Us/History/Malmstrom-History/; "Malmstrom Air Force Base," *Wikipedia*, en.wikipedia.org/wiki/Malmstrom_Air_Force_Base; "341st Missile Wing – History," U.S. Air Force, malmstrom.af.mil/About-Us/Fact-Sheets/Display/Article/346869/341st-missile-wing/; and U.S. Air Force Biography: Major General William S. Harrell, af.mil/About-Us/Biographies/Display/Article/106826/.

[168] The LGM-30 Minuteman intercontinental ballistic missile was the first three-stage, solid-fuel ICBM in the U.S. strategic arsenal. Conceived in the late 1950s and deployed in the early 1960s, the Minuteman introduced rapid launch and unprecedented survivability through its network of hardened underground silos. Each missile was housed in a vertical cylindrical launch facility, sealed by a blast door and connected via reinforced cabling to a buried launch control center staffed around the clock by two-officer crews. This system became the backbone of America's nuclear deterrent during the Cold War and remains in service today in modernized Minuteman III form. See *LGM-30G Minuteman III Fact Sheet*, U.S. Air Force, malmstrom.af.mil/About-Us/Fact-Sheets/Display/Article/1454920/lgm-30g-minuteman-iii/; "Maintaining Malmstrom's Launch Control Centers," 341st Missile Wing Public Affairs (2019); and "341st Missile Wing LGM-30 Minuteman Missile Launch Sites," *Wikipedia*, en.wikipedia.org/wiki/341st_Missile_Wing_LGM-30_Minuteman_missile_launch_sites.

[169] During the Cuban Missile Crisis of October 1962, the 341st Strategic Missile Wing at Malmstrom Air Force Base brought the first Minuteman ICBMs in the nation to operational alert status. On October 27, 1962, Launch Facility Alpha-06 was activated near

Neihart, Montana, under direct order from Strategic Air Command, marking the first real-world test of the Minuteman system. The crisis accelerated deployment and modernization across the U.S. ICBM force, transforming Malmstrom into a cornerstone of nuclear deterrence through the remainder of the Cold War. See Troy A. Hallsell, Ph.D., *"Missile Crisis Spurs Untested ICBM Usage,"* 341st Missile Wing History Office (Malmstrom AFB, Oct. 26 2022); Senior Airman Shelby Thurman, *"Sixty Years Later, America's 'Ace in the Hole' Still Stands Guard,"* Air Force Global Strike Command Public Affairs (Oct. 28 2022); and *"Malmstrom Air Force Base – History,"* U.S. Air Force, malmstrom.af.mil/About-Us/History/Malmstrom-History/.

[170] Official records confirm that on March 16 1967, all ten missiles of Echo Flight at Malmstrom AFB simultaneously entered *"No-Go"* status, defined by the U.S. Air Force as a condition in which a missile cannot respond to launch or inhibit commands and is therefore non-operational. See U.S. Air Force, *Declassified Malmstrom AFB UFO Incident Report* (1967), The Black Vault Archive, documents.theblackvault.com/documents/ufos/malmstromufo.pdf; and *Air Force Instruction 91-114: Safety Rules for the Intercontinental Ballistic Missile System* (Department of the Air Force, 6 January 2023, Incorporating Change 1 – 3 December 2024).

[171] The declassified U.S. Air Force *Report of Engineering Investigation of Echo Flight Incident* (Malmstrom AFB, 16 March 1967) records that all ten missiles in Echo Flight shut down simultaneously, each showing Voice Reporting Signal Assembly (VRSA) fault indications on channels 9 and 12—both corresponding to the missile guidance and logic systems. See U.S. Air Force, *Declassified Malmstrom AFB UFO Incident Report* (1967), The Black Vault Archive, documents.theblackvault.com/documents/ufos/malmstromufo.pdf; and *Minuteman Weapon System History and Description* (Boeing / Air Force Ballistic Systems Division, ca. 1970s), Section III: Operational Ground Equipment.

[172] U.S. Air Force, *Report of Engineering Investigation of Echo Flight Incident* (Malmstrom AFB, 16–23 March 1967) and associated correspondence (SAC, Boeing, Autonetics/Autonetics Division of North American Aviation, and OOAMA), declassified under FOIA; archived via The Black Vault, documents.theblackvault.com/documents/ufos/malmstromufo.pdf. The report states that all ten Echo Flight missiles experienced near-simultaneous shutdowns with VRSA fault indications on guidance/logic channels 9 and 12; notes no evidence of sabotage or adverse weather; records restoration of missiles to alert status by maintenance actions without corrective hardware replacement noted; and discusses an external "noise"/EMI hypothesis while finding no single definitive cause.

[173] U.S. Air Force, *Report of Engineering Investigation of Echo Flight Incident* (Malmstrom AFB, March 1967), declassified under FOIA; archived via The Black Vault, documents.theblackvault.com/documents/ufos/malmstromufo.pdf. The report attributes the simultaneous missile shutdowns to an unknown external pulse, electrical or electromagnetic in nature, while noting the absence of radar or visual anomalies and explicitly ruling out aerial objects or unidentified intrusions. It records that all missiles were restored to alert status after corrective testing, with no recurrence reported.

[174] Reports of a second 1967 incident, allegedly affecting Oscar Flight at Malmstrom AFB, derive primarily from later testimonies by former missileers Lt. Robert Salas, Capt. Frederick Meiwald, and maintenance personnel such as Henry B. Jamison, as compiled in Robert Hastings, *UFOs and Nukes: Extraordinary Encounters at Nuclear Weapons Sites* (AuthorHouse, 2008), pp. 172–181; "1967 UFO Incident Still Mystifies Man," *Great Falls Tribune*, Apr. 9 1995; "Close Encounter Prompts Nuclear Shutdown," *Independent Record* (Helena, Mont.), May 1 1997; and "How They Learned to Stop Worrying and Love the Bomb,"

*Missoula Independent*, May 8, 1997.
No comparable official Air Force or Strategic Air Command documentation has been declassified for the alleged Oscar Flight shutdown.


[175] Affidavits of Robert L. Salas and Robert C. Jamison, *Malmstrom UFO Testimonials* (DocumentCloud, 2010); Frederick Meiwald referenced therein. See also "History — Robert L. Salas, USAF Class of 1964," *U.S. Air Force Academy Class Histories*, usafaclasses.org/1964/Salas_Robert/history.htm; Bill Kaufmann, "Seminal Montana UFO Events Draw Pentagon's Attention," *Calgary Herald*, Feb 23 2023.
For representative skeptical analyses, see Brian Dunning, "The Day the UFO Deactivated the Nukes," *Skeptoid Podcast* #842 (July 26, 2022); and Mick West, "Great Expectations: UFOs in Congress," Skeptical Inquirer (May 25, 2022).
Collectively, these sources confirm each man's service at Malmstrom AFB (1967) and public testimony regarding the alleged Oscar Flight event, while illustrating the broad spectrum of interpretation—from proponents such as Salas and Hastings to skeptical reviewers questioning memory accuracy and data gaps.


[176] Captain Robert L. Salas, *History — USAF Class of 1964*, U.S. Air Force Academy Class Histories, usafaclasses.org/1964/Salas_Robert/history.htm. See also "About the Author," *Tantor Media*, tantor.com/author/robert-salas.html; and Mike Harris, "The Truth Is Out There: Ojai Author Robert Salas to Discuss UFOs and Nuclear Technology," *VC Reporter*, June 14, 2012. These sources confirm Salas's active-duty service from 1964–1971 with training at Chanute and Vandenberg Air Force Bases and subsequent assignment to Malmstrom AFB as a Minuteman launch officer. Following his military service, he worked for Martin Marietta Aerospace and Rockwell International on early Space Shuttle proposals before joining the Federal Aviation Administration.

[177] Robert Salas, as quoted in *The Los Angeles Times*, "Robert Salas," December 29, 2005, p. 113, www.newspapers.com/image/239754149/. See also Andy Bloxham, "Aliens Landed and Interfered with Missiles, US Pilots Claim," *Irish Independent*, April 13, 2010, newspapers.com/article/irish-independent-aliens-landed-and-inte/169102498/. These reports describe Salas's testimony that, while on duty at Malmstrom AFB on March 24, 1967, UAP sightings by security personnel coincided with the simultaneous shutdown of ten Minuteman missiles at Oscar Flight.

[178] U.S. Air Force, "Response to UFO Inquiries," Freedom of Information Act release archive, Office of Public Affairs, 2000s. No declassified records acknowledge a March 24, 1967, Oscar Flight shutdown linked to UAP activity. Salas has repeatedly asserted that documentation was suppressed or removed from official channels, as noted in *The Missoulian*, "Robert Salas," March 27, 2006 (p. A3), newspapers.com/article/the-missoulian-robert-salas/169100151/.

[179] *North American Aerospace Defense Command, NORAD Command Director's Log*, 24 October – 15 November 1975, NORAD/CONAD, Peterson AFB, Colorado Springs (Declassified via FOIA 1992); see also *Strategic Air Command 8th Air Force Report* (Declassified 1979; re-released 1992). These documents record repeated radar and visual detections of unidentified aerial objects over multiple U.S. Air Force nuclear installations—Loring AFB (ME), Wurtsmith AFB (MI), Minot AFB (ND), and Malmstrom AFB (MT)—prompting security alerts, NORAD coordination, and F-106 scrambles. See also Barry Greenwood and Lawrence Fawcett, "Wurtsmith and Others," in *Clear Intent* (1984), and NICAP archival reprint (nicap.org/docs/norad95-101.pdf); *The New York Times*, "U.F.O. Files: The Untold Story," October 14 1979, p. E7; and Tyler Rogoway, "The Bizarre Mystery of Unexplained Aerial Incursions over Loring Air Force Base," *The War Zone*, March 23 2020. These cross-corroborated records confirm that the 1975 NORAD/SAC

log series documented unidentified aerial incursions at nuclear facilities across the northern tier of the United States.

[180] *North American Aerospace Defense Command, NORAD Command Director's Log*, 24 October – 15 November 1975, NORAD/CONAD, Peterson AFB, Colorado Springs (Declassified via FOIA 1992); and *Strategic Air Command 8th Air Force Report* (Declassified 1979; re-released 1992). These documents record a sequence of events between 7 and 10 November 1975 in which personnel at Malmstrom AFB reported multiple unidentified aerial objects—bright orange-gold lights and structured forms—hovering near launch facilities and missile fields. Security and radar teams were dispatched, and interceptors, including F-106 aircraft, were scrambled from Malmstrom and other northern-tier bases. Logs note that objects extinguished their lights upon the approach of interceptors, re-illuminated after withdrawal, and repeatedly eluded radar contact despite confirmed visual tracking. Parallel reports from Loring AFB (ME), Wurtsmith AFB (MI), Minot AFB (ND), and Canadian Forces Station Falconbridge describe nearly identical behavior and timing. See also Barry Greenwood and Lawrence Fawcett, "Wurtsmith and Others," in *Clear Intent* (1984), 41–56, and NICAP archival reprint (nicap.org/docs/norad95-101.pdf); *The New York Times*, "U.F.O. Files: The Untold Story," October 14, 1979, p. E7; and Tyler Rogoway, "The Bizarre Mystery of Unexplained Aerial Incursions over Loring Air Force Base," *The War Zone*, March 23, 2020. Together, these declassified records confirm coordinated military responses to unidentified aerial activity at multiple U.S. and Canadian nuclear facilities during late 1975.

[181] *North American Aerospace Defense Command, NORAD Command Director's Log*, 24 October – 15 November 1975, NORAD/CONAD, Peterson AFB, Colorado Springs. Final page signed by Col. Terrence C. James, USAF Director of Administration, referencing an additional NORAD document "forwarded for review for possible downgrade and

release" but "withheld on the grounds that it was properly and currently classified and exempt from disclosure under Public Law 90-23, 5 USC 552(b)(1)." (Declassified via FOIA, 1992.)

[183] All documentation referenced to this point originates from official U.S. military or government sources, either declassified through the Freedom of Information Act or released directly by the agencies themselves. Primary materials were accessed through *The Black Vault* document archive (theblackvault.com/documentarchive/), which hosts authenticated FOIA releases from the Department of Defense, NORAD, Strategic Air Command, and related agencies. Additional records were verified via government repositories, including the FBI Vault (vault.fbi.gov/), the National Archives and Records Administration (archives.gov/research/topics/uaps), the CIA FOIA Reading Room (cia.gov/readingroom/), the NSA FOIA portal (nsa.gov/Helpful-Links/NSA-FOIA/Frequently-Requested-Information/Unidentified-Flying-Objects-UFOs/), the U.S. Air Force *Secrets Declassified* archive (secretsdeclassified.af.mil/Historical-Documents/Unidentified-Flying-Objects/), the Department of Defense All-domain Anomaly Resolution Office (aaro.mil/UAP-Records/), and the Office of the Director of National Intelligence (dni.gov/files/ODNI/documents/assessments/Prelimary-Assessment-UAP-20210625.pdf). These constitute the publicly available corpus of verified UAP-related government communications and reports.

[184] *Freedom of Information Act of 1966*, Pub. L. 89-487, 80 Stat. 250 (enacted July 4, 1966; effective July 4, 1967). See President Lyndon B. Johnson, "Statement by the President upon Signing S. 1160, the Freedom of Information Bill," *The White House Office of the Press Secretary*, July 4, 1966; and Representative Donald Rumsfeld, "Press Release: Rumsfeld Commends President for Signing Freedom of Information Bill," July 7, 1966. Both reproduced by the National Security Archive, *FOIA at 40* collection (Documents 37 and 40), nsarchive2.gwu.edu/NSAEBB/NSAEBB194/

[185] *Biographical Directory of the United States Congress*, "John Emerson Moss (M001035)," U.S. Congress, https://bioguide.congress.gov/search/bio/M001035; see also "John E. Moss," *Wikipedia*, last modified 2024, wikipedia.org/wiki/John_E._Moss. Moss represented California's 3rd Congressional District from 1953 to 1978, chaired the Special Subcommittee on Government Information and the Foreign Operations and Government Information Subcommittee of the House Government Operations Committee, and served on committees overseeing atomic energy, water, power, and ocean resources. His legislative record includes authorship of the Freedom of Information Act and extensive work in consumer protection and government transparency.

[186] U.S. Fish and Wildlife Service, "Nuclear Power Plants and Aquatic Resources," *U.S. Fish and Wildlife Service* (accessed 2024), fws.gov/node/265255; and *International Atomic Energy Agency*, "Water-Cooled Reactors," *IAEA Topics*, iaea.org/topics/water-cooled-reactors (accessed 2023). These authoritative sources confirm that most commercial nuclear power plants are located along lakes, rivers, or coastlines to draw cooling water for reactor systems, reflecting standard nuclear industry siting practice.

[187] Kevin D. Crowley and John F. Ahearne, "Managing the Environmental Legacy of U.S. Nuclear-Weapons Production," *American Scientist* 90, no. 6 (November–December 2002): 514–527, DOI 10.1511/2002.39.514; see also U.S. Department of Energy, *Linking Legacies: Connecting the Cold War Nuclear Weapons Production Processes to Their Environmental Consequences* (Washington, DC: DOE Office of Environmental Management, 1997). These studies document long-term contamination of soil and groundwater surrounding weapons-production facilities—including Hanford, Oak Ridge, Savannah River, and the Idaho National Laboratory—where radioactive and chemical waste entered nearby river systems and aquifers.

188 Adam Janos, "Why Have There Been So Many UFO Sightings Near Nuclear Facilities?" *History.com* (A&E Television Networks, June 21 2019; updated May 28, 2025), history.com/articles/ufos-near-nuclear-facilities-uss-roosevelt-rendlesham; and "UFO Reports and Atomic Sites," *Wikipedia*, last modified 2025, en.wikipedia.org/wiki/UFO_reports_and_atomic_sites. These corroborating sources summarize declassified military and intelligence records documenting repeated UAP activity near nuclear power, weapons, and research sites worldwide—from Los Alamos, Oak Ridge, and Malmstrom AFB to Rendlesham Forest and contemporary naval strike groups—and note official recognition of this pattern in U.S. and allied defense reporting.

189 United States Air Force, *Project Blue Book* case file, "Edwards Air Force Base, California, 7 October 1965" (Project 10073 Record; AFFTC correspondence; LAADS tape catalog; radar-scope photographs), declassified 1965–1966; transcripts and document set reproduced by NICAP, "UFO Alert at Edwards Air Force Base" (2006–2008), nicap.org/edwards65dir.htm; see also "Actual Documents from Blue Book Files" (17 pp.), nicap.org/docs/edwards_docs.pdf, and "Corrected Tape Transcript," nicap.org/docs/edwards_trans_corr.htm. These materials record seven to twelve luminous objects observed over Edwards AFB for approximately four and a half hours by tower personnel and other base witnesses, with intermittent radar confirmation at multiple sites; an F-106 interceptor was scrambled from George AFB at 1209Z, and LAADS/ARADCOM recorded the incident on voice tapes while radar scope cameras carried unidentified tracks (U090–U105).

190 United States Air Force, *Headquarters 90th Strategic Missile Wing, Francis E. Warren Air Force Base, Wyoming—Report to AFSC (FTD): "Unidentified Flying Objects (UFO)"* (31 July–2 August 1965), 4 pp., signed by Col. Donald W. Johnson, Commander, 809 Combat Support Gp; declassified *Project Blue Book* file. The report details 148 objects

observed over three nights by 143 missile-site personnel near Cheyenne, Wyoming, including Minuteman launch control crews. Descriptions note round or cigar-shaped lights, color changes, and hovering or erratic motion under clear weather conditions, with no radar or aircraft correlation. See also Robert Hastings, "Yet Another Nuclear Missile Launch Officer Talks About UFOs at F. E. Warren AFB," *UFOs and Nukes* (2013), ufohastings.com/articles/yet-another-nuclear-missile-launch-officer-talks-about-ufos-at-f-e-warren-afb

[191] Robert Hastings, "UFO Sightings at ICBM Sites: Ellsworth AFB, 1966," *Nuclear Connection Project* (RR0.org, 2006; updated 2008), rro.org/time/2/0/0/6/02/24/Hastings_UfosightingsAtIcbmSites/Ellsworth_1966/; and Hastings, "Three Former U.S. Air Force ICBM Launch Officers Speak Out About UFOs," *UFOs and Nukes* (2010), ufohastings.com/articles/three-former-u-s-air-force-icbm-launch-officers-speak-out-about-ufos. Both sources describe corroborating testimony by former 44th Missile Maintenance Squadron technicians and launch officers at Ellsworth AFB, South Dakota, concerning a 1966 incident in which multiple Minuteman missiles experienced power and alert-status failures while an unidentified luminous object was observed near Launch Facilities Juliet-3 and Juliet-5.

[192] Barry Greenwood and Lawrence Fawcett, *Clear Intent: The Government Coverup of the UFO Experience* (1984), chap. 2, "Intrusions at Loring," 16–26, NICAP reprint, nicap.org/books/Clear-Intent/clear-intent.pdf; and North American Aerospace Defense Command, *NORAD Command Director's Log, 24 October – 15 November 1975*, NORAD/CONAD, Peterson AFB, Colorado Springs (Declassified via FOIA 1992). These documents record repeated radar and visual detections of unidentified aerial objects over Loring AFB (ME) beginning on 27 October 1975, prompting security alerts and attempts by U.S. Air Force aircraft to intercept them. The sightings recurred over several nights and triggered parallel incidents that same season at Wurtsmith, Minot, and Malmstrom AFBs.

[193] Annette Cary, "Tri-Cities One of Hottest UFO Sighting Spots in Washington State," *The Daily Chronicle* (Aug. 19, 2022), chronline.com/stories/tri-cities-one-of-hottest-ufo-sighting-spots-in-washington-state,298544, chronicling early reports of unidentified aerial phenomena over the Hanford Nuclear Reservation in 1944–45; Adam Janos, "Why Have There Been So Many UFO Sightings Near Nuclear Facilities?" *History.com* (A&E Television Networks, June 21 2019; updated May 28 2025), history.com/articles/ufos-near-nuclear-facilities-uss-roosevelt-rendlesham, documenting UAP incidents near Los Alamos, Sandia, and Oak Ridge laboratories during the Manhattan Project and early atomic-testing period; and *Los Alamos National Laboratory FOIA Requests at the National Security Research Center* (lanl.gov, accessed 2025), lanl.gov/media/publications/the-vault/1124-foia-requests-at-the-national-security-research-center, confirming release of declassified material related to early atomic research sites. These corroborating sources establish that recorded aerial incursions near U.S. nuclear development and research facilities date back to the 1940s.

[194] U.S. Nuclear Regulatory Commission, letter to Lawrence Fawcett, Nov. 21, 1984, FOIA-84-854, *Freedom of Information Act Correspondence on the Indian Point Nuclear Power Plant* (ML21179A021), nrc.gov/docs/ML2117/ML21179A021.pdf; *The Japan News*, "Fukushima Lab Releases Images of 'Highly Likely' UFOs," June 26, 2022, japannews.yomiuri.co.jp/society/general-news/20220626-41030/; *news.ORF.at*, "Erneut Unbekannte Flugobjekte über französischem AKW," Jan. 4 2015, newsv2.orf.at/stories/2259911/; Central Intelligence Agency, "Flying Saucers Reported over Belgian Congo Uranium Mines," declassified press file (1952), cia.gov/readingroom/docs/DOC_0000015463.pdf; and Robert Weissman, "Why Do UFO Sightings Keep Happening Near Nuclear Sites?" *Vice*, June 3, 2021, vice.com/en/article/why-do-ufo-sightings-keep-happening-near-nuclear-sites/. These international and U.S. reports illustrate persistent UAP activity in proximity to civilian and

military nuclear facilities, from Indian Point to Fukushima and French reactors, confirming the pattern's global scope in the modern era.

[195] Dawn Stover, "Double Dread: UFOs and Nuclear War," *Bulletin of the Atomic Scientists* (June 4, 2019), thebulletin.org/2019/06/double-dread-ufos-and-nuclear-war/; Adam Janos, "Why Have There Been So Many UFO Sightings Near Nuclear Facilities?" *History.com* (June 21 2019; updated May 28 2025), history.com/articles/ufos-near-nuclear-facilities-uss-roosevelt-rendlesham; *The Cold War and UFOs*, The National Archives (May 23, 2019), media.nationalarchives.gov.uk/index.php/cold-war-ufos/; Stephen Bruehl and Beatriz Villarroel, "Transients in the Palomar Observatory Sky Survey (POSS-I) May Be Associated with Nuclear Testing and Reports of Unidentified Anomalous Phenomena," *Scientific Reports* 15 (2025): 34125, doi.org/10.1038/s41598-025-21620-3; *UFO Sightings at Weapons Testing Site, Woomera*, *National Archives of Australia* (Department of Supply File on the Weapons Research Establishment, Report of J.J.A. Hanlon, July 24, 1960), naa.gov.au/students-and-teachers/student-research-portal/learning-resource/themes/war/defence-equipment-and-weapons/ufo-sightings-weapons-testing-site-woomera. These sources document statistically significant or historically reported correlations between nuclear activity and unidentified aerial phenomena—from Cold War radar incidents and declassified MOD archives to the 2025 *Scientific Reports* study identifying temporal associations between nuclear tests and UAP clusters. Together, they establish that recorded UAP activity has repeatedly coincided with moments of heightened nuclear tension, including the Cuban Missile Crisis.

[196] Robert L. Salas, "Affidavit of Robert L. Salas," *Malmstrom UFO Testimonies* (DocumentCloud, Sept. 27, 2010), documentcloud.org/documents/9329-malmstrom-ufo-testimonials/; Robert C. Jamison and Dwynne C. Arneson, affidavits in the same collection, detailing missile shutdowns and security responses at

Malmstrom AFB in March 1967; *A Narrative of UFO Events at Minot Air Force Base* (Zenodo, 2023), documenting radar and visual correlations during the 1968 Minot incident; and the *Byelokoroviche Incident Report* (Soviet Strategic Rocket Forces Archive, 1982, summarized in declassified materials) describing temporary system inhibition at a nuclear missile site near Byelokoroviche, Ukraine.
Together, these accounts provide consistent testimony of unidentified aerial phenomena coinciding with launch-system malfunctions or alarms at both U.S. and Soviet nuclear facilities. These patterns are later echoed in official reviews, which acknowledge that such events remain of defense interest.

[197] *Preliminary Assessment: Unidentified Aerial Phenomena*, Office of the Director of National Intelligence (June 25, 2021), dni.gov/files/ODNI/documents/assessments/Prelimary-Assessment-UAP-20210625.pdf; David Vergun, "DoD Examining Unidentified Anomalous Phenomena," *Department of Defense News* (Nov. 14, 2024), defense.gov/News/News-Stories/Article/Article/3965403/dod-examining-unidentified-anomalous-phenomena/; and Luis Martinez, "Congressman Shows Never-Before-Seen Video at Military UFO Hearing," *ABC News* (Sept. 9, 2025), abcnews.go.com/Politics/congressman-shows-video-military-ufo-hearing/story?id=125413475.
These sources confirm that military and intelligence agencies—including the ODNI, DoD, and AARO—formally acknowledge ongoing investigation of UAP incidents, standardized reporting across branches, and the release of related data to Congress and the public. Collectively, they establish that government recognition of unidentified aerial phenomena has progressed from anecdotal reports to structured analysis within national-security frameworks.

[198] Grant Phillips, "UFOs and Nukes — Robert Salas of Ojai Recounts Shutdown of Nuclear Missiles," *Ojai Valley News* (July 25, 2025), ojaivalleynews.com/news/ufos-and-nukes-robert-salas-of-ojai-

recounts-shutdown-of-nuclear-missiles;
Anamica Singh, "Aliens Have Been Watching Over U.S. Nuclear
Missile Bases for Decades, UFO Expert Claims," *WION News* (Nov 4
2024), wionews.com/science/aliens-have-been-watching-over-us-
nuclear-missile-bases-for-decades-ufo-expert-claims-773184;
Alejandro Rojas, "Why Do UFO Sightings Keep Happening Near
Nuclear Sites?" *VICE* (June 2021), vice.com/en/article/why-do-ufo-
sightings-keep-happening-near-nuclear-sites.

Salas and Hastings describe repeated UAP interference at nuclear
installations as non-hostile demonstrations of control, while Dr.
Jensine Andresen interprets the same behavior as an effort "to help us
avoid our own self-destruction … an intelligence that recognizes the
depth of human creativity and wants to preserve it." Together, these
perspectives characterize the activity as a deterrent or warning aimed
at preventing catastrophic misuse of atomic weapons.

## CHAPTER SIX

[199] The vast majority of Earth's deep-ocean environments remain
unseen and unexplored by direct human observation. Even with the
steady progress of modern seafloor mapping, explorers have visually
surveyed less than 0.001 percent of the deep-ocean floor—an area
roughly comparable to the size of Rhode Island.
See National Oceanic and Atmospheric Administration (NOAA), *How
much of the ocean has been explored?* Ocean Explorer (June 2025),
oceanexplorer.noaa.gov/ocean-fact/explored/

[200] Human history is marked by a succession of frontiers once deemed
unreachable, from eradicating smallpox and decoding the human
genome to harnessing atomic energy and venturing into space and
the hadal depths. Each breakthrough redefined what humanity
considered possible, transforming limits into launching points. See
Centers for Disease Control and Prevention, "Smallpox Eradication: A
Model of Success" (CDC Museum STEM Lessons, 2025),

cdc.gov/museum/pdf/cdcm-pha-stem-lesson-smallpox-eradication-lesson.pdf; Joan Vernikos, "Human Exploration of Space: Why, Where, What For?" *Hippokratia* 12, Suppl. 1 (2008): 6–9, pmc.ncbi.nlm.nih.gov/articles/PMC2577404/; *POWER Magazine*, "History of Power: The Evolution of the Electric Generation Industry" (Oct. 1 2022), powermag.com/history-of-power-the-evolution-of-the-electric-generation-industry/; and National Human Genome Research Institute, "Human Genome Project Fact Sheet" (Genome.gov, 2025), genome.gov/about-genomics/educational-resources/fact-sheets/human-genome-project.

[201] Multiple lines of geological, isotopic, and biological evidence indicate that deep-sea environmental conditions—perpetual darkness, near-freezing temperatures, and immense hydrostatic pressure—have remained comparatively stable throughout the Phanerozoic, spanning roughly 500 million years. Variations in deep-sea temperature and seawater oxygen isotopes suggest that, despite major surface climate shifts, abyssal environments have maintained relative thermal and chemical continuity. See Eelco J. Rohling et al., "Comparison and Synthesis of Sea-Level and Deep-Sea Temperature Variations Over the Past 500 Million Years," *Reviews of Geophysics* 60, no. 3 (2022): e2022RG000775, doi.org/10.1029/2022RG000775; Yosef Ryb and John M. Eiler, "Oxygen Isotope Composition of the Phanerozoic Ocean and a Possible Solution to the Dolomite Problem," *Proceedings of the National Academy of Sciences* 115, no. 26 (2018): 6602–6607, doi.org/10.1073/pnas.1719681115; and Shao'e Sun et al., "Mitogenomics Provides New Insights into the Phylogenetic Relationships and Evolutionary History of Deep-Sea Sea Stars (Asteroidea)," *Scientific Reports* 12 (2022): 4656, doi.org/10.1038/s41598-022-08644-9.

[202] Across multiple domains of inquiry—microbiology, oceanography, and astrobiology—scientists continue to uncover thriving ecosystems in places once assumed barren or biologically inert. Extremophiles

flourish in high-pressure trenches, acidic vents, subglacial lakes, and radiation-saturated rock, demonstrating life's astonishing reach and resilience. See P.H. Rampelotto, "Extremophiles and Extreme Environments," *Life* 3, no. 3 (2013): 482–485, doi.org/10.3390/life3030482; Schmidt Ocean Institute, "Life Found a Way" (2020), schmidtocean.org/cruise-log-post/life-found-a-way/; SciTechDaily, "Earth's Oldest Living Organisms Discovered Trapped in 2-Billion-Year-Old Rock" (2024), scitechdaily.com/earths-oldest-living-organisms-discovered-trapped-in-2-billion-year-old-rock/; NASA Astrobiology, "Life in the Extreme: Surviving Beneath a Glacier, Part II" (2025), astrobiology.nasa.gov/news/life-in-the-extreme-surviving-beneath-a-glacier-part-ii/; and Anna Davison, "The Most Extreme Life-Forms in the Universe, Part II," *New Scientist* (26 June 2008), newscientist.com/article/dn14209-the-most-extreme-life-forms-in-the-universe-part-ii/.

[203] A growing interdisciplinary literature in comparative cognition and behavioral neuroscience demonstrates that many nonhuman species display advanced cognitive abilities once thought to be uniquely human. Studies document tool manufacture, abstract reasoning, social learning, and self-awareness across birds, mammals, and cephalopods, eroding long-held assumptions of human exclusivity. See Lesley J. Rogers, "More Evidence of Complex Cognition in Nonhuman Species," *Animal Sentience* 23(20) (2019), wellbeingintlstudiesrepository.org/animsent/vol3/iss23/20/; Michael Colombo and Damian Scarf, "Are There Differences in 'Intelligence' Between Nonhuman Species? The Role of Contextual Variables," *Frontiers in Psychology* 11 (2020): 2072, doi.org/10.3389/fpsyg.2020.02072; Bianca Zidar et al., "Animal Cognition: Bridging Behavioural and Neurobiological Perspectives," *Frontiers in Veterinary Science* 12 (2025): 1541615, doi.org/10.3389/fvets.2025.1541615; and Heather C. McGonigle et al., "Animal Intelligence Revisited," *Animals* 13 (2023): 1165, doi.org/10.3390/animals13111665.

[204] Ongoing discoveries continue to overturn long-held assumptions about how life adapts and evolves, revealing unexpected pathways of innovation and resilience. From microbial cooperation to urban evolution and symbiotic defense, these findings underscore evolution's inherent creativity and unpredictability. See Eleanor C. Beans, "Cities Serve as Testbeds for Evolutionary Change," *Royal Society Open Biology* (2017), pmc.ncbi.nlm.nih.gov/articles/PMC5434096/; Elizabeth L. Branscome et al., "The Evolution of Cooperation and Metabolic Innovation in Microbial Communities," *Proceedings of the National Academy of Sciences* 116 (2019): 18285–18294, pnas.org/doi/10.1073/pnas.1820852116; Takanori Nishino et al., "Defensive Fungal Symbiosis on Insect Hindlegs," *bioRxiv* (2024), doi.org/10.1101/2024.03.25.586038; and Sara Frueh, "Evolution's Surprises," *National Academies* (Aug. 6, 2020), nationalacademies.org/news/2020/08/evolutions-surprises.

[205] Geological reconstructions and marine geochemical records indicate that deep-marine basins have been a continuous component of Earth's biosphere since at least the mid-Paleozoic, even though individual seafloor basins are cyclically destroyed and renewed through plate-tectonic spreading. This turnover does not interrupt the broader ecological continuity of abyssal environments. See Shuzhong Shen et al., "A Long-Term Record of Early to Mid-Paleozoic Marine Redox Change," *Science Advances* 7, no. 18 (2021): eabf4382, science.org/doi/10.1126/sciadv.abf4382; Dietmar Müller et al., "Oceanic Basin Evolution and Plate-Tectonic Cycles Since 500 Ma," *Earth-Science Reviews* 218 (2021): 103691, doi.org/10.1016/j.earscirev.2021.103691; and Huaiyu Zhang et al., "Global Reconstruction of Oceanic Basin Formation and Subduction History Since the Paleozoic," *Science Advances* 11 (2025): eaw5878, doi.org/10.1126/sciadv.adw5878.

[206] Fossil and genetic evidence demonstrate that numerous deep-sea lineages—ranging from echinoderms and foraminifera to early

bilaterian worms and modern coelacanths—have persisted across hundreds of millions of years, surviving multiple mass-extinction events and maintaining continuity through profound shifts in ocean chemistry and climate. Recent molecular analyses have likewise traced extant taxa such as xenophyophores and black corals to ancient divergence points extending to the Paleozoic and even Precambrian eras. See Smithsonian National Museum of Natural History, "Foraminifera," *Ocean Portal* (2025), ocean.si.edu/ocean-life/plankton/foraminifera; University of California San Diego, "About Echinoderms," *Echinoderm Tree of Life* (2025), echinotol.ucsd.edu/about-echinoderms/; Scott Evans et al., "Discovery of Ikaria wariootia, an Early Bilaterian," *Proceedings of the National Academy of Sciences* 117 (2020): 10160–10166, doi.org/10.1073/pnas.2001045117; Maria Schweitzer et al., "Remarkably Preserved Annelid Worms from the La Meseta Formation (Eocene, Seymour Island)," *Palaeontology* 48 (2005): 107–122, doi.org/10.1111/j.1475-4983.2004.00440.x; Fabrice Bassoullet et al., "Migration Controls Extinction and Survival Patterns of Foraminifers during the Permian–Triassic Crisis in South China," *Comptes Rendus Palevol* 4 (2005): 545–552, doi.org/10.1016/j.crpv.2005.02.005; Jeremy Horowitz et al., "Bathymetric Evolution of Black Corals through Deep Time," *Proceedings of the Royal Society B* 290 (2023): 20231107, doi.org/10.1098/rspb.2023.1107; and J. Schmidt et al., "New Insights into the Evolutionary History of Xenophyophores," *Frontiers in Marine Science* 12 (2025): 1582660, doi.org/10.3389/fmars.2025.1582660.

[207] Deep-ocean environments are comparatively insulated from catastrophic planetary disturbances, including asteroid impacts and rapid climatic oscillations, due to their extreme depth, limited light, and thermal inertia. Evidence from Paleoarchean impact strata shows that shallow-water phototrophs were devastated by meteorite-induced heating and tsunamis, whereas life in deeper basins persisted with minimal disruption. See Nadja Drabon et al., "Effect of a Giant Meteorite Impact on Paleoarchean Surface Environments and Life,"

*Proceedings of the National Academy of Sciences* 121, no. 44 (2024): e2408721121, doi.org/10.1073/pnas.2408721121. See also Rebecca Lindsey and LuAnn Dahlman, "Climate Change: Ocean Heat Content," *NOAA Climate.gov* (June 26, 2025), climate.gov, summarizing data indicating that > 90 percent of excess planetary heat is retained in upper ocean layers while deep-water strata remain largely thermally stable.

[208] Geological and isotopic evidence indicates that density-driven deep-ocean circulation—analogous to modern thermohaline flow—has likely operated since at least the Paleozoic. Paleoclimate reconstructions show that even during the warm Cenozoic, deep-water formation and global overturning persisted, maintaining long-term connectivity among ocean basins. This system, known as thermohaline circulation, continues to link surface and abyssal layers in a planetary-scale loop driven by temperature and salinity gradients. See Mitchell Lyle, "Could Early Cenozoic Thermohaline Circulation Have Warmed the Poles?" *Paleoceanography* 12, no. 2 (1997): 161–167, doi.org/10.1029/96PA03330; N. J. Shackleton et al., "Evolution of Atlantic Thermohaline Circulation: Early Oligocene Onset of Deep-Water Production in the North Atlantic," *Geology* 34, no. 6 (2006): 441–444, doi.org/10.1130/G22427.1; Stefan Rahmstorf, "Thermohaline Ocean Circulation: A Brief Fact Sheet," *Potsdam Institute for Climate Impact Research* (2006), pik-potsdam.de/~stefan/thc_fact_sheet.html; and J. D. H. Nance et al., "Persistent Ocean Circulation Since the Mid-Paleozoic," *Science Advances* 11, no. 3 (2025): eadw5878, doi.org/10.1126/sciadv.adw5878.

[209] In January 2025, an iceberg roughly the size of Chicago—designated A-84—broke from the George VI Ice Shelf on the Antarctic Peninsula, exposing a 209-square-mile expanse of previously inaccessible seafloor. The international team aboard Schmidt Ocean Institute's research vessel *Falkor (too)* conducted the first exploration of the site and reported a vibrant benthic ecosystem of sponges, corals, and

icefish. Co-chief scientist Dr. Patricia Esquete remarked, "We didn't expect to find such a beautiful, thriving ecosystem." See Schmidt Ocean Institute, "*Thriving Antarctic Ecosystems Found in Wake of Recently Detached Iceberg*," press release, March 20, 2025, schmidtocean.org/thriving-antarctic-ecosystems-found-in-wake-of-recently-detached-iceberg/; and Sarah Kuta, "*A Chicago-Sized Iceberg Broke Off From Antarctica, Revealing a Hidden Ecosystem Never Seen Before*," Smithsonian Magazine, March 21 2025, smithsonianmag.com/smart-news/a-chicago-sized-iceberg-broke-off-from-antarctica-revealing-a-hidden-ecosystem-never-seen-before-180986291/.

[210] Contemporary research in animal cognition and philosophy identifies anthropocentrism not merely as a cultural attitude but as a cognitive bias—one that treats human modes of perception and reasoning as the default standard for all life. This bias fosters what comparative psychologist Gordon Burghardt calls a "cognitive dissonance between human exceptionalism and biological continuity," narrowing our ability to recognize alternative forms of awareness. As philosopher Carolina Scotto notes, such bias "assumes that human ways of experiencing, conceiving, and thinking provide the gold standard for understanding the behavior of non-human animals." See Gordon M. Burghardt, "*Anthropocentrism as Cognitive Dissonance in Animal Research*," Animal Sentience 7, no. 28 (2022): 1–6, wellbeingintlstudiesrepository.org; and Carolina Scotto, "*The Anthropocentric Bias in Animal Cognition*," ArtefaCToS 13, no. 1 (2024): 85–116, doi.org/10.14201/art2024.31800.

[211] Research over the past two decades has revealed thriving microbial ecosystems beneath more than a kilometer of Antarctic ice. Microbiological sampling from subglacial Lake Vostok demonstrated viable bacterial communities in a dark, high-pressure, chemically powered environment. Subsequent expeditions to Lakes Whillans and Mercer uncovered abundant bacteria and archaea, as well as the

preserved remains of crustaceans and tardigrades, proving that even these isolated, nutrient-poor systems sustain complex biospheres. See Martin J. Siegert et al., *"Physical, Chemical and Biological Processes in Lake Vostok and Other Antarctic Subglacial Lakes,"* Nature 414 (2001): 603–609, doi.org/10.1038/35104550; Douglas Fox, *"Life Found 800 Meters Down in Antarctic Subglacial Lake,"* Scientific American, January 17 2020, scientificamerican.com/article/life-found-800-meters-down-in-antarctic-subglacial-lake/; and Meilan Solly, *"Trove of Tiny Ancient Animal Remains Recovered From Depths of Antarctic Ice,"* Smithsonian Magazine, January 22, 2019, smithsonianmag.com/smart-news/trove-tiny-ancient-animal-remains-recovered-depths-antarctic-ice-180971316/.

[212] In 2019, researchers from the Japan Agency for Marine-Earth Science and Technology (JAMSTEC) and partner institutions identified metabolically active microbial cells within basaltic rock cores recovered from 2.5 kilometers beneath the seafloor of the South Pacific Gyre. These microbes, preserved in 33.5–104 million-year-old oceanic crust, demonstrate that even Earth's most nutrient-poor marine region sustains slow-living life. Subsequent analyses confirmed aerobic bacterial populations surviving under extreme energy limitation, in what oceanographers describe as the planet's most isolated marine environment. See Yohey Suzuki et al., *"Deep Microbial Proliferation at the Basalt Interface in 33.5–104 Million-Year-Old Oceanic Crust,"* Nature Communications Biology 3 (2020): 136, doi.org/10.1038/s42003-020-0860-1; Henning A. Müller et al., *"The Ultra-Slow Microbial Life in the South Pacific Gyre Sediments,"* Frontiers in Microbiology 12 (2021): 785743, doi.org/10.3389/fmicb.2021.785743; and Peter Dockrill, *"There's a 'Desert' in the Middle of the Pacific, and We Finally Know What Lives There,"* ScienceAlert, July 3, 2019, sciencealert.com/theres-a-desert-in-the-middle-of-the-pacific-and-we-finally-know-what-lives-there.

[213] Despite being one of the driest and most irradiated regions on Earth, the Atacama Desert supports resilient microbial communities adapted to extreme desiccation and radiation. These organisms persist through metabolic pathways that scavenge trace atmospheric gases—hydrogen, carbon monoxide, and nitrogen oxides—for energy and carbon. Because its solar intensity rivals that of Venus, NASA and ESA researchers use the Atacama as a terrestrial analogue for planetary habitability studies. See Dirk Schulze-Makuch et al., "*Transitory Microbial Habitat in the Hyperarid Atacama Desert*," *Proceedings of the National Academy of Sciences* 115, no. 11 (2018): 2670–2675, doi.org/10.1073/pnas.1714341115; C. Wierzchos et al., "*Atmospheric Gas Scavenging Supports Microbial Life in the Hyperarid Core of the Atacama Desert*," *Proceedings of the Royal Society B* 291 (2024): wrae062, doi.org/10.1098/rspb.2024.0062; and Jennifer Nalewicki, "*Chile's Atacama Desert Is the Sunniest Spot on Earth, Catching as Many Rays as Venus*," *Live Science*, July 21, 2023, livescience.com/planet-earth/chiles-atacama-desert-is-the-sunniest-spot-on-earth-catching-as-many-rays-as-venus.

[214] Recent expeditions to the Atacama Trench—one of the planet's deepest and most isolated hadal environments—have revealed a wealth of previously unknown life. Investigations by the Schmidt Ocean Institute and partners uncovered "living fossils" and diverse crustaceans filmed by the *SuBastian* ROV. At the same time, subsequent studies described new amphipods and decapods unique to Chilean waters. In 2024, researchers from the Woods Hole Oceanographic Institution identified *Dulcibella camanchaca*, the first large active predatory amphipod from these depths, underscoring the trench's exceptional endemism and ecological richness. See Woods Hole Oceanographic Institution, "*Woods Hole Oceanographic Institution and Partners Discover New Ocean Predator in the Atacama Trench*," press release, December 9 2024, whoi.edu/press-room/news-release/dulcibella-camanchaca/; Guillermo L. Guzmán et al., "*Discoveries from the Atacama Trench: New Species of Decapods Emerge*

*from the Abyss,"* *Marine Biodiversity and Biogeography Poster,* July 2025, researchgate.net/publication/393917783; Schmidt Ocean Institute, *"Living Fossils of the Atacama Trench,"* schmidtocean.org/cruise/living-fossils-of-the-atacama-trench/; and Peter Dockrill, *"How We Found a Beautiful New Species of Snailfish Deep Beneath the Sea,"* *Popular Science,* September 13, 2018, popsci.com/new-species-snailfish-deep-trenches/.

[215] Studies over the past decade have confirmed that the atmosphere itself hosts abundant and sometimes metabolically active microbial life. Bacteria and archaea have been detected in the lower and middle troposphere—often within ice-bearing clouds—where they endure freezing temperatures and intense ultraviolet bombardment. Airborne populations comprise up to 20 percent of particles in the 0.25–1 μm range, with genomic evidence suggesting ongoing metabolic processes even at high altitude. See Natasha DeLeon-Rodriguez et al., *"Microbiome of the Upper Troposphere: Species Composition and Prevalence, Effects of Tropical Storms, and Atmospheric Implications,"* *Proceedings of the National Academy of Sciences* 110, no. 7 (2013): 2575–2580, doi.org/10.1073/pnas.1212089110; David J. Smith et al., *"Aerobiology: The Global Transport and Viability of Airborne Microorganisms,"* *Frontiers in Microbiology* 7 (2016): 772, doi.org/10.3389/fmicb.2016.00772; and Tristan A. Caro et al., *"Ultraviolet Light Measurements (280–400 nm) Acquired from Stratospheric Balloon Flight to Assess Influence on Bioaerosols,"* *Aerobiologia* 35 (2019): 247–259, doi.org/10.1007/s10453-019-09597-9.

[216] Investigations of the continental deep biosphere have revealed thriving microbial ecosystems extending as far as 5 kilometers beneath Earth's surface. The Deep Carbon Observatory's decade-long global survey estimates that this subterranean biosphere contains between 16.5 and 25 billion tons of microbial biomass—roughly 70 percent of all Earth's bacteria and archaea—inhabiting high-pressure, low-energy environments within boreholes and mines worldwide.

Complementary studies demonstrate that these communities persist through chemolithotrophic metabolisms, forming complex ecosystems in rock-hosted fluids far below the reach of sunlight. See Karen Lloyd et al., Deep Carbon Observatory, quoted in Eric Betz, *"Scientists Discover Staggering Amount of Life Deep Below Earth's Surface,"* *Discover Magazine* (Dec. 10, 2018), astronomy.com/science/scientists-discover-staggering-amount-of-life-deep-below-earths-surface/; Robert M. Hazen et al., *"The Deep Terrestrial Biosphere: An Exploration of Microbial Diversity and Function,"* *Frontiers in Microbiology* 12 (2021): 658988, doi.org/10.3389/fmicb.2021.658988; and Patrick H. Thieringer, Alexander S. Honeyman, and John R. Spear, *"Spatial and Temporal Constraints on the Composition of Microbial Communities in Subsurface Boreholes of the Edgar Experimental Mine,"* *Microbiology Spectrum* 9, no. 3 (2021): e00631-21, doi.org/10.1128/spectrum.00631-21.

[217] Microbial communities have been documented surviving and even flourishing in environments saturated with ionizing radiation, including nuclear waste storage pools. Species such as *Deinococcus radiodurans*, *Arthrobacter*, and *Pseudomonas* not only tolerate extreme radiation but also actively contribute to bioremediation, sequestering and detoxifying radionuclides such as uranium and cesium. These findings extend the known boundaries of habitability and suggest promising applications for nuclear site cleanup. See Brian Hobman et al., *"Radiation-Resistant Microorganisms and Their Role in Environmental Bioremediation,"* *Microorganisms* 11, no. 8 (2023): 1871, doi.org/10.3390/microorganisms11081871; and Y. Zhu et al., *"Radiation-Resistant Microbes and Their Environmental Applications: A Review,"* *Biotechnology for Environment* 1 (2024): 8, doi.org/10.1186/s44314-024-00008-z.

[218] Recent comparative-cognition syntheses demonstrate that cognitive capacities once regarded as uniquely human—tool use, planning, communication, emotional inference, problem-solving, and social learning—are widespread across taxa. See Frans B. M. de Waal

and Sarah F. Brosnan, "Evolution of Cognition: The Comparative Approach," *Frontiers in Psychology* 11 (2020): 2072, doi.org/10.3389/fpsyg.2020.02072; Juliane Bräuer et al., "Old and New Approaches to Animal Cognition: There Is Not 'One Cognition,'" *Journal of Intelligence* 8 (2020): 28, doi.org/10.3390/jintelligence8030028; and Jennifer Vonk et al., "Animal Cognition: Recent Progress and Emerging Themes," *Trends in Cognitive Sciences* 24, no. 9 (2020): 715–728, doi.org/10.1016/j.tics.2020.06.007.

[219] In 1960, Jane Goodall began her seminal field study of wild chimpanzees at Gombe Stream National Park in Tanzania, where she documented that chimpanzees not only used tools but also modified them for specific purposes such as termite fishing. This discovery overturned the prevailing scientific belief that tool use was a defining trait of humans alone, marking a watershed moment in primatology and comparative cognition. See Jane Goodall Institute, *"Gombe Stream Research Centre"* and *"Gombe 60: Six Decades of Discovery, Innovation & Hope,"* janegoodall.org (2020); and Jane Goodall, *"My Chimpanzees Have Settled the Argument for Once and for All,"* National Geographic (December 1963), nationalgeographic.com/magazine/article/jane-goodall-original-story-chimpanzees-still-astonishes.

[220] Following Jane Goodall's first observation of chimpanzee tool use in 1960, her mentor Louis Leakey famously wrote, "Now we must redefine tool, redefine man, or accept chimpanzees as human." The line was part of a telegram sent to Goodall after she reported that Gombe chimpanzees used and modified twigs to extract termites. This observation overturned long-standing definitions of humanity's uniqueness. See Renuka Surujnarain, *"Now We Must Redefine Man, or Accept Chimpanzees as Humans?"* Jane Goodall Institute News (July 24, 2019), news.janegoodall.org/2019/07/24/now-we-must-redefine-man-or-accept-chimpanzees-ashumans/; and *National Geographic Education,*

*"Jane Goodall"* (2023),
education.nationalgeographic.org/resource/jane-goodall/.

[221] Goodall's findings were initially dismissed as overly sentimental, yet her data ultimately compelled the scientific establishment to revise its understanding of the human–animal boundary. Her decades of fieldwork demonstrated that chimpanzees possess social, emotional, and cognitive sophistication once thought to be uniquely human. As a result, the concept of "man" as defined by tool use and culture was fundamentally reexamined. See Dimitris Xygalatas, *"Jane Goodall Changed How We See Ourselves,"* Psychology Today (Oct. 2, 2025), psychologytoday.com/us/blog/the-ritual-mind/202510/jane-goodall-changed-how-we-see-ourselves; and Carly Cassella, *"How Jane Goodall Changed Our View of Chimps (And Humans) Forever,"* ScienceAlert (Oct. 2, 2025), sciencealert.com/how-jane-goodall-changed-our-view-of-chimps-and-humans-forever.

[222] Since Jane Goodall's discovery of tool use among chimpanzees, documentation of nonhuman toolmaking has become so widespread that it now features in educational media for all ages. Works such as Jennifer Swanson's *Animal Toolkit: How Animals Use Tools* (Houghton Mifflin Harcourt, 2022) and Mike Barfield's *Curious Creatures Working with Tools* (Flying Eye Books, 2024) present these behaviors to young readers, while documentaries like PBS's *Nature: Primates — "Secrets of Survival"* (2020) and *A Murder of Crows* (2010) showcase tool use among capuchins, corvids, and other species. See Junior Library Guild, *Animal Toolkit: How Animals Use Tools* (2022), juniorlibraryguild.com/animal-toolkit-how-animals-use-tools-9780358244448j; Flying Eye Books, *Curious Creatures Working with Tools* (2024), flyingeyebooks.com/book/curious-creatures-working-with-tools/; PBS *Nature*, *"Primates: Secrets of Survival"* (Season 39, Episode 3, 2020), pbs.org/wnet/nature/about-primates-secrets-survival/23112/; and PBS *Nature*, *"A Murder of Crows"* (Season 29, Episode 2, 2010), pbs.org/wnet/nature/a-murder-of-crows-introduction/5838/.

[223] Tool use has been documented across diverse taxa—from primates and marine mammals to birds. Capuchin monkeys employ stones to crack nuts, sea otters use rock "anvils" to open mussels, and New Caledonian crows craft hooked tools from twigs to extract insects. These behaviors, once considered uniquely human, now form part of the growing field of animal archaeology. See Daniel Potter, *"Using Archaeology to Uncover Sea Otters' Historical Habitats," Monterey Bay Aquarium Stories* (Mar. 15, 2019), montereybayaquarium.org/stories/using-archaeology-to-uncover-sea-otters-historical-habitats.

[224] Humpback whales (*Megaptera novaeangliae*) create elaborate "bubble-net" structures that function as foraging tools—circular curtains of air designed to corral and concentrate krill and small fish. Recent tag and drone observations reveal that whales regulate the number, size, and spacing of these rings to maximize prey density without increasing energetic cost. See A. Szabo et al., *"Solitary Humpback Whales Manufacture Bubble-Nets as Tools to Increase Prey Intake," Royal Society Open Science* 11 (2024): 240328, doi.org/10.1098/rsos.240328.

[225] Certain assassin bugs (*Reduviidae: Gorareduvius* sp.) deliberately scrape resin from spinifex grass (*Triodia bitextura*) and apply it to their forelegs, turning their bodies into adhesive traps that enhance prey capture. Controlled trials show that resin-equipped individuals enjoy up to a 33 percent higher success rate in securing prey than resin-deprived counterparts—one of the few verified examples of stereotyped tool use in insects. See Fernando G. Soley and Marie E. Herberstein, *"Assassin Bugs Enhance Prey Capture with a Sticky Resin," Biology Letters* 19 (2023): 20220608, doi.org/10.1098/rsbl.2022.0608.

[226] The evolution of opposable-thumb dexterity roughly two million years ago conferred early *Homo* species a decisive advantage in tool production and cultural development. At the same time, modern

research shows that cephalopods achieve comparable precision through wholly different neural architectures. See Eberhard Karls Universität Tübingen, *"Thumb Dexterity Helped Spark the Development of Human Culture,"* press release (Jan 28, 2021), summarizing Katerina Harvati et al., *Current Biology* 31 (2021): DOI 10.1016/j.cub.2020.12.041; and Cassady S. Olson et al., *"Neuronal Segmentation in Cephalopod Arms,"* *Nature Communications* 16 (2025): 443, DOI 10.1038/s41467-024-55475-5.

[227] Research in comparative cognition demonstrates that corvids such as ravens and scrub jays exhibit "mental time travel," recalling past experiences to plan for future needs and altering their behavior when observed by potential competitors. These abilities—once believed uniquely human—show that nonhuman animals can anticipate, strategize, and act on inferred intent. See Nicola S. Clayton and Anthony Dickinson, "Episodic-Like Memory During Cache Recovery by Scrub Jays," *Nature* 395 (1998): 272–274; and Thomas R. Raby, D. M. Alexis, A. Dickinson, and N. S. Clayton, "Planning for the Future by Western Scrub-Jays," *Nature* 445 (2007): 919–921, pmc.ncbi.nlm.nih.gov/articles/PMC2830244/.

[228] Controlled experiments show that chimpanzees differentiate between a human who is unwilling to share food and one who is physically unable to do so. This ability suggests a rudimentary awareness of agency and intention—an understanding of other minds, once thought exclusive to humans. See Jan M. Engelmann et al., "Chimpanzees Consider Freedom of Choice in Their Evaluation of Social Action," *Biology Letters* 18 (2022): 20210502, doi.org/10.1098/rsbl.2021.0502.

[229] Evidence for mirror self-recognition now spans multiple animal lineages once thought incapable of self-awareness. Beyond humans and great apes, successful or strongly suggestive results have been recorded in dolphins, elephants, magpies, certain fish such as the

cleaner wrasse, and even ants. See Amanda Pachniewska, "*List of Animals That Have Passed the Mirror Test*," *Animal Cognition* (April 15, 2015), animalcognition.org/2015/04/15/list-of-animals-that-have-passed-the-mirror-test/; and Yanyu Lei, "*Sociality and Self-Awareness in Animals*," *Frontiers in Psychology* 13 (2023): 1065638, doi.org/10.3389/fpsyg.2022.1065638.

[230] Comparative communication research demonstrates that symbolic and language-like expression is not confined to humans. Enculturated great apes, dolphins, and parrots have each shown the capacity to use arbitrary symbols or learned referents to represent objects, actions, and categories—meeting many definitions of proto-linguistic communication. See Heidi Lyn and Jennie L. Christopher, "How Environment Can Reveal Semantic Capacities in Nonhuman Animals," *Animal Behavior and Cognition* 7, no. 2 (2020): 159–167, doi.org/10.26451/abc.07.02.10.2020; and Erica A. Cartmill, "Overcoming Bias in the Comparison of Human Language and Animal Communication," *Proceedings of the National Academy of Sciences* 120, no. 47 (2023): e2218799120, doi.org/10.1073/pnas.2218799120.

[231] Field and laboratory research now confirms that many animal societies transmit learned behaviors culturally across generations. Tool use, hunting strategies, vocal dialects, and social customs in species ranging from chimpanzees to whales and birds are maintained through imitation, teaching, and social learning—constituting non-genetic inheritance systems akin to culture. See Luke Rendell, Rachel L. Kendal et al., "Social Learning Strategies: Bridge-Building Between Fields," *Philosophical Transactions of the Royal Society B* 377, no. 1843 (2022): 20200308, doi.org/10.1098/rstb.2020.0308; and Andrew Whiten et al., "The Emergence of Collective Knowledge and Cumulative Culture in Animals, Humans and Machines," *Philosophical Transactions of the Royal Society B* 376 (2021): 20200268, doi.org/10.1098/rstb.2020.0268.

[232] Advanced cognitive traits once considered uniquely human—empathy, altruism, metacognition, and awareness of one's own knowledge gaps—are now well documented across a range of animal species. Comparative studies reveal evidence of emotional contagion and helping behavior in mammals and birds, inequity aversion and cooperation in primates, and uncertainty monitoring and information-seeking in apes, dolphins, and corvids. See Frans B. M. de Waal and Stephanie D. Preston, "Mammalian Empathy: Behavioural Manifestations and Neural Basis," *Nature Reviews Neuroscience* 18 (2017): 498–509; Friederike Range and Sarah Brosnan, "Altruism in Animals and Classification: A View," *Frontiers in Psychology* 6 (2015): 1465; Luke J. Hernandez-Lallement et al., "Prosocial Behavior in Animals: The Influence of Social Context," *Frontiers in Behavioral Neuroscience* 9 (2015): 254; and Jonathan Birch, Alexandra K. Schnell, and Nicola S. Clayton, "Dimensions of Animal Consciousness," *Trends in Cognitive Sciences* 24, no. 10 (2020): 789–801, doi.org/10.1016/j.tics.2020.07.007.

[233] In controlled experiments, individual honeybees were trained to discriminate "greater than" and "less than" using visual numerosity cues and were later tested with novel combinations, including the absence of any elements. The bees consistently selected zero over one and performed significantly above chance, demonstrating an understanding that an empty set lies at the lower end of the numerical continuum—an ability previously observed only in vertebrates. See Scarlett R. Howard et al., "Numerical Ordering of Zero in Honey Bees," *Science* 360, no. 6393 (2018): 1124–1126, doi.org/10.1126/science.aar4975.

[234] "Life, uh... finds a way." The line, delivered by Dr. Ian Malcolm (portrayed by Jeff Goldblum) in *Jurassic Park* (dir. Steven Spielberg, 1993; based on Michael Crichton's 1990 novel of the same name), has since become a cultural shorthand for biological adaptability and

evolutionary inevitability. See *Jurassic Park*, IMDb Quotes Database, imdb.com/title/tt0107290/quotes/?item=qt0463040.

[235] The environmental and physiological parameters that sustain *Homo sapiens*—oxygen, liquid water, narrow thermal and pressure ranges—represent only a small subset of the conditions in which life can exist. Extremophilic organisms thrive in habitats long assumed sterile, from boiling hydrothermal vents and acid pools to subglacial lakes and hadal trenches. At the same time, astrobiological research continues to expand the known boundaries of habitability. See OpenStax, "Requirements for Human Life," *Anatomy and Physiology I* (Lumen Learning, 2025), courses.lumenlearning.com/suny-ap1/chapter/requirements-for-human-life-and-homeostasis/; P. H. Rampelotto, "Extremophiles and Extreme Environments," *Life* 3, no. 3 (2013): 482–485, doi.org/10.3390/life3030482; and Katherine L. C. Bell et al., "Reconsidering the Limits of Life," *Science Advances* 11, no. 22 (2025): eads5698, doi.org/10.1126/sciadv.ads5698.

[236] Many microorganisms are chemolithoautotrophs, deriving energy from the oxidation of inorganic substrates—such as iron, sulfur, or thiosulfate—rather than from organic carbon. These metabolisms, long considered incompatible with life, sustain diverse microbial ecosystems from soils to hydrothermal vents. See Jun Ye et al., "Chemolithotrophic Processes in the Bacterial Communities on the Surface of Mineral-Enriched Biochars," *The ISME Journal* 11 (2017): 1087–1101, doi.org/10.1038/ismej.2016.187.

[237] Numerous taxa on Earth exhibit anhydrobiosis—the ability to survive near-total desiccation for extended periods. Certain nematodes, tardigrades, rotifers, and microbial species can suspend metabolism entirely and later revive when rehydrated. Even in hyper-arid deserts and polar regions, microorganisms persist on trace atmospheric moisture or infrequent deliquescent brines. See Lois M. Crowe and John H. Crowe, "Life Without Water," NASA Technical

Memorandum 100471 (1989); Andrew Stevenson et al., "Microbial Life in the Hyperarid Core of the Atacama Desert," *Microorganisms* 10, no. 3 (2022): 432, doi.org/10.3390/microorganisms10030432; and Malcolm Potts, "Desiccation Tolerance of Prokaryotes," *Philosophical Transactions of the Royal Society B* 346, no. 1318 (1994): 3–10, doi.org/10.1098/rstb.1994.0123.

[238] Deep-sea hydrothermal vents were first discovered in 1977 near the Galápagos Rift, revealing entire ecosystems thriving independently of sunlight. Within these mineral-rich volcanic chimneys, life endures crushing pressures, extreme temperatures, and toxic chemical gradients by harnessing chemosynthesis—the microbial conversion of reduced compounds such as hydrogen sulfide into chemical energy. These bacteria and archaea form both the base of the food web and symbiotic partnerships with larger fauna, from tube worms to mussels and shrimp. See National Oceanic and Atmospheric Administration (NOAA), *Hydrothermal Vents Fact Sheet* (2022), oceanexplorer.noaa.gov; Woods Hole Oceanographic Institution, "Exploring Hydrothermal Vents: Vent Biology" (2023), whoi.edu; and Breea Govenar, "Energy Transfer Through Food Webs at Hydrothermal Vents: Linking the Lithosphere to the Biosphere," *Oceanography* 25, no. 1 (2012): 246–255, doi.org/10.5670/oceanog.2012.23.

[239] The giant tubeworm *Riftia pachyptila*, a keystone of hydrothermal vent ecosystems, lacks a mouth and digestive tract. Instead, it relies entirely on chemosynthetic bacterial symbionts housed within a specialized internal organ called the trophosome. These bacteria oxidize hydrogen sulfide and fix carbon dioxide, providing all of the host's nutritional needs. See Tanja Hinzke et al., "Host–Microbe Interactions in the Chemosynthetic *Riftia pachyptila* Symbiosis," *mBio* 10, no. 6 (2019): e02243-19, doi.org/10.1128/mBio.02243-19.

[240] Deep-sea mussels and clams of the genus *Bathymodiolus* have evolved massive, bacteria-filled gills that function as internal habitats for chemosynthetic symbionts. These cavernous lamellae can be up to twenty times larger than those of shallow-water relatives. They may contain more than a trillion symbiotic bacteria in a single individual, comparable to the bacterial abundance found in over two hundred liters of seawater. The symbionts, primarily sulfur- and methane-oxidizing Gammaproteobacteria, provide the host's nutrition and drive much of the biomass production at hydrothermal vents. See Daniel L. Distel et al., "The Dual Symbiosis of the Deep-Sea Mussel *Bathymodiolus thermophilus*," *Proceedings of the National Academy of Sciences* 91 (1994): 9642–9646, doi.org/10.1073/pnas.91.20.9642; Nicolas Gros et al., "A Mussel's Life Around Deep-Sea Hydrothermal Vents," *Frontiers for Young Minds* 7 (2019): 76, doi.org/10.3389/frym.2019.00076; Andreas Ponnudurai et al., "Metabolic and Physiological Interdependencies in the *Bathymodiolus* Symbiosis," *Frontiers in Marine Science* 5 (2018): 282, doi.org/10.3389/fmars.2018.00282; and Bérénice Piquet et al., "'There and Back Again': Ultrastructural Changes in the Gills of *Bathymodiolus* Vent Mussels During Symbiont Loss," *Frontiers in Marine Science* 9 (2022): 968331, doi.org/10.3389/fmars.2022.968331.

[241] In 2023 (an event widely reported through 2024), an international research team aboard the Schmidt Ocean Institute's *Falkor (too)* discovered an entirely new ecosystem beneath hydrothermal vents at a well-studied undersea volcano on the East Pacific Rise, off Central America. Using the remotely operated vehicle ROV SuBastian, the scientists overturned sections of volcanic crust, revealing cave systems teeming with giant tubeworms, snails, and chemosynthetic bacteria thriving in 25 °C water at a depth of 2,500 meters. This was the first confirmation that vent animals can inhabit sub-seafloor cavities,

forming interconnected habitats both above and below the ocean floor. See Schmidt Ocean Institute, *"Scientists Discover New Ecosystem Underneath Hydrothermal Vents,"* press release, August 8, 2023, schmidtocean.org; and Royal Netherlands Institute for Sea Research (NIOZ), "Evidence of Hydrothermal Vent Animals in Volcanic Caves Beneath the Ocean Floor," news release, August 8, 2023, nioz.nl.

[242] The line "There are more things in heaven and earth, Horatio, / Than are dreamt of in your philosophy" appears in *Hamlet*, Act I, Scene 5, spoken by Prince Hamlet to his confidant Horatio. See William Shakespeare, *Hamlet, Prince of Denmark*, arranged by Charles Kean (London: Bradbury and Evans, 1859), Project Gutenberg eText no. 27761, gutenberg.org/files/27761/27761-h/27761-h.htm.

[243] Evidence from geomicrobiology and subsurface microbiology confirms that microbial life inhabits microscopic pores and fractures within solid rock, kilometers beneath Earth's surface. These lithotrophic communities derive energy from hydrogen and sulfur compounds generated by radiolysis and water–rock interactions, surviving in isolation for millions of years. Recent studies have documented indigenous microbial cells colonizing mineral-filled veins in 2-billion-year-old mafic rock, demonstrating long-term habitability in geologically stable crust. See *Microbiology Resource of the Month: Rock-Inhabiting Microbes*, American Society for Microbiology (October 2021), asm.org/articles/2021/october/microbiology-resource-of-the-month-rock-inhabiting; Tullis C. Onstott et al., "Two Miles Underground, Strange Bacteria Are Found Thriving," *Princeton University News* (October 20 2006); and Yohey Suzuki et al., "Subsurface Microbial Colonization at Mineral-Filled Veins in 2-Billion-Year-Old Mafic Rock from the Bushveld Igneous Complex, South Africa," *Microbial Ecology* 87 (2024): 116, doi.org/10.1007/s00248-024-02434-8.

[244] The bacterium *Deinococcus radiodurans* remains the best-characterized example of an organism capable of withstanding ionizing radiation at doses hundreds to thousands of times greater than what would destroy human DNA. Laboratory experiments have shown that it can survive acute exposures exceeding 5,000 grays (Gy), approximately 1,000 times the lethal dose for humans, through a combination of efficient DNA repair and protection of its proteome from oxidative damage. See Anita Krisko and Miroslav Radman, "Biology of Extreme Radiation Resistance: The Way of *Deinococcus radiodurans*," *Cold Spring Harbor Perspectives in Biology* 5, no. 12 (2013): a012765, doi.org/10.1101/cshperspect.a012765; and Aaron Gronstal, "Life in the Extreme: Radiation," *NASA Astrobiology* (February 10, 2023), astrobiology.nasa.gov/news/life-in-the-extreme-radiation/.

[245] Acidophiles inhabit environments so acidic that they would destroy the cellular structures of most other organisms. These extremophiles exhibit remarkable biochemical adaptations allowing them to thrive at pH levels approaching zero, metabolizing metals and sulfur compounds in conditions equivalent to battery acid. Their metabolic capabilities have been harnessed in biomining, where species such as *Acidithiobacillus ferrooxidans* and *Leptospirillum ferriphilum* catalyze the oxidative dissolution of metal sulfides, facilitating environmentally efficient extraction of copper, nickel, and other valuable elements. See Elizabeth L. J. Watkin, Ivan Nancucheo, and Axel Schippers, "Editorial: Acidophile Microbiology: From Extreme Environments to Biotechnological Applications," *Frontiers in Microbiology* 15 (2024): 1454559, doi.org/10.3389/fmicb.2024.1454559.

[246] Alkaliphilic microorganisms thrive at pH levels that would rupture the membranes of most cells, employing biochemical strategies such as acidic polymers and specialized transport systems to stabilize their cellular structures. Their negatively charged surface polymers and modified lipid compositions create protective barriers that maintain internal pH homeostasis even in environments with

pH values exceeding 10. See Koki Horikoshi, "Alkaliphiles: Some Applications of Their Products for Biotechnology," *Microbiology and Molecular Biology Reviews* 63, no. 4 (1999): 735–750, doi.org/10.1128/mmbr.63.4.735-750.1999.

[247] Once considered biological deserts, the anoxic and oxygen minimum zones of the ocean are now known to host dynamic microbial ecosystems. Recent oceanographic and genomic studies demonstrate that these "dead zones" support thriving communities of bacteria and archaea engaged in complex nitrogen and sulfur cycling. Aerobic and anaerobic metabolisms often coexist at nanomolar oxygen levels, with facultative microbes switching between respiration modes to maximize survival. These communities drive major biogeochemical transformations—including denitrification, anammox, and chemoautotrophic carbon fixation—revealing that anoxia does not equate to sterility but to hidden productivity. See Osvaldo Ulloa et al., "Microbial Oceanography of Anoxic Oxygen Minimum Zones," *Proceedings of the National Academy of Sciences* 109, no. 40 (2012): 15996–16003, doi.org/10.1073/pnas.1205009109; Ana M. M. B. Sousa et al., "Anoxic Marine Zones—Microbial Hotspots for Novel Biogeochemical Interactions," *Frontiers in Microbiology* 12 (2021): 748961, doi.org/10.3389/fmicb.2021.748961; and Emily J. Zakem et al., "Stable Aerobic and Anaerobic Coexistence in Anoxic Marine Zones," *ISME Journal* 14 (2020): 288–301, doi.org/10.1038/s41396-019-0523-8.

[248] The sacoglossan sea slug *Elysia chlorotica* exemplifies the rare phenomenon of kleptoplasty—retaining functional chloroplasts from consumed algae and using them for photosynthesis. These "stolen" plastids are incorporated into specialized digestive tissues where they remain photosynthetically active for extended periods, allowing the slug to generate energy directly from sunlight. The process blurs the line between animal and plant physiology, with kleptoplastic species such as *E. chlorotica* and *E. viridis* exhibiting measurable oxygen

production and nutrient assimilation from photosynthates. See
Sidney K. Pierce et al., "The Making of a Photosynthetic Animal:
Kleptoplasty in Sacoglossan Sea Slugs," *PLoS Biology* 20, no. 11 (2022):
e3001857, doi.org/10.1371/journal.pbio.3001857; and Paulo Cartaxana
et al., "Kleptoplast Photosynthesis Is Nutritionally Relevant in the
Sea Slug *Elysia viridis*," *Scientific Reports* 7 (2017): 7714,
doi.org/10.1038/s41598-017-08002-0.

[249] Bdelloid rotifers are microscopic freshwater invertebrates found
worldwide, remarkable for their ability to withstand extremes that
would destroy most animal life. They can survive complete
desiccation at any stage of their life cycle, enter a dormant state,
disperse aerially as dried cysts, and rehydrate to resume activity and
reproduction. Experiments have demonstrated that *Adineta* species
revived from Late Pleistocene Siberian permafrost—radiocarbon-
dated to approximately 24,000 years BP—remained viable and capable
of parthenogenetic reproduction after thawing. Bdelloids are
traditionally considered entirely female, reproducing via obligate
parthenogenesis, though genomic evidence now suggests rare or
facultative sexual events in some species. Their DNA is known to
fragment during desiccation and reassemble upon rehydration. This
process enables horizontal gene transfer from bacteria, fungi, and
plants and contributes to their extraordinary radiation resistance and
rapid adaptive potential. See Elena R. Gladyshev and Matthew
Meselson, "Extreme Resistance of Bdelloid Rotifers to Ionizing
Radiation," *Proceedings of the National Academy of Sciences* 105, no. 13
(2008): 5139–5144, pnas.org/doi/10.1073/pnas.0800966105; Lyubov
Shmakova et al., "A Living Bdelloid Rotifer from 24,000-Year-Old
Arctic Permafrost," *Current Biology* 31, no. 15 (2021): R697–R713,
sciencedirect.com/science/article/pii/S0960982221006242; Veronika
N. Laine, Timothy B. Sackton, and Matthew Meselson, "Genomic
Signature of Sexual Reproduction in the Bdelloid Rotifer
*Macrotrachella quadricornifera*," *Genetics* 220, no. 2 (2022): iyab221,
pmc.ncbi.nlm.nih.gov/articles/PMC9208647/; Veronika Laine et al.,

"Gene Exchange and Desiccation Tolerance Shape Bdelloid Rotifer Evolution," *Nature Communications* 15 (2024): 49919, nature.com/articles/s41467-024-49919-1; and Matthew Meselson et al., "Bdelloid Rotifers as Models for Horizontal Gene Transfer and Desiccation Adaptation," *PLOS Biology* 20, no. 11 (2022): e3001857, journals.plos.org/plosbiology/article?id=10.1371/journal.pbio.200483 0.

[250] The concept that every species inhabits its own perceptual world originates with Jakob von Uexküll's theory of the *Umwelt*—the subjective "self-world" formed by an organism's sensory and motor capacities. Uexküll argued that animals do not experience a single shared objective reality but rather construct distinct experiential worlds through their species-specific organs of perception and action. Contemporary perceptual ecology and cognitive ethology have extended this principle to demonstrate that divergent sensory modalities—such as echolocation, polarization vision, or magnetoreception—create fundamentally different phenomenal realities across taxa. See Jakob von Uexküll, "Jakob von Uexküll: Umwelt," *The Philosopher* (1923): thephilosopher1923.org/post/jakob-von-uexkull-umwelt; and Jui-Pi Chien, "Of Animals and Men: A Study of Umwelt in Uexküll, Cassirer, and Heidegger," *Concentric: Literary and Cultural Studies* 32, no. 1 (2006): 57–79, concentric-literature.url.tw/issues/Animals/3.pdf.

[251] Quantitative studies of canine olfaction demonstrate that dogs possess between 220 and 300 million olfactory receptors—orders of magnitude greater than the roughly 5 to 6 million found in humans—and that their olfactory epithelium is proportionally far larger, allowing odor detection 10,000 to 100,000 times more acute than human capacity. Neural imaging confirms that a dog's brain allocates nearly 40 times more cortical area to scent processing than ours, supporting this extraordinary sensitivity. See Lucia Lazarowski and Marc Bekoff, "Canine Olfaction Science in Service of Humanity—

Canine Detection Capabilities and Applications," *Frontiers in Veterinary Science* 5 (2018): 56, frontiersin.org/journals/veterinary-science/articles/10.3389/fvets.2018.00056/full.

[252] Numerous species perceive wavelengths of light invisible to humans. Birds, insects, fish, and reptiles commonly detect ultraviolet light, while some snakes and other vertebrates sense infrared radiation as thermal energy. These expanded visual ranges allow animals to navigate, hunt, and communicate in perceptual domains completely outside human experience—UV patterns on flower petals guide bees, reindeer detect lichens against UV-reflective snow, and pit vipers use heat-sensing pits to locate prey in darkness. See Ed Yong, "How Science Came To See Ultraviolet Light In Animals," *Science Friday* (June 24, 2022), sciencefriday.com/articles/ultraviolet-light-animals/; Tom Mangan, "Animals See a World That's Completely Invisible to Our Eyes," *All About Vision* (October 25, 2021), allaboutvision.com/eye-care/pets-animals/how-animals-see/; and H. D. Wolpert, "Life Lessons," *OE Magazine* (February 2002), spie.org/news/life-lessons?highlight=x2416&ArticleID=x25379.

[253] The electric eel (*Electrophorus electricus*) is capable of generating electrical discharges exceeding 600 volts—among the highest recorded in any known animal—sufficient to stun or kill prey and potentially harm humans. In addition to these high-voltage strikes, eels produce low-voltage pulses for electrolocation and social signaling, allowing them to navigate, sense prey, and coordinate movement in murky environments. They actively "ping" their surroundings with weak electric fields to form spatial maps and then amplify discharge strength during predatory attacks. This mechanism doubles the field intensity by curling their bodies to position positive and negative poles on opposite sides of their target. See William G. R. Crampton et al., "Review of the Gymnotiform Electric Fish Genus *Electrophorus*," *Neotropical Ichthyology* 7, no. 4 (2009): 693–698,

scielo.br/j/ni/a/rkRxs6S8TQDxZDRCPNgd9My/?lang=en; Kenneth C.
Catania, "The Shocking Predatory Strike of the Electric Eel," *Nature
Communications* 10 (2019): 11690,
pmc.ncbi.nlm.nih.gov/articles/PMC6736962/; and Kenneth C. Catania,
"Electric Eels Concentrate Their Electric Field to Induce Involuntary
Fatigue in Struggling Prey," *Current Biology* 25, no. 22 (2015): 2889–
2898, cell.com/current-biology/fulltext/S0960-9822(15)01147-1.

[254] Magnetotactic bacteria are aquatic prokaryotes that biomineralize
intracellular chains of magnetic iron minerals—typically magnetite
($Fe_3O_4$) or greigite ($Fe_3S_4$)—enclosed in lipid-bilayer organelles called
magnetosomes. These magnetosome chains function as microscopic
compass needles, aligning the cells with Earth's magnetic field and
guiding their movement along geomagnetic lines toward optimal
oxygen or redox zones in sediments and water columns. The
phenomenon, termed *magnetotaxis*, was first described in 1963 by
Salvatore Bellini and independently confirmed in 1975 by Richard P.
Blakemore, whose *Science* paper provided the first micrographic
evidence of these magnetosomes. Subsequent genomic and structural
analyses established magnetotaxis as a monophyletic adaptation
with deep evolutionary origins. See Pranami Goswami et al.,
"Magnetotactic Bacteria and Magnetofossils: Ecology, Evolution and
Environmental Implications," *NPJ Biofilms and Microbiomes* 8 (2022):
43, nature.com/articles/s41522-022-00304-0; and Christopher T.
Lefèvre and Dennis A. Bazylinski, "Ecology, Diversity, and Evolution
of Magnetotactic Bacteria," *Microbiology and Molecular Biology Reviews*
77 (2013): 497–526, mmbr.asm.org/content/77/3/497.

[255] Certain birds appear able to perceive Earth's magnetic field visually
through specialized photoreceptors in their eyes. Cryptochrome
proteins in the retina form light-dependent radical pairs whose spin
states are influenced by geomagnetic fields, producing visual patterns
aligned with magnetic lines. This quantum-biophysical process,
confirmed in migratory species such as the European robin, provides a

plausible mechanism for the extraordinary navigational precision observed in long-distance migration. See Jingjing Xu et al., "Magnetic Sensitivity of Cryptochrome 4 from a Migratory Songbird," *Nature Communications* (2021), europepmc.org/article/med/34163056; and Roswitha Wiltschko and Wolfgang Wiltschko, "Magnetoreception in Birds," *Journal of the Royal Society Interface* 16 (2019): 20190295, royalsocietypublishing.org/doi/10.1098/rsif.2019.0295.

[256] The platypus (*Ornithorhynchus anatinus*) is one of nature's few electroreceptive mammals. Its bill contains roughly 40,000 electroreceptors arranged in longitudinal rows, each sensitive to the faint electric fields generated by the muscle contractions of prey. During dives, the platypus closes its eyes, ears, and nostrils, navigating and hunting solely through these receptors. Behavioral experiments have shown that it can determine both the distance and direction of prey by detecting differences in the arrival times of electrical and mechanical signals. Electroreception in the platypus was first confirmed during the 1980s, and subsequent studies mapped the sensory fields and cortical integration of its electrosensory and mechanosensory systems. A comparable ability was later identified in the Guiana dolphin (*Sotalia guianensis*), which uses specialized vibrissal crypts on its rostrum to detect weak electric fields in turbid waters, making these two species the only known electroreceptive mammals. See Uwe Proske and Ed Gregory, "Electrolocation in the Platypus—Some Speculations," *Comparative Biochemistry and Physiology Part A: Molecular & Integrative Physiology* 136, no. 4 (2003): 821–825, sciencedirect.com/science/article/abs/pii/S1095643303001600; Paul R. Manger, Rita Collins, and John D. Pettigrew, "The Development of the Electroreceptors of the Platypus (*Ornithorhynchus anatinus*)," *Philosophical Transactions of the Royal Society B* (1998), researchgate.net/publication/13565632_The_development_of_the_ele ctroreceptors_of_the_platypus_Ornithorhynchus_anatinus; University of Western Australia, "Platypus Electroreception" (2015), uwa.edu.au/study/-/media/Faculties/Science/Docs/Platypus-

electroreception.pdf; and Nicole U. Czech-Damal et al.,
"Electroreception in the Guiana Dolphin (*Sotalia guianensis*)*,"
*Proceedings of the Royal Society B* 279 (2012): 663–668,
pmc.ncbi.nlm.nih.gov/articles/PMC3248726/pdf/rspb20111127.pdf.

## CHAPTER SEVEN

[257] The Aguadilla UAP Incident occurred at approximately 9:20 p.m.
Atlantic Standard Time on April 25, 2013 (01:20 UTC on April 26) when
a U.S. Customs and Border Protection (CBP) De Havilland Canada
DHC-8 turboprop equipped with a Wescam MX-15D thermal-imaging
system captured infrared footage of unidentified aerial targets near
Rafael Hernández Airport in Aguadilla, Puerto Rico. The Department
of Defense's All-domain Anomaly Resolution Office (AARO)
confirmed the encounter, identifying the data source as infrared video
recorded by a CBP aircraft and noting that the timestamp discrepancy
between reports (April 25 vs. April 26) arises from conversion between
local Atlantic Standard Time and Coordinated Universal Time (UTC).
The Scientific Coalition for UAP Studies (SCU) independently analyzed
the same footage, verifying the aircraft type, sensor metadata, and
flight path, and documenting the reported airspace intrusion over the
runway. See U.S. Department of Defense, *All-domain Anomaly
Resolution Office Case Resolution: "The Puerto Rico Object"* (20 Mar 2025),
aaro.mil/UAP-Cases/UAP-Case-Resolution-Reports; and Scientific
Coalition for UAP Studies, *2013 Aguadilla Puerto Rico UAP Incident
Report: A Detailed Analysis* (2015), explorescu.org/post/2013-aguadilla-
puerto-rico-uap-incident-report-a-detailed-analysis.

[258] The Wescam MX-15D is a fully digital, high-definition, multi-sensor
targeting and reconnaissance system designed for medium-altitude
surveillance, search and rescue, border patrol, and precision targeting.
It is employed across multiple branches of the U.S. government,
including Customs and Border Protection, the Coast Guard, the Navy,
and on unmanned aerial systems such as the MQ-9

Reaper. The MX-15D supports simultaneous imaging and lasing payloads—thermal, daylight, and low-light cameras, providing continuous 24-hour operation—with a cooled mid-wave infrared imager capable of resolutions up to 1280 × 1024 pixels. Its laser rangefinder and designator systems achieve effective ranges of up to 20 kilometers with range resolution to within ±2 meters under standard atmospheric conditions. See L3Harris Technologies, *Wescam MX-15D Airborne Targeting and Designating* (2020), l3harris.com/all-capabilities/wescam-mx-15d-airborne-targeting-and-designating; *Electro-Optical/Infrared Systems Brochure MX-15D-0501AB* (2019), l3harris.com/sites/default/files/2020-07/ims_eo_brochure_wescam_MX-15D-0501AB-Brochure.pdf; and *Product Data Sheet: MX-15D* (March 2018), militarysystems-tech.com/sites/militarysystems/files/supplier_docs/PDS-MX-15D-March-2018.pdf.

[259] By the mid-2000s, CMOS (Complementary Metal-Oxide Semiconductor) sensors began to overtake CCD (Charge-Coupled Device) technology as the dominant standard in digital imaging. This shift accelerated after 2007, when manufacturers such as Nikon and Sony introduced CMOS-based professional cameras—the Nikon D3 and Sony α700—that marked a decisive transition point. CMOS sensors offered faster readout speeds, lower power consumption, and reduced heat generation compared to CCDs, leading to their adoption across nearly all imaging platforms. See Matt Williams, "What Is the Difference Between a CCD and CMOS Camera Sensor?" *PetaPixel* (Aug. 4, 2021), petapixel.com/what-is-ccd-cmos-sensor/.

[260] In 2013, CCD (Charge-Coupled Device) sensors continued to dominate defense and aerospace imaging applications, maintaining advantages in low-light performance, thermal integration, and global-shutter precision. CCDs produced cleaner, distortion-free images in high-velocity tracking and high-contrast environments such as shorelines and open water, where CMOS (Complementary Metal-

Oxide Semiconductor) sensors of the time still exhibited rolling-shutter artifacts and reduced dynamic range. These characteristics made CCDs the preferred choice for long-range reconnaissance and infrared integration throughout the early 2010s. See Adimec Advanced Image Systems, "CCD vs. CMOS Image Sensors in Defense Cameras" (2020), adimec.com/ccd-vs-cmos-image-sensors-in-defense-cameras/; Ted Pella, Inc., "CCD and CMOS Technology" (2025), tedpella.com/cameras_html/ccd_cmos.aspx; FRAMOS Imaging Systems, "[Links] Digital Cameras in Unmanned Aerial Vehicles (UAV) for Military and Commercial Uses" (2018), framos.com/news/links-digital-cameras-in-unmanned-aerial-vehicles-uav-for-military-and-commercial-uses/; and Dave Litwiller, "CCD vs. CMOS: Facts and Fiction," *Photonics Spectra* (Jan. 2001), dalsa.com/shared/content/Photonics_Spectra_CCDvsCMOS_Litwiller.pdf.

[261] The 2013 Aguadilla UAP incident was recorded by a U.S. Customs and Border Protection DHC-8 aircraft using a Wescam MX-15D thermal imaging system. According to the *Scientific Coalition for UAP Studies (SCU)* technical analysis, tower personnel first observed the object as a reddish-pink light approaching from the ocean before it extinguished its visible illumination and was thereafter tracked solely by its infrared signature. The SCU report documents the object's low-altitude flight at estimated speeds up to 120 mph (193 km/h) through residential areas, its entry into the ocean without visible splash or deceleration, and its apparent division into two distinct heat sources upon re-emergence. Time-stamped thermal data from the MX-15D were cross-referenced with radar tracks from a long-range array located approximately 145 km east-southeast of Aguadilla at an elevation of 1,042 m (3,417 ft). That radar system, operating under a transponder code reserved for military and special-operations use, recorded 50 primary (non-transponder) strikes at a mean velocity of 270 km/h (168 mph) before the object vanished below its 122 m (400 ft) minimum detection threshold—immediately before appearing in the

FLIR sequence. A subsequent *All-domain Anomaly Resolution Office (AARO)* case report confirmed the authenticity of the CBP-sourced video and the radar correlation. Still, it concluded that the object was "most likely" a sky lantern. See *Scientific Coalition for UAP Studies, 2013 Aguadilla Puerto Rico UAP Incident Report: A Detailed Analysis* (2015), explorescu.org/post/2013-aguadilla-puerto-rico-uap-incident-report-a-detailed-analysis; and U.S. Department of Defense, *All-domain Anomaly Resolution Office Case Resolution: "The Puerto Rico Object"* (20 Mar 2025), aaro.mil/UAP-Cases/UAP-Case-Resolution-Reports.

[262] On October 21, 2013, the Scientific Coalition for UAP Studies (SCU) obtained the infrared video of the Aguadilla incident through a vetted intermediary who was in direct contact with a primary witness. According to the SCU's account, the intermediary was not present during the event but was verified as credible and signed a nondisclosure agreement along with other participants to protect witness identities and data integrity. In 2015, the SCU released a 161-page technical analysis authored by a multidisciplinary team of scientists, engineers, and former defense analysts. The report documents the chain of custody, data provenance, and methodological procedures used in the study. See *Scientific Coalition for UAP Studies, 2013 Aguadilla Puerto Rico UAP Incident Report: A Detailed Analysis* (2015), explorescu.org/post/2013-aguadilla-puerto-rico-uap-incident-report-a-detailed-analysis.

[263] The Scientific Coalition for UAP Studies' analysis of the Aguadilla infrared video found that the recorded object displayed no visible wings, rotors, exhaust, or other means of propulsion. Frame-by-frame and motion-vector studies showed that it moved independently of prevailing winds, executing abrupt accelerations and stationary hovers. Thermal analysis documented its entry into the ocean without deceleration or splash, and its subsequent re-emergence, behavior inconsistent with that of any known natural or artificial object. The report concluded that common explanations, such as

balloon, bird, drone, or insect, were inconsistent with the data. See *Scientific Coalition for UAP Studies, 2013, Aguadilla, Puerto Rico, UAP Incident Report: A Detailed Analysis* (2015), explorescu.org/post/2013-aguadilla-puerto-rico-uap-incident-report-a-detailed-analysis.

[264] The All-domain Anomaly Resolution Office (AARO) confirmed receipt of requests from civilian research groups, including the Scientific Coalition for UAP Studies (SCU), seeking the full technical dataset from the Aguadilla case. In its correspondence, AARO stated that it was "not yet prepared to fully respond" but might do so in the future. In its final summary, AARO concluded that the object recorded by the CBP DHC-8's infrared system was "most likely a sky lantern released during a nearby wedding celebration." The report describes such lanterns as small, heat-lifted paper balloons commonly used in festive events. See U.S. Department of Defense, *All-domain Anomaly Resolution Office Case Resolution: "The Puerto Rico Object"* (20 Mar 2025), aaro.mil/UAP-Cases/UAP-Case-Resolution-Reports.

[265] Puerto Rico's earliest comprehensive fire-safety regulations were established under the *Fire Prevention Code* (Reglamento No. 4048), approved October 2, 1990, pursuant to Act No. 43 of June 21, 1988. This code, enforced by the Puerto Rico Fire Department, broadly restricted the use of open-flame and airborne ignition sources. In 2011, the *Puerto Rico Fire Code* adopted the 2009 *International Fire Code* through Reglamento 7950, which explicitly prohibited the release of untethered sky lanterns (§ 308.1.6.3). That prohibition remained in effect during the 2013 Aguadilla incident and was later reaffirmed by Law 202-2015, which formalized criminal penalties for the release of sky lanterns. See *the Fire Prevention Code of Puerto Rico (1990), app.estado.gobierno.pr/ReglamentosOnLine/Reglamentos/4048ING.pdf; and the Puerto Rico Fire Code 2011* (Reglamento 7950, adopting 2009 IFC § 308.1.6.3).

[266] The Scientific Coalition for UAP Studies (SCU) performed the first comprehensive technical review of the 2013 Aguadilla incident, analyzing the Department of Homeland Security thermal footage, radar correlations, and meteorological data. Their findings documented object velocities ranging from 80 to 120 mph—far exceeding local wind speeds—and confirmed motion patterns inconsistent with any wind-borne object. Thermal analysis revealed a uniform infrared signature with no central hot spot, flicker, or thermal plume, thereby eliminating open-flame sources, such as sky lanterns. The report further noted the object's entry into and exit from the ocean without visible splash or deceleration, continued thermal visibility in "black-hot" FLIR mode, and an absence of image artifacts, frame loss, or sensor anomalies. Conventional explanations, including birds, drones, balloons, and plasma phenomena, were ruled out, and frame-by-frame analysis recorded the object dividing into two independent, thermally coherent bodies. See *Scientific Coalition for UAP Studies, "Aguadilla, Puerto Rico UAP 2013: A Detailed Analysis"* (2015), explorescu.org/post/2013-aguadilla-puerto-rico-uap-incident-report-a-detailed-analysis.

[267] The SCU report acknowledges that radar accuracy in the Aguadilla case was limited to approximately one-eighth of a mile, citing FAA radar performance standards. The authors emphasize that these data were used only to verify temporal and spatial correlation between the CBP DHC-8 aircraft and the unidentified target, not to calculate a precise three-dimensional trajectory. The radar evidence thus serves as corroboration of the aircraft's flight path and the object's coincident presence, not as a primary tracking source. See *Scientific Coalition for UAP Studies, "Aguadilla, Puerto Rico UAP 2013: A Detailed Analysis"* (2015), pp. 34–38, explorescu.org/post/2013-aguadilla-puerto-rico-uap-incident-report-a-detailed-analysis.

[268] The SCU report acknowledges that the team lacked access to the original DHS video file. Their analysis was conducted on a high-

resolution copy publicly released online, believed to be a direct digital duplicate of the original CBP footage. The authors were transparent about this limitation and described the steps taken to verify that no evidence of alteration, frame loss, or data corruption was present in the analyzed copy. See *Scientific Coalition for UAP Studies, "Aguadilla, Puerto Rico UAP 2013: A Detailed Analysis"* (2015), pp. 6–7, explorescu.org/post/2013-aguadilla-puerto-rico-uap-incident-report-a-detailed-analysis.

[269] The Scientific Coalition for UAP Studies (SCU) analyzed a non-original, YouTube-derived copy of the DHS video and therefore limited its work to *relative* temperature/contrast and motion analysis. Within that constraint, the team reports that the object's behavior remains consistent across multiple zoom levels and polarity/contrast modes; they explicitly model aircraft motion and parallax, concluding the target's lateral movement and speed are not explained by wind drift (while noting range/speed uncertainties). Their method compares the object's thermal signature against varied scene elements (e.g., roads, asphalt, open pastures, livestock) rather than any single reference. The frame sequence documenting the apparent "split" is presented alongside a discussion of potential IR artifacts the team argues would more typically appear as flicker, dropout, or uneven contrast—not a symmetric bifurcation with stable motion and a signature. See Scientific Coalition for UAP Studies, *2013 Aguadilla, Puerto Rico UAP Incident: A Detailed Analysis* (2015), explorescu.org/post/2013-aguadilla-puerto-rico-uap-incident-report-a-detailed-analysis.

[270] The Scientific Coalition for UAP Studies (SCU) publicly confirmed in April 2025 that it had *formally submitted a series of technical questions to the All-domain Anomaly Resolution Office (AARO)* concerning the latter's "sky lantern" hypothesis for the 2013 Aguadilla video. AARO acknowledged receipt of the request but stated it was "not currently prepared to provide answers" and might do so in a future report. SCU's

statement outlines specific data and methodological clarifications sought—including reconstructed flight-path coordinates, radar sources, and control-tower logs—underscoring the group's ongoing effort to obtain official release of the underlying data. See Scientific Coalition for UAP Studies, *SCU Announcement on the AARO Investigation of the Aguadilla Video* (April 28, 2025), explorescu.org/post/scu-announcement-on-the-aaro-investigation-of-the-aguadilla-video.

[271] The All-domain Anomaly Resolution Office (AARO) categorized the 2013 Aguadilla, Puerto Rico incident as *"Explained—Conventional Object (Sky Lantern)"* in its public case-resolution summary. Still, the accompanying narrative specifies that this is a tentative explanation rather than a confirmed identification. The report notes that "data limitations prevent full characterization of the object" and that AARO is "not yet prepared to provide additional details." Thus, while the case is marked "resolved," no definitive determination has been made regarding the object's precise nature. See All-domain Anomaly Resolution Office, *Puerto Rico UAP Case Resolution Report* (2024), aaro.mil/UAP-Cases/UAP-Case-Resolution-Reports/.

[272] The Scientific Coalition for UAP Studies (SCU) formally disputed the All-domain Anomaly Resolution Office's "sky-lantern" classification of the 2013 Aguadilla video in its public statement of April 2025. SCU reiterated that its own 2015 technical report had already ruled out balloons, birds, lanterns, and other prosaic explanations based on infrared, radar, and meteorological data, and affirmed that "the case remains unresolved pending full data release." See Scientific Coalition for UAP Studies, *SCU Announcement on the AARO Investigation of the Aguadilla Video* (April 28, 2025), explorescu.org/post/scu-announcement-on-the-aaro-investigation-of-the-aguadilla-video; and *2013 Aguadilla, Puerto Rico UAP Incident: A Detailed Analysis* (2015), explorescu.org/post/2013-aguadilla-puerto-rico-uap-incident-report-a-detailed-analysis.

273 In medical terminology, an *idiopathic* condition refers to one whose cause is unknown after all other explanations have been ruled out. Physicians use the term "diagnosis by exclusion" to describe this process of eliminating known causes until only the unexplained remains. See Cleveland Clinic, *Idiopathic: Definition & Characteristics* (reviewed May 13, 2025), my.clevelandclinic.org/health/articles/idiopathic.

274 The Lockheed F-117A Nighthawk, the world's first operational stealth aircraft, was developed under the classified *Have Blue* program and achieved its first flight on June 18, 1981. It followed Lockheed's experimental *Have Blue* demonstrator (first flown in 1977) and incorporated faceted surfaces and radar-absorbent coatings that dramatically reduced radar cross-section. The F-117A entered operational service with the 4450th Tactical Group in 1983 and remained secret until 1988. See Lockheed Martin, *F-117 Nighthawk* (History & Overview), lockheedmartin.com/en-us/news/features/history/f-117.html; *Operation Nighthawk Landing Fast Facts* (2019), lockheedmartin.com/content/dam/lockheed-martin/aero/documents/F-117/ONL%20Fast%20Facts%20Final.pdf; and National Museum of the United States Air Force, *Lockheed F-117A Nighthawk Fact Sheet*, nationalmuseum.af.mil/Visit/Museum-Exhibits/Fact-Sheets/Display/Article/198056/lockheed-f-117a-nighthawk/.

275 RAF test pilot Squadron Leader Dave Southwood, who evaluated the F-117 under deep secrecy at Groom Lake and Tonopah in 1986, later recalled his first sight of the aircraft: "It looked impossible, something out of science fiction." See Steve Davies, *Unveiling the Secrets: Flying the F-117 Stealth Fighter at Area 51* (Mar 25 2025), 10percenttrue.com/post/into-the-black-evaluating-the-stealth-fighter.

[276] The composite name *Pelagomorph noetica* derives from classical Greek roots: *pelag-* (from *pélagos*, "sea" or "deep sea"); *morph-* (from *morphē*, "form" or "shape"); and *noēsis* ("mind," "intellect," or "understanding"). Together, they literally translate as "a deep-sea form possessed of intellect." See Merriam-Webster Dictionary entries for *pelag-* ("from Greek pelagos, sea"), merriam-webster.com/dictionary/pelag-; *morph* ("from Greek morphē, form, shape"), merriam-webster.com/dictionary/morph; and *noesis* ("from Greek noēsis, from noein to perceive, think"), merriam-webster.com/dictionary/noesis.

[277] Analyses of both historical and contemporary UAP databases reveal a persistent correlation between unidentified aerial or submersible phenomena and large bodies of water. The International Business Times reported in 2025 that the Enigma app recorded over 9,000 "Unidentified Submersible Object" (USO) sightings within ten miles of U.S. coastlines, with hundreds describing objects plunging into or emerging from the sea without disturbance. Enigma Labs' own USO collection notes that such objects are frequently observed crossing the air–water boundary with "transmedium capabilities," moving at extraordinary speeds without visible splashes or wakes. Earlier reports from *Wired* similarly cite declassified Russian Navy records describing anomalous underwater contacts during the Cold War, further reinforcing the historical continuity of water-linked encounters. See *Wired*, "Russian Navy Declassifies Cold War Close Encounters" (July 23, 2009), wired.com/2009/07/russian-navy-declassifies-cold-war-close-encounters/; Annalyn Zoglmann, "What Are USOs? Tracker Detects 9,000 Unexplained Objects Beneath U.S. Waters" (*IBTimes UK*, Oct 27, 2025); and Enigma Labs, *Unidentified Submerged Objects (USOs)*, enigmalabs.io/collection/1ac6fede-9cbe-49aa-8b44-169fd90b9e33.

[278] Numerous documented UAP encounters describe seamless traversal between air and water, with no visible deceleration, splash,

or wake—behavior broadly categorized as "transmedium." The 2013 Aguadilla, Puerto Rico report by the Scientific Coalition for UAP Studies provides one of the most detailed analyses, showing an object that "sliced into the water with little or no splash," maintained comparable velocity underwater (~82 mph) and in the air, and later re-emerged while retaining thermal integrity. Complementary peer-reviewed research explores the electrodynamic interactions between plasmoid-type UAPs and conductive marine environments, suggesting that charged plasma structures may be naturally drawn to oceanic interfaces where air and water exchange charge. These reports collectively illustrate a recurring observational pattern of craft or luminous bodies crossing the air–water boundary without causing hydrodynamic disturbance. See Scientific Coalition for UAP Studies, *Aguadilla Puerto Rico UAP Report: A Detailed Analysis* (2015), explorescu.org/post/2013-aguadilla-puerto-rico-uap-incident-report-a-detailed-analysis; Joseph, R. et al., *Unifying Scalar Fields and Field Repulsion: A New Approach to Understanding Unidentified Anomalous Phenomena and Trans-Medium Travel* (2024), researchgate.net/publication/389503379_Unifying_Scalar_Fields_and _Field_Repulsion_A_New_Approach_to_Understanding_Unidentifie d_Anomalous_Phenomena_and_Trans-Medium_Travel; and Joseph, R. et al., *Unidentified Anomalous Phenomena, Extraterrestrial Life, Plasmoids, Shape Shifters, Replicons, Thunderstorms, Lightning, Hallucinations, Aircraft Disasters, Ocean Sightings* (2024), researchgate.net/publication/383034675_Unidentified_Anomalous_P henomena_Extraterrestrial_Life_Plasmoids_Shape_Shifters_Replico ns_Thunderstorms_Lightning_Hallucinations_Aircraft_Disasters_Oc ean_Sightings.

[279] Numerous accounts over the past eight decades describe persistent UAP and USO activity in the waters surrounding the Channel Islands—particularly near Santa Catalina, Santa Cruz, and San Clemente Islands. Reports span both military and civilian witnesses and include the 1953 Lockheed sighting by Kelly Johnson, the 1962

Avalon fishing-boat incident, and the modern Navy "drone swarm" encounters of 2019. See *The Catalina Islander*, "Mysterious Island: UFO Action Galore" (2022), thecatalinaislander.com/mysterious-island-ufo-action-galore/; *Ojai Valley News*, "USO Hot Spots: Underwater Mysteries from Lake Casitas to Channel Islands" by Grant Phillips (July 31 2025), ojaivalleynews.com/news/county/uso-hot-spots-underwater-mysteries-from-lake-casitas-to-channel-islands/article_9c411ce1-e209-487f-ae6f-024c3d85bb66.html; and Marik von Rennenkampff, "UFOs, the Channel Islands and the Navy's 'Drone Swarm' Mystery," *The Hill* (Jan 5 2022), thehill.com/opinion/national-security/588223-ufos-the-channel-islands-and-the-navys-drone-swarm-mystery/.

[280] Ronald Reagan publicly recounted two separate UFO sightings in California. The first, dating to the early 1950s, reportedly occurred while Reagan and his wife Nancy were driving to actor William Holden's home near the Pacific coast; actress Lucille Ball recalled that the couple arrived "all out of breath and so excited" after witnessing what they described as a flying saucer. See Mitch Horowitz, *"Ronald Reagan and the Occultist: The Amazing Story of the Thinker Behind His Sunny Optimism,"* Salon (Jan 5, 2014), salon.com/2014/01/05/ronald_reagan_and_the_occultist_the_amazing_story_of_the_thinker_behind_his_sunny_optimism; and *Chicago Tribune*, "Lucille Ball Was More Than Just a Comedienne" (June 9, 1991), chicagotribune.com/1991/06/09/lucille-ball-was-more-than-just-a/.
The second incident occurred in 1974 during a Cessna Citation flight to Bakersfield, witnessed by Reagan, pilot Bill Paynter, and two security officers. The object "went up at a 45-degree angle at a high rate of speed," accelerating "like a hot rod." Reagan later told *Wall Street Journal* reporter Norman C. Miller that it "went straight up into the heavens." See *HowStuffWorks*, *"Ronald Reagan Sees a UFO"* (Apr 16 2024), science.howstuffworks.com/space/aliens-ufos/ronald-reagan-ufo.htm; and Garrett M. Graff, *"Which Presidents Have Seen UFOs? Yep, It's More Than One,"* *Politico Magazine* (Nov 17, 2023),

politico.com/news/magazine/2023/11/17/us-presidents-ufo-obsession-00127519. Both events describe aerial objects exhibiting extraordinary acceleration and non-ballistic flight behavior.

[281] On July 15, 2019, personnel aboard *USS Omaha* (LCS-12) detected and tracked as many as 14 unidentified radar contacts over several hours off the coast of San Diego. The Independence-class *Omaha*—built for high-speed littoral and open-ocean operations—is equipped with advanced sensors, including the SAAB AN/SPS-77 Sea GIRAFFE 3D radar, AN/KAX-2 electro-optical/FLIR system, and Northrop Grumman Integrated Combat Management System. During the 2019 encounter, multiple onboard systems recorded the anomalous targets, including infrared video captured from the ship's Combat Information Center showing a spherical object moving above the water before descending and disappearing beneath the surface. Radar data reportedly indicated velocities reaching approximately 138 knots (159 mph) and abrupt directional changes inconsistent with any known aircraft. After the video was released publicly, the Department of Defense confirmed its authenticity and that it had been reviewed by the UAP Task Force, while declining to identify the object. See NBC News, "Leaked Navy Video Appears to Show UFO off California" (2021), nbcnews.com/news/us-news/leaked-navy-video-appears-show-u-f-o-california-coast-n1267688/; The Debrief, "Pentagon Confirms Leaked Video Showing 'Transmedium' UFO is Authentic" (2021), thedebrief.org/pentagon-confirms-leaked-video-showing-transmedium-ufo-is-authentic/; Global News, "Leaked Video Appears to Show UFO Plunging Under Water off California" (2021), globalnews.ca/news/7871671/ufo-video-water-san-diego-california-2019-omaha/; SeaForces.org, "Independence-Class LCS (Littoral Combat Ship) — U.S. Navy/Austal Fact File" (2023), seaforces.org/usnships/lcs/Independence-class.htm.

[282] Point Loma in San Diego is home to Submarine Squadron 11, a U.S. Navy command composed of four Los Angeles-class fast-attack

submarines—*USS Alexandria (SSN 757), USS Santa Fe (SSN 763), USS Scranton (SSN 756), and USS Greeneville (SSN 772)*—each powered by nuclear reactors. See Commander, Submarine Squadron 11 (United States Navy), *About COMSUBRON 11* (2024), csp.navy.mil/css11/About/.

[283] Naval Air Station North Island, located on the northern end of Coronado, has long served as a central Pacific Fleet aviation hub and is identified in public records as a non-strategic nuclear weapons depot. Investigative reporting during the Navy's late-1990s nuclear homeport expansion detailed the base's role in storing radioactive and mixed hazardous waste associated with nuclear-powered carriers. See The Center for Land Use Interpretation, *North Island Naval Air Station* (2024), clui.org/ludb/site/north-island-naval-air-station; Phyllis Orrick, "North Island: Key Element in the Navy's Nuclear Fleet," *San Diego Reader* (Jan. 15 1998), sandiegoreader.com/news/1998/jan/15/special-glow/.

[284] During environmental remediation at the decommissioned Long Beach Naval Station, Navy and state health officials documented the presence of radioactive contaminants, including Radium-226 and Strontium-90, in soil and groundwater. These radionuclides were first detected in the early 2000s and were confirmed to exceed remediation goals in subsequent Naval Facilities Engineering Systems Command reports released through 2023. See Melissa Chan, "Shipyard Veterans May Have Been Exposed to Cancer-Causing Radioactive Materials. The Navy Has Not Told Them," *NBC News* (Jan. 27, 2024), nbcnews.com/news/us-news/shipyard-veterans-may-exposed-cancer-causing-radioactive-materials-nav-rcna129810/; U.S. Nuclear Regulatory Commission, *Long Beach Naval Complex Environmental Site Data Summary, Sites 1 & 2* (2022), nrc.gov/docs/ML2222/ML22227A091.pdf.

[285] The San Onofre Nuclear Generating Station (SONGS), situated on the Marine Corps Base Camp Pendleton near San Clemente, remains a defueled facility that continues to store spent nuclear fuel in dry cask canisters within its Independent Spent Fuel Storage Installation, located just inland from the Pacific Ocean. The site is approximately 44 miles from Santa Catalina Island. See U.S. Nuclear Regulatory Commission, *San Onofre – Unit 1* (updated Mar. 9, 2023), nrc.gov/info-finder/decommissioning/power-reactor/san-onofre-unit-1/; County of San Diego Office of Emergency Services, *SONGS Facts and Preparedness* (2024), sandiegocounty.gov/content/sdc/oes/emergency_management/oes_j l_SONGS.html.

[286] The Boiling Nuclear Superheater (BONUS) reactor, located northwest of Rincón and approximately 17 miles south of Aguadilla, Puerto Rico, was constructed in 1960 as a prototype power plant to test the integral boiling-superheating concept. It operated experimentally between 1964 and 1968 before being decommissioned and entombed in concrete. Long-term stewardship of the site is maintained by the U.S. Department of Energy Office of Legacy Management under the Defense Decontamination and Decommissioning Program. See U.S. Department of Energy, *BONUS, Puerto Rico, Decommissioned Reactor Site Fact Sheet* (2024), energy.gov/lm/articles/bonus-puerto-rico-decommissioned-reactor-site-fact-sheet.

[287] Rafael Hernández Airport (BQN), located in Aguadilla, Puerto Rico, occupies the site of the former Ramey Air Force Base, which was closed between 1971 and 1973 and subsequently converted for civilian use. See *Rafael Hernández Airport*, Wikipedia (accessed 2025), en.wikipedia.org/wiki/Rafael_Hern%C3%A1ndez_Airport.

[288] Ramey Air Force Base hosted one of the earliest U.S. Navy undersea surveillance installations, commissioned in 1954 as Naval

Facility Ramey AFB under Project Caesar, the classified cover name for the Sound Surveillance System (SOSUS). The station—later redesignated NAVFAC Punta Borinquen—was part of the network that evolved into the Integrated Undersea Surveillance System (IUSS). See Commander, Submarine Force, U.S. Pacific Fleet, "NOPF Whidbey Island Changes Name to Theater Undersea Surveillance Command Pacific" (Sept. 30 2022), csp.navy.mil/Media/News-Admin/Article/3182737/nopf-whidbey-island-changes-name-to-theater-undersea-surveillance-command-pacif/; Ernest Castillo III et al., *SOSUS/IUSS: Monitoring the World's Oceans—System Log 1949–20XX* (IUSS-CAESAR Alumni Association, rev. 2023), iusscaa.org/sosus_iuss_log_1949_20xx.pdf.

[289] Official U.S. Air Force memoranda cited in a Department of Veterans Affairs case confirm that Ramey Air Force Base was designated a "nuclear-capable unit," with technicians in the 1950s and 1960s servicing early unsealed-pit nuclear systems for Strategic Air Command aircraft. The Air Force Safety Center later reconstructed recorded ionizing-radiation doses for those personnel in connection with VA compensation claims. See Department of Veterans Affairs, *Board of Veterans' Appeals*, Citation Nr. 1215233 (Apr. 26, 2012), va.gov/vetapp12/Files2/1215233.txt.pdf; Norris & Arkin, *U.S. Nuclear Weapons Deployments, 1945–1977* (Nautilus Institute Report No. 306, 1998), nautilus.org/wp-content/uploads/2015/04/306.pdf; U.S. Army Corps of Engineers, *Former Ramey Air Force Base: Comprehensive Work Summary* (2008), usace.contentdm.oclc.org/digital/collection/p16021coll7/id/6293/.

[290] Veteran accounts describe a 1964 emergency transfer of Mk-28 thermonuclear bombs at Ramey Air Force Base, during which Air Police and munitions specialists escorted the weapons from a special-weapons maintenance facility to transport aircraft—an event later recalled as "the day Puerto Rico almost disappeared." These testimonies also reference plutonium pit cleaning and the disposal of

contaminated residue in an area known as the Caliche Pit, consistent with the presence of the 21st Munitions Maintenance Squadron and the 72d Bombardment Wing, then stationed at Ramey. See *Military Truth*, "Verification of Ramey AFB Incident (Contact Info Redacted)" (2018), militarytruth.org/wp-content/uploads/2018/02/Verification-of-Ramey-AFB-Incident-contact-info-redacted.pdf; Don Chapin, Capt. USAF (ret.), "A Near Miss" (*Military Truth*, 2018), militarytruth.org/a-near-miss/; *Wikipedia*, "North American F-100 Super Sabre" (accessed 2025), en.wikipedia.org/wiki/North_American_F-100_Super_Sabre; U.S. Army Corps of Engineers, *Former Ramey Air Force Base: Comprehensive Work Summary* (2008), usace.contentdm.oclc.org/digital/collection/p16021coll7/id/6293/.

[291] Ramey Air Force Base's nuclear and surveillance operations formed part of a broader Cold War triad linking air, sea, and space missions. The 72d Bombardment Wing, based at Ramey, is listed among participating units in *Operation Argus* (1958), a series of high-altitude nuclear tests conducted by Joint Task Force 88 in the South Atlantic, and the base simultaneously supported the Navy's first SOSUS undersea-surveillance facility commissioned there in 1954. See Defense Nuclear Agency, *Operation Argus: A Historical Summary, August–September 1958* (Report DNA 6039F, 1958), dtra.mil/Portals/61/Documents/NTPR/2-Hist_Rpt_Atm/1958_DNA_6039F.pdf; Commander, Submarine Force, U.S. Pacific Fleet, "NOPF Whidbey Island Changes Name to Theater Undersea Surveillance Command Pacific" (Sept. 30 2022), csp.navy.mil/Media/News-Admin/Article/3182737/nopf-whidbey-island-changes-name-to-theater-undersea-surveillance-command-pacif/; Ernest Castillo III et al., *SOSUS/IUSS: Monitoring the World's Oceans—System Log 1949–20XX* (IUSS-CAESAR Alumni Association, rev. 2023), iusscaa.org/sosus_iuss_log_1949_20xx.pdf.

[292] The Puerto Rico Trench lies just north of the island and represents the deepest point in the Atlantic Ocean, reaching an estimated depth

of about 8,400 meters (8.4 kilometers or 5.5 miles). It marks the boundary where the North American plate subducts beneath the Caribbean plate. See Alan J. Jamieson, Henri-Germain Delauze, and Michel Nargeolet, "Exploration of the Puerto Rico Trench in the Mid-Twentieth Century," *Deep Sea Research Part I: Oceanographic Research Papers* 163 (2020), sciencedirect.com/science/article/abs/pii/S0160932720300363.

[293] Geological studies led by the U.S. Geological Survey confirm that the Puerto Rico Trench region has produced megathrust earthquakes exceeding magnitude 8.0, including a prehistoric event estimated at magnitude 8.7 or greater that generated tsunamis capable of reshaping the island's northwestern coastline. These findings are based on paleotsunami deposits dated to 1470-1530 CE, discovered in mangrove cores from East Bajura, Puerto Rico. See U.S. Geological Survey, "USGS-Led Team Discovers Pre-Columbian Tsunami Deposits in Puerto Rico" (2023), usgs.gov/centers/pcmsc/news/usgs-led-team-discovers-pre-columbian-tsunami-deposits-puerto-rico.

[294] NASA's Gravity Recovery and Climate Experiment (GRACE) mission identified significant variations in Earth's gravitational field associated with the Puerto Rico Trench and Lesser Antilles region, where the anomaly reflects extreme density contrasts generated by subducting lithosphere. Earlier geophysical analyses found that the trench's pronounced negative free-air gravity anomaly, first modeled by Peter Molnar, results from a dense, descending slab of Atlantic lithosphere whose excess mass causes the crust to bend downward. These mass concentrations locally distort the Earth's gravity field and influence orbital path measurements from satellites, such as GRACE. See Peter Molnar, "Gravity Anomalies and the Origin of the Puerto Rico Trench," *Geophysical Journal of the Royal Astronomical Society* 51, no. 3 (1977), academic.oup.com/gji/article/51/3/701/774225; NASA, *Gravity by GRACE* (2001),

eospso.nasa.gov/sites/default/files/publications/2001_Gravity_by_GR
ACE.pdf.

[295] The Bermuda Triangle—also known as the Devil's Triangle—is
commonly described as the area bounded by Miami, Bermuda, and
Puerto Rico, an expanse of ocean long associated in popular culture
with navigational anomalies and unexplained disappearances. The
boundaries, size, and number of recorded incidents vary by source,
but most geographical references identify Puerto Rico as the southern
vertex of the triangle. See Bill Norrinton, "The Geography of the
Bermuda Triangle," University of California, Santa Barbara,
Department of Geography (2018), legacy.geog.ucsb.edu/the-
geography-of-the-bermuda-triangle/.

[296] Numerous official and analytical reports describe electromagnetic
and radio-frequency interference associated with unidentified aerial
phenomena. A declassified U.S. Customs and Border Protection file
cites a 1977 Finland incident involving military personnel and radio-
frequency jamming during UAP exposure and notes that such
encounters have been observed to cause temporary malfunction of
electrical and electronic systems—language originating in the U.K.
Ministry of Defence's *Project Condign* report. U.S. Defense Intelligence
Agency analyses likewise document anomalous field effects on
biological tissues and technical equipment. At the same time, later
aggregate reviews identify electromagnetic-spectrum distortion and
sensor interference as recurring observables in UAP data.
Contemporary assessments by NASA and independent researchers
continue to examine these patterns in radar and optical
instrumentation. See U.S. Customs and Border Protection, *Records
Pertaining to Unidentified Aerial Phenomenon* (FOIA release, 2021),
documents3.theblackvault.com/documents/cbp/Records%20pertaini
ng%20to%20Unidentified%20Aerial%20Phenomenon.pdf; U.S.
Defense Intelligence Agency, *Defense Intelligence Reference Document:
Anomalous and Subacute Field Effects on Human Biological Tissues* (2010),

dia.mil/FOIA/FOIA-Electronic-Reading-Room/FileId/170026/; Deep Knowledge Ventures, *UAP Analytical Frameworks* (2023), analytics.dkv.global/EI/UAP-Analytical-Frameworks.pdf; David Robson and Zaria Gorvett, "The Weird Incidents Piquing NASA's Interest," *BBC Future* (July 26, 2023), bbc.com/future/article/20230726-the-weird-incidents-piquing-nasas-interest.

[297] Numerous historical and contemporary reports describe interference by unidentified aerial phenomena with complex technical and defensive systems. Documented cases include the 1967 Malmstrom Air Force Base incident, where missile-launch platforms and targeting panels were simultaneously disabled, as well as later U.S. Air Force testimony describing alarm activations and security-system malfunctions near nuclear sites. Pilots have reported temporary loss of avionics, radar, radio contact, and engine function in the vicinity of unidentified objects, while civilian witnesses and early intelligence files record failures of automobile engines, headlamps, and photographic equipment. Recent congressional testimony and investigative journalism further document incursions by unknown craft near strategic installations and associated radar disruptions. See Thomas Novelly, "Air Force Veterans Who Are UFO True Believers Return to Newly Attentive Washington," *Military.com* (Oct. 19, 2021), military.com/daily-news/2021/10/19/air-force-veterans-who-are-ufo-true-believers-return-newly-attentive-washington.html; National Aviation Reporting Center on Anomalous Phenomena, *Unidentified Aerial Phenomena: Aviation Safety and Official Denial* (2021), narcap.org/blog/uap-aviation-safety-and-official-denial; *Unidentified Anomalous Phenomena: Implications on National Security, Public Safety, and Government Transparency*, Hearing before the U.S. House Committee on Oversight and Accountability, 118th Congress, July 26 2023, congress.gov/event/118th-congress/house-event/116282/text; Central Intelligence Agency, *Flying Saucers: UFO Reports* (FOIA release, ca. 1978), cia.gov/readingroom/document/cia-rdp81r00560r000100010002-9; Michael Shellenberger, *Written*

*Testimony before the House Oversight Committee on Unidentified Anomalous Phenomena* (Nov. 2024), oversight.house.gov/wp-content/uploads/2024/11/Written-Testimony-Shellenberger.pdf.

[298] Analysis of U.S. government imagery and human-intelligence data compiled in the 2024 congressional report *Technologies of Unknown Origin and Non-Human Intelligence* shows recurring geometric consistency across UAP observations. Shapes most frequently recorded between 1991 and 2022 include spheres or orbs, discs or saucers, ovals or "tic-tacs," triangles, boomerangs or arrowheads, and irregular or "organic" forms such as jellyfish- or brain-like structures. These morphologies and light formations were observed in multiple theaters under U.S. Central, Indo-Pacific, and Northern Commands, often in groupings or structured formations. See U.S. Congress, *Technologies of Unknown Origin and Non-Human Intelligence: Report to Congress* (Nov. 13, 2024), congress.gov/118/meeting/house/117721/documents/HHRG-118-GO12-20241113-SD003.pdf.

[299] Multiple radar–visual correlation studies and declassified intelligence summaries describe unidentified aerial objects that disappeared instantaneously from both visual observation and radar contact. In several well-documented cases, targets tracked simultaneously by radar and eyewitnesses—such as those during the 1952 Washington National sightings and the Haneda Air Force Base incident—were observed to vanish from screens without trace or loss of instrument function. See Central Intelligence Agency, *The National Investigations Committee on Aerial Phenomena (NICAP) Report* (declassified FOIA release, ca. 1978), cia.gov/readingroom/document/cia-rdp81r00560r000100010001-0; Central Intelligence Agency, *Flying Saucers: UFO Reports* (FOIA release, ca. 1978), cia.gov/readingroom/document/cia-rdp81r00560r000100010002-9.

## CHAPTER EIGHT

[300] The term *Unidentified Aerial Phenomena* first appeared in official government usage within a classified report by the United Kingdom Ministry of Defence titled *Unidentified Aerial Phenomena in the UK Air Defence Region: Volume 2 – Information on Associated Natural & Man-Made Phenomena* (Scientific & Technical Memorandum No. 55/2/00, Defence Intelligence Analysis Staff, December 2000; declassified 2006). Commonly known as the *Project Condign Report*, this study marked the formal shift from the label 'UFO' to 'UAP'. See United Kingdom Ministry of Defence, *Unidentified Aerial Phenomena in the UK Air Defence Region*, archive.org/details/condign-vol-2-1-258/Condign_Vol_2_1-258/.

[301] While the expression *Unidentified Aerial Phenomena* predates 2000 in some unclassified U.S. usage, the first verified appearance of the term in a classified government report occurred in the United Kingdom Ministry of Defence study *Unidentified Aerial Phenomena in the UK Air Defence Region* (Project Condign, 2000). Subsequent linguistic analyses and usage references indicate that UAP and UFO continued to be used interchangeably for many years, with the renewed preference for UAP gaining traction as a means to avoid the cultural stigma associated with "UFO." See Timothy L. Taylor, "UAP: A UFO Reset Made in the USA," *ResearchGate* (2024), researchgate.net/publication/390370311_UAP_A_UFO_RESET_MADE_IN_THE_USA; Dictionary.com, "What Does 'UAP' Mean?" (2021), dictionary.com/e/acronyms/uap/; and Defense Technical Information Center, *Project Blue Book Special Report No. 14* (1968 reprint), apps.dtic.mil/sti/tr/pdf/AD0680977.pdf.

[302] The social and institutional stigma surrounding UFO and UAP witnesses has been documented since the mid-20th century, when government investigations such as Project Blue Book and the Condon Report characterized most sightings as misperceptions or

psychological aberrations. Scholars like Greg Eghigian, PhD, have noted that "outsiders make judgments not about the sighting, but about the reliability and even the character of the person reporting the sighting," leading to decades of marginalization of credible observers. Contemporary analyses confirm that this stigma persists even among professionals and academics who report anomalous experiences. See Greg Eghigian, *The History and Psychology of UFOs*, Hogan Assessments (2023), hoganassessments.com/blog/history-psychology-ufos/; Tatyana Woodall, "US Government UFO Investigations Could Combat Stigma," *Popular Science* (June 17, 2022), popsci.com/science/space/aliens-evidence-us-government/; and University of Virginia News, "Despite Stigma, UFO Survey Finds 19% of Academics Say They've Had Strange Sightings" (2024), news.virginia.edu/content/despite-stigma-ufo-survey-finds-19-academics-say-theyve-had-strange-sightings.

[303] The Pentagon's shift from "aerial" to "anomalous" phenomena was explicitly motivated by a data-driven effort to improve systematic analysis across air, sea, and space domains. As Undersecretary of Defense for Intelligence Ronald S. Moultrie stated in 2022, the terminology was expanded to include "submerged and trans-medium objects," emphasizing improved data collection rather than semantics. See Brandi Vincent, *Pentagon Changes 'UAP' Terminology as It Looks to Investigate Unexplainable Sightings Across All Domains*, *DefenseScoop* (December 19, 2022), defensescoop.com/2022/12/19/pentagon-changes-uap-terminology-as-it-looks-to-investigate-unexplainable-sightings-across-all-domains/; *What Are UAPs, and Why Do UFOs Have a New Name? CBS News* (June 14, 2023), cbsnews.com/news/what-are-uaps-unexplained-aerial-phenomenon-ufos-new-name/.

[304] The term *trans-medium* refers to objects or phenomena capable of moving through more than one physical environment—such as air, sea, and space—without apparent loss of speed, stability, or integrity.

The idea has gained visibility through modern UAP investigations and Navy encounters, such as the 2004 Nimitz incident, in which radar and eyewitness data suggested movement between the atmosphere and the ocean without visible disruption. See *History.com*, "UFO Sightings: Speed, Appearance, and Movement" (2023), history.com/articles/ufo-sightings-speed-appearance-movement.

[305] The blobfish (*Psychrolutes marcidus*) inhabits depths of roughly 600–1,200 meters (2,000–4,000 feet) off the coasts of Australia and New Zealand, where high hydrostatic pressure preserves its natural structure. When brought rapidly to the surface, the loss of that pressure causes its gelatinous tissues to collapse, producing the distorted, sagging form that made it a viral symbol of "ugly" marine life. See Koh Ewe, "Blobfish: 'World's Ugliest Animal' Is NZ's Fish of the Year," *BBC News* (March 19, 2025), bbc.com/news/articles/col125d05970; Franz Lidz, "Behold the Blobfish," *Smithsonian Magazine* (November 2015), smithsonianmag.com/science-nature/behold-the-blobfish-180956967/.

[306] Many barophilic and deep-sea organisms experience lethal or extreme physiological stress when removed from their native pressure environments. Rapid decompression causes gases in their bodies to expand and tissues to rupture, while pressure-dependent cellular structures and enzymes fail to function under surface conditions. See Horst Felbeck, *Voyager: "Some Animals Live Deep in the Ocean at Great Depths and Pressure. How Do You Study Them? Won't They Die When You Remove Them from Their Native Environment?"* Scripps Institution of Oceanography (Aug. 1, 2011), scripps.ucsd.edu/news/voyager-some-animals-live-deep-ocean-great-depths-and-pressure-how-do-you-study-them-wont-they; See also Annie Machordom et al., "In-situ Studies on Hadal Fauna: Pressure Effects on Deep-Sea Microbiomes and Macrofauna," *Frontiers in Marine Science* (2024), pmc.ncbi.nlm.nih.gov/articles/PMC11593575/.

[307] Tardigrades (*Tardigrada*) represent one of the most resilient known animal phyla, capable of surviving conditions ranging from near-absolute zero to the vacuum of space. Their survival is mediated through *cryptobiosis*, particularly the *tun* state—an ametabolic form induced by desiccation or freezing in which metabolism drops to less than 0.01% of normal and body water to around 1–3%. In this state, tardigrades vitrify their cellular components through intrinsically disordered proteins (CAHS, SAHS) and protective sugars such as trehalose, creating a glass-like matrix that stabilizes DNA and membranes until rehydration revives the organism. See Thomas C. Boothby and Bob Goldstein, "How Tardigrades Work Their Magic," *American Scientist* (2018), americanscientist.org/article/tardigrades; A. Kunieda et al., "Tardigrade-unique disordered proteins protect from desiccation," *Communications Biology* 7 (2024): 364, nature.com/articles/s42003-024-06643-2; and Lauren Robertson, "Everything You Need (and Want) to Know About Tardigrades," *Front Line Genomics* (Oct. 18, 2022), frontlinegenomics.com/everything-you-need-and-want-to-know-about-tardigrades/.

[308] Many deep-sea animals employ gelatinous or hydrogel-like tissues with minimal gas content to withstand extreme hydrostatic pressure. Hadal amphipods and snailfish, for instance, rely on highly hydrated matrices that provide buoyancy and mechanical integrity through viscosity and cohesion rather than rigid skeletal support. This adaptation minimizes compressible spaces, preventing structural collapse or implosion when transitioning between depths. See M.E. Gerringer et al., "Distribution, Composition, and Functions of Gelatinous Tissues in Deep-Sea Fishes," *Royal Society Open Science* 5, no. 171063 (2017), royalsocietypublishing.org/doi/10.1098/rsos.171063; Susanne Wedlich, *"How Jelly-like Bodies Help Sea Creatures Survive Extreme Conditions," Big Think* (Mar. 1, 2023), bigthink.com/life/slime-jelly-ocean-extreme-conditions/; and Jacob R. Winnikoff et al., "Homeocurvature Adaptation of Phospholipids to Pressure in Deep-

Sea Invertebrates," *Science* 384, no. 6703 (2024): 1482–1488, science.org/doi/10.1126/science.adm7607.

[309] *Viscoelastic materials* exhibit both elastic (energy-storing) and viscous (energy-dissipating) properties. Their mechanical response can be described by two moduli: the *storage modulus* (E′), which measures the material's ability to store energy and recover shape, and the *loss modulus* (E″), which reflects internal friction and energy dissipation as heat. Examples include polymers, rubbers, and biological tissues such as articular cartilage, which deform like fluids under sustained stress yet rebound elastically once the stress is removed. See Bernard M. Lawless et al., "Viscoelasticity of Articular Cartilage: Analysing the Effect of Induced Stress and the Restraint of Bone in a Dynamic Environment," *Journal of the Mechanical Behavior of Biomedical Materials* 75 (2017): 293–301, sciencedirect.com/science/article/pii/S1751616117303284; and G. Timpone et al., "A Non-Destructive Methodology for the Viscoelastic Characterization of Polymers: Toward the Identification of the Time–Temperature Superposition Shift Law," *Sensors* 23, no. 22 (2023): 9213, mdpi.com/1424-8220/23/22/9213.

[310] *Hemocyanin* is a copper-based oxygen-transport protein found in the hemolymph of marine arthropods and mollusks. It functions similarly to hemoglobin but relies on copper ions rather than iron, binding and releasing oxygen through a reversible oxidation–reduction process that gives the protein its characteristic blue coloration. See Ruiyang Ji et al., "*A Comprehensive Review on Hemocyanin from Marine Products: Structure, Functions, Its Implications for the Food Industry and Beyond*," *International Journal of Biological Macromolecules* 269, Pt 1 (June 2024): 132041, sciencedirect.com/science/article/abs/pii/S0141813024028460.

[311] All aircraft rely on some form of propulsion to generate thrust sufficient to overcome drag and maintain flight. The three principal

categories of flight propulsion are *reciprocating (piston) engines* that drive propellers, *turbine or jet engines* that compress and combust air for thrust, and *rocket engines* that carry both fuel and oxidizer to expel high-velocity exhaust gases. Each system operates in accordance with Newton's Third Law of Motion—every action producing an equal and opposite reaction. See J. Gordon Leishman, *"Fundamentals of Propulsion Systems,"* in *Introduction to Aerospace Flight Vehicles* (Embry-Riddle Aeronautical University Press, 2025), eaglepubs.erau.edu/introductiontoaerospaceflightvehicles/chapter/introduction-to-propulsion-systems/.

[312] Natural aerial locomotion occurs in both powered and unpowered forms. Gliding and parachuting represent unpowered flight, while powered flight—driven by muscular wings—has evolved at least four times independently in the animal kingdom: among insects, pterosaurs, birds, and bats. Additional specialized modes include *soaring*, in which large birds and some extinct pterosaurs exploit rising air currents, such as thermals, ridge lift, wave lift, and convergence zones, to maintain altitude with minimal energy expenditure. Ballooning spiders achieve passive aerial dispersal by releasing silk threads that catch the wind, allowing them to travel on air currents. See *"Flying and Gliding Animals," Wikipedia* (rev. Oct. 2024), en.wikipedia.org/wiki/Flying_and_gliding_animals; and *"Vertebrate Flight: Introduction,"* University of California Museum of Paleontology (UCMP), ucmp.berkeley.edu/vertebrates/flight/flightintro.html.

[313] Numerous pilot and sensor reports describe Unidentified Aerial Phenomena (UAP) performing maneuvers that appear to defy known aerodynamic and propulsion principles—exhibiting instantaneous accelerations, right-angle turns, and sudden stops without visible thrust or aerodynamic surfaces. The National Aviation Reporting Center on Anomalous Phenomena (NARCAP) characterizes these behaviors as inconsistent with any known mechanical or biological means of flight, citing recurrent patterns of "non-inertial motion"

observed in Balls of Light, Discs, Cylinders, and Spheres. See Ted Roe, *"Flight Dynamics of Unidentified Aerial Phenomena, UAP,"* NARCAP *Technical Report 20* (October 2019), narcap.org/blog/flightdynamicsofuap. See also Alan C. Holt, *"Field Resonance Propulsion Concept,"* NASA Technical Memorandum 80961 (1979), ntrs.nasa.gov/api/citations/19800010907/downloads/19800010907.pdf.

[314] Physicist Robert "Bob" Lazar first appeared publicly in November 1989 on *KLAS-TV* in Las Vegas, interviewed by journalist George Knapp. He claimed to have worked in a secret facility known as S-4 near Area 51, where he allegedly studied nine recovered extraterrestrial craft and an "anti-gravity reactor" powered by a super-heavy element he called *Element 115*. Lazar described the material as stable, gravity-amplifying, and capable of folding space-time to produce instantaneous propulsion. He asserted that this element was not naturally found on Earth, whereas later scientific discoveries confirmed the existence of a distinct synthetic element—Moscovium (115)—with properties unlike those initially claimed. Lazar's testimony remains both influential and disputed, drawing attention through subsequent coverage by George Knapp, Jeremy Corbell's *Bob Lazar: Area 51 & Flying Saucers* (2018), and Lazar's own *Dreamland: An Autobiography* (2019). See *"Bob Lazar, UFO Hoaxster,"* HowStuffWorks (Nov 22, 2023), science.howstuffworks.com/space/aliens-ufos/bob-lazar.htm; and *"Bob Lazar,"* Wikipedia (rev. Oct 2024), en.wikipedia.org/wiki/Bob_Lazar.

[315] Moscovium (*Mc*, element 115) is a highly radioactive, synthetic element first confirmed in 2015 by the Joint Institute for Nuclear Research (Dubna, Russia) in collaboration with Lawrence Livermore and Oak Ridge National Laboratories. Only a few atoms have ever been produced, each decaying in fractions of a second into lighter elements such as nihonium. Current research demonstrates that

Moscovium is unstable, extremely short-lived, and exhibits no known gravitational or exotic energy properties. Theoretical stabilization of heavier isotopes remains speculative. See GSI Helmholtz Centre for Heavy Ion Research, *"The Periodic Table Just Got Wilder: Scientists Unveil the Secrets of the Heaviest Element Ever – Moscovium,"* *SciTechDaily* (Nov 20, 2024), scitechdaily.com/the-periodic-table-just-got-wilder-scientists-unveil-the-secrets-of-the-heaviest-element-ever-moscovium/; and *"Moscovium – Element Information, Properties and Uses,"* *Royal Society of Chemistry Periodic Table* (2025), periodic-table.rsc.org/element/115/moscovium.

[316] Copernicium (*Cn*, element 112) is a synthetic, radioactive element produced through heavy-ion fusion reactions and first identified in 1996 by physicist Sigurd Hofmann and colleagues at the GSI Helmholtz Centre for Heavy Ion Research in Darmstadt, Germany. Although most isotopes of copernicium decay within milliseconds, theoretical models propose that the heavier isotopes *Copernicium-291* and *Copernicium-293* could possess half-lives approaching 1,200 years, possibly existing in the so-called "island of stability." See *"Copernicium (Cn),"* *EBSCO Research Starters* (2023), ebsco.com/research-starters/chemistry/copernicium-cn.

[317] The concept of the *Alcubierre Drive* was first proposed by physicist Miguel Alcubierre in *"The Warp Drive: Hyper-Fast Travel Within General Relativity,"* *Classical and Quantum Gravity* 11 (1994): L73–L77. Alcubierre demonstrated mathematically that, within general relativity, it is possible to achieve apparent faster-than-light motion by contracting space-time ahead of a vessel and expanding it behind, forming a localized "warp bubble." The model would not violate relativity because the craft itself remains locally subluminal; however, it requires regions of negative energy density—"exotic matter"—and would generate extreme radiation levels, as later noted in follow-up analyses such as L. H. Ford and M. J. Pfenning, *Classical and Quantum Gravity* 14 (1997): 1743–1751. See Miguel Alcubierre, *"The Warp Drive:*

Hyper-Fast Travel Within General Relativity," *Classical and Quantum Gravity* 11 (1994): L73–L77, DOI: 10.1088/0264-9381/11/5/001; available via ResearchGate at researchgate.net/publication/1963139_The_Warp_Drive_Hyper-fast_Travel_Within_General_Relativity.

[318] Research and experimental efforts exploring advanced or speculative propulsion have examined electromagnetic and plasma-based thrust systems, high-frequency oscillation, and hybrid concepts that merge electrodynamic and plasma physics principles. NASA's Innovative Advanced Concepts (NIAC) program and related work by Eagleworks Laboratories have investigated *ambient plasma wave propulsion* and *magnetohydrodynamic (MHD) thrusters*. Meanwhile, the Defense Systems Information Analysis Center (DSIAC) reviews the lineage of plasma thrusters, tracing it from early ion engines to contemporary electric and Hall-effect designs. Such studies, alongside theoretical work on "Q-thrusters" and warp-field interferometry, represent ongoing attempts to expand the limits of propulsion physics beyond chemical or nuclear methods. See *Space Travel Aided by Plasma Thrusters: Past, Present and Future* (DSIAC, 2023), dsiac.dtic.mil/articles/space-travel-aided-by-plasma-thrusters-past-present-and-future/; and Harold "Sonny" White, *The Potential for Ambient Plasma Wave Propulsion* (NASA, 2023), nasa.gov/general/the-potential-for-ambient-plasma-wave-propulsion/.

[319] Emerging materials research supports the possibility of using metamaterial or plasma-based shielding systems capable of manipulating electromagnetic or gravitational fields. John Slough's NASA-funded plasma-shield concept proposed surrounding a spacecraft with a charged hydrogen cloud confined by superconducting wire mesh, generating a magnetic field strong enough to deflect cosmic radiation—essentially a *protective bubble* in space. Similar innovations in "smart" and self-healing materials have been reviewed in recent aerospace studies, which highlight

metamaterials, carbon nanotube composites, and multifunctional nanostructures for active radiation shielding and structural adaptability in extreme environments. See David Shiga, *"Plasma Bubble Could Protect Astronauts on Mars Trip,"* New Scientist (July 17, 2006), newscientist.com/article/dn9567-plasma-bubble-could-protect-astronauts-on-mars-trip/; and L. Pernigoni and A. M. Grande, *"Advantages and Challenges of Novel Materials for Future Space Applications,"* Frontiers in Space Technologies 4 (2023): 1253419, doi: 10.3389/frspt.2023.1253419.

[320] Stimuli-responsive hydrogels demonstrate the capacity to alter their density, porosity, or volume in response to external physical fields, providing an empirical analogue for hypothetical biological matrices capable of self-regulated buoyancy or environmental adaptation. Acharya et al. summarize that DNA-based hydrogels can reversibly expand, contract, and deform in response to triggers such as temperature, light, magnetic, and electric fields, allowing for modulation of cross-link density and phase transitions. These mechanisms underpin current research into electromagnetic and magneto-responsive hydrogel systems that mimic the dynamic properties of living tissue and demonstrate controlled deformation under applied fields. See Rumi Acharya et al., *"Physical Stimuli-Responsive DNA Hydrogels: Design, Fabrication Strategies, and Biomedical Applications,"* Journal of Nanobiotechnology 23 (2025): 233, pmc.ncbi.nlm.nih.gov/articles/PMC11929200/; and Anusua Bhowmik et al., *"Recent Advancements in Electro- and Magneto-Responsive Hydrogels: From Fundamentals to Biomedical Applications,"* Frontiers in Bioengineering and Biotechnology 12 (2024): 11214942, pmc.ncbi.nlm.nih.gov/articles/PMC11214942/.

[321] Thermoresponsive hydrogels and related smart materials demonstrate the capacity to rapidly absorb or release heat in response to environmental stimuli such as temperature changes. These systems adjust molecular alignment or phase state to achieve

efficient thermal exchange, a property widely studied for use in energy storage and biomedical applications. See *M. H. Jabbar et al.*, "Design of Thermoresponsive and Thermo-Regulating Hydrogels: A Review of Materials for Advanced Energy and Biomedical Applications," *Materials Advances* (2024), ScienceDirect, sciencedirect.com/science/article/pii/S2666123324001016.

[322] Localized heat gradients can alter air density and create buoyant convection currents, a principle explained by standard thermodynamic laws governing energy transfer and equilibrium between systems and surroundings. See *Oxford University Department of Physics, Basics of Thermodynamics* (2023), physics.ox.ac.uk/system/files/file_attachments/basic_thermo.pdf.

[323] Many marine species possess electrogenic or electroreceptive capabilities, including elasmobranchs such as sharks, rays, and skates, which detect weak bioelectric fields from prey or environmental cues through specialized organs like the ampullae of Lorenzini. These fields are typically in the microvolt-per-centimeter range, far below anthropogenic levels but biologically significant. See *Zoë L. Hutchison et al.*, "Anthropogenic Electromagnetic Fields (EMF) Influence the Behaviour of Bottom-Dwelling Marine Species," *Scientific Reports* 10 (2020): 4219, nature.com/articles/s41598-020-60793-x; and Jason Bittel, "The Shocking Ways Wild Animals Use Electricity," *National Geographic* (2023), nationalgeographic.com/animals/article/how-wild-animals-use-electricity.

[324] Geomagnetic and ionospheric gradients represent measurable variations in Earth's magnetic and plasma-field environments that can fluctuate dramatically under solar or geomagnetic disturbances. These gradients—often quantified in millimeters per kilometer—are shaped by solar activity and magnetospheric coupling, with especially strong perturbations during geomagnetic storms. The high-latitude ionosphere, in particular, behaves as a dynamic, nonequilibrium

medium that links the magnetosphere and upper atmosphere through plasma instabilities, electric field variations, and charged-particle precipitation. See Yixin Zhang et al., "A Study on the Characteristics of the Ionospheric Gradient under Geomagnetic Perturbations," *Sensors* 20 (2020): 1805, pmc.ncbi.nlm.nih.gov/articles/PMC7181138/; and M. J. Keskinen, "The Structure of the High-Latitude Ionosphere and Magnetosphere," *Johns Hopkins APL Technical Digest* 5 no. 2 (1984): 153–158, secwww.jhuapl.edu/techdigest/content/techdigest/pdf/V05-N02/05-02-Keskinen.pdf.

[325] In theoretical and experimental physics, magnetohydrodynamic (MHD) and Lorentz-force propulsion systems demonstrate how conductive fluids or plasmas can generate thrust by interacting with magnetic and electric fields. This principle provides a physical analogue for biological electromagnetic propulsion. The Lorentz-force propeller concept describes momentum transfer through electromagnetic field gradients, while magnetohydrodynamic thrusters achieve propulsion by accelerating conductive fluid via Lorentz forces. See J. Moreau, "Magnetohydrodynamics Propulsion: A Global Approach of an Inner DC Thruster," *Energy Conversion and Management* 40, no. 17 (1999): 1835–1846, sciencedirect.com/science/article/abs/pii/S0196890499000722; and Yingying Hu et al., "Principle and Characteristic of Lorentz Force Propeller," *Proceedings of the IEEE ICIEA* (2008), researchgate.net/publication/228765779_Principle_and_Characteristic_of_Lorentz_Force_Propeller.

[326] Magnetic levitation, or *mag-lev*, arises from the Meissner effect—an electromagnetic interaction in which a superconductor expels magnetic flux and levitates above a magnetic field through repulsive field coupling. This phenomenon, first observed in superconductors cooled below their critical temperature, forms the physical foundation for frictionless transport and electromagnetic lift systems. See Elie

W'ishe Sorongane, "The Classical Description of the Meissner Effect: Theory and Applications," *Open Journal of Applied Sciences* 13 (2023): 275–287, scirp.org/journal/paperinformation?paperid=123514.

[327] Marine magnetic anomalies—variations in Earth's magnetic field intensity recorded within the oceanic crust—are produced as newly formed basalt cools and locks in the prevailing geomagnetic polarity. These anomalies occur symmetrically on either side of spreading centers and persist along fracture zones and trench systems, including subduction margins like the Puerto Rico Trench, which lies within a geologically active area of crustal magnetization and flux distortion. See Bruce Peter Luyendyk, *"Marine Magnetic Anomalies,"* Encyclopaedia *Britannica* (2024), britannica.com/science/oceanic-crust/Marine-magnetic-anomalies; and R. A. R. McKay et al., *Magnetic Anomaly Patterns in Deep-Ocean Basins*, Naval Ocean Research and Development Activity Report ADA139602 (1983), apps.dtic.mil/sti/tr/pdf/ADA139602.pdf.

[328] Electrohydrodynamics (EHD), also known as ionic air propulsion, describes the generation of airflow by accelerating charged particles using high-voltage electric fields. The effect was first noted in early electrostatic experiments and later formalized by Townsend in 1914, laying the groundwork for 20th-century studies in electrostatic and corona discharge phenomena. Only in the 21st century has EHD been adapted for practical applications, such as ionic lifters and drones, made possible by lighter power systems and high-efficiency converters. Recent research confirms the viability of EHD-based drones and thrust devices that operate without moving parts, relying entirely on ionized air momentum transfer. See Rushikesh Patil, "Design of Power Supply and EHD System for Ion-Propulsion Drone," *Bohr International Journal of Engineering* 2, no. 3 (2022): 15–29, journals.bohrpub.com/index.php/bije/article/download/511/4739; and Hongtao Liu et al., "Investigation of an Electrohydrodynamic Thruster Using a Miniaturized Corona Discharge," *International Journal of Heat*

*and Mass Transfer* 223 (2024): 124–139,
sciencedirect.com/science/article/pii/S0017931024007269.

[329] Recent interdisciplinary research has shown that swarming insects can measurably influence atmospheric electric fields, generating weak but detectable ion clouds. Field studies conducted by biologists at the University of Bristol demonstrated that honeybee swarms can alter atmospheric charge by 100–1,000 volts per meter, comparable in magnitude to storm-cloud effects when scaled to large locust swarms. These findings reveal that collective biological activity can contribute to atmospheric electricity. See Cell Press, *"Insects Contribute to Atmospheric Electricity,"* *ScienceDaily* (October 24, 2022), sciencedaily.com/releases/2022/10/221024131048.htm; and Darren Incorvaia, *"Insect Swarms Might Generate as Much Electric Charge as Storm Clouds,"* *Science News* (October 31, 2022), sciencenews.org/article/insect-swarms-electric-charge-static-electricity-storm-clouds-honeybees.

[330] Electric eels (*Electrophorus voltai*) are capable of delivering discharges of up to 860 volts, the highest known voltage produced by a living organism. These pulses, generated by thousands of electrocyte cells, are used for navigation, communication, and stunning prey. The Atlantic torpedo ray (*Torpedo nobiliana*) likewise generates strong electrical discharges, reaching 170–220 volts via paired electric organs used to stun prey and deter predators. See Katherine J. Wu, "Newly Described Species of Electric Eel Serves Up Shocks of 860 Volts," *NOVA | PBS* (September 10, 2019), pbs.org/wgbh/nova/article/electric-eel-species/; and "Atlantic Torpedo Ray," *EBSCO Research Starters* (2024), ebsco.com/research-starters/science/atlantic-torpedo-ray.

[331] The electric eel (*Electrophorus electricus*) generates power through stacked electrocyte cells—thin, excitable membranes arranged in long series that together produce potential differences exceeding 600

volts. Recent work has demonstrated that similar stacked hydrogel systems can mimic these biological architectures, generating over 100 volts from ion gradients across alternating cation- and anion-selective membranes. These layered structures exemplify how organized electrolyte sheets can generate powerful, synchronized discharges, providing a model for bioelectric layering in nature and a potential source of inspiration for synthetic systems. See Thomas B. H. Schroeder et al., "An Electric Eel-Inspired Soft Power Source from Stacked Hydrogels," *Nature* 552 (2017): 214–218, pmc.ncbi.nlm.nih.gov/articles/PMC6436395/.

[332] Across documented aviation encounters, numerous reports describe UAP executing "instantaneous" accelerations and reversals, abrupt stops, and hovering with no visible propulsion or aerodynamic structures. A comprehensive NARCAP technical report catalogs pilot accounts of objects performing sudden full stops, hovering silently, and then accelerating beyond visual tracking—motions inconsistent with inertia-limited flight dynamics. Parallel analyses from the Alternative Propulsion Engineering Conference highlight similar "five observables," including positive lift without exhaust, extreme acceleration, and rapid departures following silent hovers, as in the Aguadilla incident. These recurring dynamics remain unexplained by current aeronautical models. See Ted Roe, "Flight Dynamics of Unidentified Aerial Phenomena (UAP)," *National Aviation Reporting Center on Anomalous Phenomena Technical Report 20* (October 2019), narcap.org/blog/flightdynamicsofuap; and Tim Ventura, "UAP Flight Characteristics: What the 'Five Observables' Actually Tell Us," *Alternative Propulsion Engineering Conference* (September 21, 2025), altpropulsion.com/uap-flight-characteristics-what-the-five-observables-actually-tell-us/.

[333] Deep-sea hydrothermal vent ecosystems rely on chemosynthesis—biological energy production driven by chemical gradients rather than sunlight. When reducing vent fluids rich in hydrogen sulfide,

methane, and hydrogen, these mix with oxidizing seawater, and chemolithoautotrophic microbes catalyze redox reactions that release usable energy and fix carbon dioxide into biomass. These redox-driven processes, dominated by Epsilonproteobacteria, sustain entire ecosystems independent of photosynthesis and operate under high pressure and total darkness. See Jesse McNichol et al., "Assessing Microbial Processes in Deep-Sea Hydrothermal Systems by Incubation at *In Situ* Temperature and Pressure," *Deep Sea Research Part I: Oceanographic Research Papers* 115 (2016): 221–232, sciencedirect.com/science/article/abs/pii/S0967063716300681.

[334] Laboratory plasma-physics studies show that ultraviolet and near-infrared radiation can ionize air at relatively low energies, producing localized semi-plasma regions through multiphoton and avalanche ionization. Ultraviolet excitation, in particular, initiates weakly ionized, low-temperature plasmas that heat the surrounding air, altering its pressure and density. These conditions provide a physical basis for how localized radiation could create low-pressure plasma boundaries. See Ciprian Dumitrache, Christopher M. Limbach, and Azer P. Yalin, "Threshold Characteristics of Ultraviolet and Near Infrared Nanosecond Laser-Induced Plasmas," *Physics of Plasmas* 23 (2016): 093515, pubs.aip.org/aip/pop/article/23/9/093515/319659/Threshold-characteristics-of-ultraviolet-and-near.

[335] Numerous documented UAP reports describe luminous halos or corona effects surrounding observed objects. For example, declassified UK Ministry of Defence files include a 2001 Oxfordshire sighting in which witnesses reported "a bright white halo seen heading northwest." Similar phenomena have been recorded in other cases archived by the National UFO Reporting Center and local press. See Emily Allen, "Documents Reveal UFO Sightings," *Herald Series* (May 23, 2008), heraldseries.co.uk/news/2293801.documents-reveal-ufo-sightings/; *Chorley Guardian*, "'UFO' Sighting: Chorley Man

Captures Mysterious White Halo on Video in His Living Room" (May 2023), lep.co.uk/news/people/ufo-sighting-chorley-man-captures-mysterious-white-halo-on-video-in-his-living-room-4140322; and National UFO Reporting Center case entries 179191 and 76542, nuforc.org/sighting.

[336] Research on acoustic and vibration-based propulsion demonstrates that controlled oscillations can generate measurable thrust through localized pressure-wave manipulation. Laboratory studies of ultrasonic disc transducers have shown that oscillatory vibration produces directed microjets and pressure differentials in surrounding media. Additionally, NASA field-resonance and vortex-induced vibration research further illustrates how cyclic structural motion can couple with surrounding fluid flow to create lift or impulse effects. These findings establish a physical basis for the concept of biological vibration-induced displacement. See Lin Zhang et al., "Acoustic Underwater Propulsion System Based on Ultrasonic Disc PZT Transducer," *Sensors and Actuators A: Physical* (2023), sciencedirect.com/science/article/abs/pii/S0924424723003515; Charles Dalton, *Fundamentals of Vortex-Induced Vibration* (University of Houston, Bureau of Safety and Environmental Enforcement Report 485AB, 1988), bsee.gov/sites/bsee.gov/files/tap-technical-assessment-program/485ab.pdf; and Alan C. Holt, *Field Resonance Propulsion Concept* (NASA Technical Memorandum 80961, 1979), ntrs.nasa.gov/api/citations/19800010907/downloads/19800010907.pdf.

[337] U.S. House of Representatives, *Hearing on Unidentified Anomalous Phenomena (UAP)*, Committee on Oversight and Accountability (Nov. 13 2024). The accompanying submission, "Unidentified Anomalous Phenomena: Implications on National Security, Public Safety, and Government Transparency," summarizes classified-origin imagery, infrared, and radar intelligence showing instances of UAPs "suddenly disappearing," producing "sensor dropout across multiple detection

systems," and evading simultaneous optical and FLIR tracking. See
HHRG-118-GO12-20241113-SD003,
congress.gov/118/meeting/house/117721/documents/HHRG-118-GO12-
20241113-SD003.pdf.

[338] Transparency in marine life arises primarily from refractive-index
matching and the reduction of internal light scattering—mechanisms
evolved independently among many pelagic species. Studies show
that glass squids and larval eels achieve near invisibility by
minimizing pigmentation and aligning their internal refractive index
with that of seawater, whereas other gelatinous species, such as salps
and comb jellies, approach optical parity with their environment. See
Edwin Barkdoll, "Nature's Invisibility Cloak," *Nautilus* (Nov. 16 2023),
nautil.us/natures-invisibility-cloak-442975; and Erika P., "These
Transparent Plants and Animals Are a Must See," *Science Times* (Jan. 6
2021), sciencetimes.com/articles/28940/20210106/look-transparent-
plants-animals-see.htm.

[339] Biophotonic crystals—periodic nanoscale lattices that manipulate
light through interference rather than pigment—are responsible for
much of the structural coloration observed in nature. These photonic
architectures, found in organisms ranging from beetles to butterflies,
act as optical lattices that selectively reflect or refract particular
wavelengths depending on lattice spacing and refractive index.
Inverse-opal and gyroid-type nanostructures have been documented
in butterfly wing scales, producing vibrant iridescence through
coherent scattering. The *Blue Morpho* butterfly, for example, exhibits
diamond-like photonic crystals within its wing scales that generate
its characteristic electric-blue shimmer purely through physical
nanostructure rather than chemical pigment. See Yanqiu Zhu *et al.*,
"Photonic Crystal Structures with Tunable Structure Color as
Colorimetric Sensors," *Sensors* 13, no. 4 (2013): 4192–4210,
mdpi.com/1424-8220/13/4/4192; Richard W. Corkery and Elizabeth C.
Tyrode, "On the Colour of Wing Scales in Butterflies: Iridescence and

Preferred Orientation of Single Gyroid Photonic Crystals," *Interface Focus* 7 (2017): 20160154, royalsocietypublishing.org/doi/10.1098/rsfs.2016.0154; and Vinod Saranathan et al., "Structure, Function, and Self-Assembly of Single-Network Gyroid (I4132) Photonic Crystals in Butterfly Wing Scales," *Proceedings of the National Academy of Sciences* 107 (2010): 11676–11681, pnas.org/doi/10.1073/pnas.0909616107.

[340] Quantum coherence in biology is now well documented: enzymes can exploit quantum tunneling and, under coherent conditions, sample multiple reaction pathways before resolving into the most efficient outcome; photosynthetic complexes show long-lived coherence guiding energy transfer at physiological temperatures. In classical terms, enzymes catalyze reactions by lowering activation barriers (lock-and-key/induced-fit mechanism), but in proton- or electron-transfer reactions, tunneling can dominate. This phenomenon has been reviewed across various enzyme systems and demonstrated in model and physiological contexts. Coherence and tunneling have also been probed in nucleic-acid and photosynthetic systems, supporting the broader conclusion that life has evolved to use quantum effects in specific, efficient ways. See Lorenzo Stella et al., "Quantum Biology—New Perspective on the Living World," *Biology* 13 (2024): 74, pmc.ncbi.nlm.nih.gov/articles/PMC10939336/; C. B. de Oliveira et al., "Probing Quantum Coherence in a Biological System by Means of DNA Amplification," *Journal of Photochemistry and Photobiology B* 54 (2000): 162–166, sciencedirect.com/science/article/abs/pii/S0303264700000952; Michael J. Janko et al., "Quantum Coherence and Tunneling in Biology," *International Journal of Molecular Sciences* 24 (2023): 16464, pmc.ncbi.nlm.nih.gov/articles/PMC10671017/; and G. Panitchayangkoon et al., "Long-Lived Quantum Coherence in Photosynthetic Complexes at Physiological Temperature," *PNAS* 107 (2010): 12766–12770, pnas.org/doi/10.1073/pnas.1005484107.

CHAPTER NINE

[341] Human historical knowledge captures only a minute fraction of the Earth's full narrative. Geological and evolutionary records reveal vast stretches of "deep history" that extend beyond the boundaries of human memory and written documentation, encompassing both biological and planetary processes. See Daniel Lord Smail and Andrew Shryock, "Deep History: An Integrative Approach Toward the History of Everything," *Frontiers in Psychology* 14 (2023): 1238272, pmc.ncbi.nlm.nih.gov/articles/PMC10603192/.

[342] Tardigrades are among the few known multicellular organisms that can endure direct exposure to the vacuum and radiation of space. Multiple experiments conducted aboard the ESA FOTON-M3 and NASA Endeavour missions confirmed their survival under conditions of cosmic and ultraviolet radiation, vacuum, and microgravity, making them invaluable model species in astrobiology. See Weronika Erdmann and Łukasz Kaczmarek, "Tardigrades in Space Research – Past and Future," *Origins of Life and Evolution of Biospheres* 47 (2017): 545–553, pmc.ncbi.nlm.nih.gov/articles/PMC5705745/.

[343] Monotremes—represented today only by the platypus and echidnas—occupy a singular branch of the mammalian tree, combining reptilian, avian, and mammalian traits. They lay eggs, produce milk, and bear venomous or display spurs on the hind limbs, while also possessing electroreceptors that allow detection of weak electrical fields in prey. Genomic sequencing has revealed that their chromosomes and gene organization are a mosaic of reptilian and therian features, reflecting an early divergence from other mammals during the Mesozoic. See Frank Grützner and Jennifer A. Marshall Graves, "A Platypus' Eye View of the Mammalian Genome," *Current Opinion in Genetics & Development* 14 (2004): 642–649, sciencedirect.com/science/article/abs/pii/S0959437X04001492; and Timothy F. Flannery et al., "A Review of Monotreme (*Monotremata*)

Evolution," *Alcheringa: An Australasian Journal of Palaeontology* 46 (2022): 59–74, tandfonline.com/doi/full/10.1080/03115518.2022.2025900.

[344] Certain jumping spiders of the genus *Portia* exhibit aggressive mimicry by entering other spiders' webs and imitating the precise vibration patterns of ensnared prey. This behavioral strategy deceives the resident spider into approaching, allowing *Portia* to ambush it. These complex, risk-laden tactics have been shown to involve trial-and-error learning, flexible signal generation, and feedback-driven adaptation. See R. R. Jackson and F. R. Cross, "A Cognitive Perspective on Aggressive Mimicry," *Journal of Zoology* 290 (2013): 161–171, zslpublications.onlinelibrary.wiley.com/doi/10.1111/jzo.12036.

[345] Experimental work on *Manduca sexta* has shown that associative memory can survive complete metamorphosis. Caterpillars trained to avoid a specific odor through aversive conditioning later exhibited the same avoidance as adult moths, confirming that learned information persists through the pupal stage. This finding provides the first definitive evidence that neuronal connections or synaptic patterns may persist through metamorphosis, allowing for the continuity of memory from larva to adult. See Douglas J. Blackiston, Elena Silva Casey, and Martha R. Weiss, "Retention of Memory through Metamorphosis: Can a Moth Remember What It Learned As a Caterpillar?" *PLOS ONE* 3, no. 3 (2008), doi.org/10.1371/journal.pone.0001736.

[346] Ant colonies function as highly integrated eusocial systems in which individual behavior contributes to emergent collective intelligence, often described as a form of distributed cognition or "superorganism" dynamics. Theoretical analyses have even proposed that such colonies may exhibit rudimentary forms of self-modeling or proto-consciousness arising from decentralized feedback networks. Among these eusocial groups, fungus-growing ants of the tribe

*Attini*—particularly *Atta* and *Acromyrmex*—practice true agriculture, cultivating fungal gardens in a multipartite symbiosis with bacterial mutualists and crop fungi, a system evolved over 50 million years. See Ted R. Schultz and Seán G. Brady, "Major Evolutionary Transitions in Ant Agriculture," *Proceedings of the National Academy of Sciences* 105 (2008): 5435–5440, pnas.org/doi/10.1073/pnas.0711024105; and David C. Queller and Joan E. Strassman, "Is the Ant Colony a Conscious Organism?" *Synthese* (2019), 10.1007/s11229-019-02130-y; see also J. M. Izquierdo, "Is the Ant Colony a Conscious Organism?" *Synthese* (2020): 1–18, researchgate.net/publication/385587564_IS_THE_ANT_COLONY_A_C ONSCIOUS_ORGANISM.

[347] Bombardier beetles (*Coleoptera: Carabidae: Brachinini*) defend themselves by explosively mixing hydroquinone and hydrogen peroxide in a specialized reaction chamber, producing a superheated jet of benzoquinone gases that can reach temperatures near 100 °C. The reaction is pulsed and can be aimed with remarkable precision through rapid modulation of the beetle's abdominal valves. This phenomenon has been documented through high-speed thermography and direct predator-interaction studies. See Thomas Eisner and Daniel J. Aneshansley, "Spray Aiming in the Bombardier Beetle: Photographic Evidence," *Proceedings of the National Academy of Sciences* 96 (1999): 9705–9709, pnas.org/doi/10.1073/pnas.96.17.9705; and Shinji Sugiura, "Anti-Predator Defences of a Bombardier Beetle: Is Bombing Essential for Successful Escape from Frogs?" *PeerJ* 6 (2018): e5942, doi.org/10.7717/peerj.5942.

[348] Mantis shrimp (*Stomatopoda*) deliver one of the fastest known strikes in the animal kingdom. High-speed videography and fluid dynamics analyses reveal that raptorial appendages accelerate at up to 23 m s$^{-1}$, generating cavitation bubbles that implode with shockwave force capable of stunning prey and fracturing aquarium glass. Their eyes are equally extraordinary. Species such as

*Odontodactylus scyllarus* and *Gonodactylus smithii* possess specialized midband photoreceptors that detect and analyze both linear and circularly polarized light—an ability unique among known organisms. Each eye can move independently, providing trinocular depth perception and polarization analysis within a single visual organ. These optical systems integrate signals across multiple photoreceptor types and outperform current engineered polarization sensors in dynamic range and efficiency. See S. N. Patek, "The Power of Mantis Shrimp Strikes: Interdisciplinary Impacts of an Extreme Cascade of Energy Release," *Integrative and Comparative Biology* 59 (2019): 1573–1585, academic.oup.com/icb/article/59/6/1573/5531653; Y. Chiou et al., "A Biological Graded Polarizer Provides a Basis for Animal Color Vision in the Ultraviolet," *Scientific Reports* 8 (2018): 6845, nature.com/articles/s41598-018-28004-w; and Y. Gagnon, J. F. Marshall, and N. W. Roberts, "Polarization Vision in Stomatopod Crustaceans: A Comparative Study of the Structure and Function of the Midband," *The Journal of Comparative Neurology* 527 (2019): 2479–2492, onlinelibrary.wiley.com/doi/10.1002/cne.24788.

[349] The phylum *Cnidaria* (corals, jellyfish, sea anemones, hydroids) encompasses roughly 10,000 described species and represents one of Earth's oldest animal lineages. Molecular-clock analyses place its divergence at ≥ 740 million years ago, with body-fossil evidence extending to ≈ approximately 600 Million Years Ago. Because most cnidarians possess soft, gelatinous tissues that decay rapidly, their fossil record is fragmentary and likely underestimates their true temporal range. See Ehsan Kayal et al., "Cnidarian phylogenetic relationships as revealed by mitogenomics," *BMC Evolutionary Biology* 13 (2013): 5, biomedcentral.com/1471-2148/13/5; Junhyong Park et al., "Precambrian origin and early diversification of Cnidaria inferred from molecular clocks," *Proceedings of the Royal Society B* 288 (2021): 20202939,

pmc.ncbi.nlm.nih.gov/articles/PMC7893222/; and Ehsan Kayal et al., "The Cnidarian Tree of Life: Insights from Comparative Genomics and Phylogenomics," *Genes* 12 (2021): 1072, mdpi.com/2073-4425/12/7/1072.

[350] Cnidocytes—also called cnidoblasts—are specialized ectodermal cells unique to the phylum *Cnidaria*. Each contains a Golgi-derived, pressurized capsule (the cnidocyst or nematocyst) that explosively everts a coiled tubule when triggered, injecting toxins to capture prey or defend itself. This stinging apparatus represents one of evolution's oldest and most refined cellular weapons, with origins tracing back over 600 million years. Modern imaging and biomechanical studies have shown that nematocyst discharge occurs in roughly *700 nanoseconds*, reaching accelerations of up to $5.4 \times 10^6\,g$, making it one of the fastest biological processes ever measured. See M. Peteya et al., "Integrative and Comparative Anatomy of Cnidocytes," *Integrative and Comparative Biology* 65 (2024): 661–684, academic.oup.com/icb/article/65/3/661/8161043; A. Karabulut et al., "The Architecture and Operating Mechanism of a Cnidarian Stinging Organelle," *Nature Communications* 13 (2022): 3494, nature.com/articles/s41467-022-31090-0; and T. Holstein and P. Tardent, "An Ultrahigh-Speed Analysis of Exocytosis: Nematocyst Discharge," *Science* 223 (1984): 830–833; with corroborating measurements in T. Nüchter et al., "Nanosecond-Scale Kinetics of Nematocyst Discharge," *Current Biology* 16 (2006): R316–R318.

[351] Human tolerance to acceleration varies sharply across contexts. Military and aerobatic pilots routinely endure sustained positive-Gz loads approaching *9 G*, beyond which loss of consciousness and visual "blackout" can occur as blood flow to the brain is compromised. In contrast, impact data from professional motorsports show that vehicle collisions producing *50–100 G* commonly result in significant traumatic brain injury. See *Federal Aviation Administration, Civil Aerospace Medical Institute*, "Acceleration in Aviation: G-Force" (AAM-400, 2021),

faa.gov/pilots/safety/pilotsafetybrochures/media/acceleration.pdf;
and Christopher S. Weaver et al., "An Analysis of Maximum Vehicle G
Forces and Brain Injury in Motorsports Crashes," *Medicine & Science in
Sports & Exercise* 38 (2006): 246–249, journals.lww.com/acsm-
msse/fulltext/2006/02000/an_analysis_of_maximum_vehicle_g_forc
es_and_brain.8.aspx.

[352] Nematocyst discharge operates without muscular input, powered
instead by a rapid interplay of *osmotic influx and hydrostatic pressure*
within the cnidocyst capsule. Ion dissociation—primarily of bound
$Ca^{2+}$ and protons—triggers a sudden rise in intracapsular pressure,
forcing the tubule to evert in a fraction of a microsecond. Modeling
and experimental work confirm that this process ranks among the
*fastest and most forceful cellular mechanisms known*, driven purely by
physical and chemical gradients. See Shawn C. Oppegard, Peter A.
Anderson, and David T. Eddington, "Puncture Mechanics of Cnidarian
Cnidocysts: A Natural Actuator," *Journal of Biological Engineering* 3
(2009): 17,
researchgate.net/publication/26852712_Puncture_mechanics_of_cnid
arian_cnidocysts_A_natural_actuator; and Jochen Weber et al.,
"Osmotic and Hydrodynamic Mechanisms of Nematocyst Discharge,"
*Fluids* 5 (2020): 20, mdpi.com/2311-5521/5/1/20.

[353] Cnidarians lack a centralized brain or ganglia, relying instead on a
diffuse nerve net—a decentralized mesh of neurons embedded within
the body wall. This network radiates just beneath the epidermis,
forming a full-body sensory and response system capable of
*bidirectional conduction* and *localized reflex loops.* Electrophysiological
and molecular studies confirm that cnidarian neurons transmit
signals bidirectionally, enabling sensation and action to occur
simultaneously across the organism. *In essence, perception and response
are integrated at every point along the body.* Comparative neurobiology
suggests that this architecture represents not a primitive deficit but
an alternative optimization—one that emphasizes redundancy,

flexibility, and energetic efficiency rather than hierarchical control. See C. J. P. Grimmelikhuijzen and J. A. Westfall, "The Nervous Systems of Cnidarians," in *The Nervous Systems of Invertebrates: An Evolutionary and Comparative Approach* (Birkhäuser, 1995), 7–22, link.springer.com/chapter/10.1007/978-3-0348-9219-3_2; Hiroshi Watanabe, Toshitaka Fujisawa, and Thomas W. Holstein, "Cnidarians and the Evolutionary Origin of the Nervous System," *Development, Growth & Differentiation* 51 (2009): 167–183, onlinelibrary.wiley.com/doi/10.1111/j.1440-169X.2009.01103.x; and Paul A. V. Anderson and George O. Mackie, "Do Jellyfish Have Central Nervous Systems?" *Journal of Experimental Biology* 214 (2011): 1215–1223, journals.biologists.com/jeb/article/214/8/1215/10743.

[354] Rhopalia are specialized sensory structures in cubozoan jellyfish that function as integrated components of the central nervous system. Each rhopalium contains multiple image-forming and non-image-forming eyes, as well as a statolith used for gravity detection. Neural integration within the rhopalial neuropil enables both *light and balance perception*, qualifying these organs as *primitive proto-ganglia* that coordinate visual input and locomotor control. See Anders Garm, Peter Ekström, Mathilde Boudes, and Dan-Eric Nilsson, "Rhopalia Are Integrated Parts of the Central Nervous System in Box Jellyfish," *Cell and Tissue Research* 325 (2006): 333–343, link.springer.com/article/10.1007/s00441-005-0134-8.

[355] Box jellyfish (Cubozoa) are distinguished by their cube-shaped medusae and active predatory behavior, pursuing prey rather than drifting passively. Members such as *Tripedalia cystophora* possess *up to twenty-four eyes*—including lens eyes with *retinas, corneas, and crystalline lenses*—organized within sensory complexes called rhopalia. *Despite lacking a central brain, these animals navigate obstacles, track movement, and detect prey through visually guided behaviors mediated by the rhopalial nervous system.* Recent neurophysiological work shows that *T. cystophora* performs operant associative learning, integrating visual

and mechanical stimuli within each rhopalium. See Christian Skogh, Anders Garm, Dan-Eric Nilsson, and Peter Ekström, "Bilaterally Symmetrical Rhopalial Nervous System of the Box Jellyfish *Tripedalia cystophora*," *Journal of Morphology* 267 (2006): 1391–1405, onlinelibrary.wiley.com/doi/10.1002/jmor.10472; and Jan Bielecki et al., "Associative Learning in the Box Jellyfish *Tripedalia cystophora*," *Current Biology* 33 (2023): 4150–4159, cell.com/current-biology/fulltext/S0960-9822(23)01136-3.

[356] Cnidarians retain deeply conserved genomic toolkits despite their ancient evolutionary lineage. Their genomes contain HOX genes and WNT signaling pathways, which govern body axis formation and tissue organization—mechanisms also essential in bilaterians. Genomic and transcriptomic studies reveal that cnidarians *harbor homologues of genes associated with neural development*, such as *Sox* and *BMP* family members, despite lacking a centralized brain. These findings underscore that *the molecular framework for complex body patterning and neurogenesis predates the emergence of centralized nervous systems*, tracing back to the common ancestor of all eumetazoans. See Chuya Shinzato et al., "Cnidarian Genome Evolution and the Origin of Animal Complexity," *Communications Biology* 2 (2019): 476, pmc.ncbi.nlm.nih.gov/articles/PMC6879750/; and Iva Kelava, Fabian Rentzsch, and Ulrich Technau, "Evolution of Eumetazoan Nervous Systems: Insights from Cnidarians," *Philosophical Transactions of the Royal Society B* 370 (2015): 20150065, pmc.ncbi.nlm.nih.gov/articles/PMC4650132/.

[357] The genome of the freshwater cnidarian *Hydra magnipapillata* was sequenced and found to contain approximately twenty thousand protein-coding genes—roughly equivalent to the number in humans. This landmark study revealed extensive transposable element activity and conservation of eumetazoan signaling pathways despite morphological simplicity. See Jarrod A. Chapman et al., "The Dynamic

Genome of Hydra," *Nature* 464 (2010): 592–596,
pmc.ncbi.nlm.nih.gov/articles/PMC4479502/.

[358] Many cnidarians are capable of stem-cell-powered regeneration
driven by interstitial stem cells and somatic cell reprogramming. In
*Hydractinia symbiolongicarpus*, injury triggers a cascade in which
senescent cells induce nearby tissues to generate new pluripotent
stem cells, allowing complete body reconstruction from fragments.
See Miguel Salinas-Saavedra et al., "Senescence-Induced Cellular
Reprogramming Drives Cnidarian Whole-Body Regeneration," *Cell
Reports* 42 (2023): 112687, cell.com/cell-reports/fulltext/S2211-
1247(23)00698-8.

[359] *Turritopsis dohrnii* exhibits a unique biological process known as
reverse development, in which mature medusae can revert to the
juvenile polyp stage. This transformation is triggered by stress, injury,
or senescence and involves cell transdifferentiation, regeneration, and
activation of telomere maintenance and DNA repair pathways. The
process effectively resets the organism's life cycle, earning it the
popular moniker "immortal jellyfish." See S. Piraino et al., "Reverse
Development in Cnidaria," *Canadian Journal of Zoology* 82 (2004): 1748–
1754, vliz.be/imisdocs/publications/ocrd/71191.pdf; and Yui
Matsumoto et al., "Transcriptome Characterization of Reverse
Development in *Turritopsis dohrnii* (Hydrozoa, Cnidaria)," *G3: Genes,
Genomes, Genetics* 9 (2019): 4127–4138,
pmc.ncbi.nlm.nih.gov/articles/PMC6893190/.

[360] Cnidarian genomes exhibit epigenetic flexibility driven by
mechanisms such as DNA methylation, histone modification, and
chromatin remodeling, allowing dynamic gene regulation in response
to environmental and symbiotic cues. Studies on the symbiotic sea
anemone Exaiptasia diaphana reveal that activating histone marks
(H3K4me3, H3K27ac, H3K9ac, and H3K36me3) act cooperatively with
DNA methylation to promote the expression of symbiosis-associated

genes, while repressive marks, such as H3K27me3, silence others, demonstrating a finely tuned crosstalk between epigenetic systems. See Kashif Nawaz et al., "Histone Modifications and DNA Methylation Act Cooperatively in Regulating Symbiosis Genes in the Sea Anemone *Aiptasia*," *BMC Biology* 20 (2022): 265, pmc.ncbi.nlm.nih.gov/articles/PMC9717517/; and Groves Dixon et al., "Coral DNA Methylation and Transcriptional Responses to Environmental Stress," *PLoS Genetics* 16 (2019): e1008397, journals.plos.org/plosgenetics/article?id=10.1371/journal.pgen.100839 7.

[361] Cnidarians such as *Hydra* and *Nematostella* exhibit evidence of horizontal gene transfer (HGT) from bacterial and unicellular donors, thereby acquiring novel metabolic and defensive traits. Genomic surveys have identified dozens of bacterial-derived genes in *Hydra magnipapillata*, including sugar-modifying enzymes and aerolysin-like toxins, which are likely involved in immunity and predation. Functional integration of these transferred genes has been demonstrated through their incorporation into operons and expression within cnidarian tissues, indicating adaptive co-option into host biology. See Luis Boto, "Horizontal Gene Transfer in the Acquisition of Novel Traits by Metazoans," *Proceedings of the Royal Society B* 281 (2014): 20132450, royalsocietypublishing.org/doi/epdf/10.1098/rspb.2013.2450; and Catherine E. Dana et al., "Incorporation of a Horizontally Transferred Gene into an Operon during Cnidarian Evolution," *PLoS ONE* 7 (2012): e31643, journals.plos.org/plosone/article?id=10.1371/journal.pone.0031643.

[362] Cephalopods constitute a distinct *class* within the phylum *Mollusca*, subphylum *Conchifera*, encompassing nautiluses, squids, cuttlefish, and octopuses. Modern taxonomic records recognize roughly 800 extant species and more than 11,000 extinct taxa—including ammonites and belemnites—across the group's 500-million-year

evolutionary history. See Vicente E. A. S. Lopes and Teodoro Vaske Júnior, "Creation and Maintenance Processes of Malacological Collections of Cephalopod Beaks," *Brazilian Journal of Aquatic Science and Technology* 26 (2022): 77–82, periodicos.univali.br/index.php/bjast/article/view/17226; and Fernanda C. N. Vasconcelos et al., "Cephalopods: Fascinating Mollusks of the Sea," *Frontiers in Communication* 3 (2018): 20, frontiersin.org/journals/communication/articles/10.3389/fcomm.2018.00020/full.

[363] Cephalopods demonstrate advanced cognitive and behavioral complexity supported by robust neuroanatomical architecture and learning capacity. Experimental studies confirm that octopuses, cuttlefish, and squid exhibit long- and short-term memory, operant learning, spatial mapping, and problem-solving abilities rivaling those of some vertebrates. Individual recognition of humans has been observed in captive octopuses, along with goal-directed behavior and the flexible use of learned strategies. See Alex K. Schnell and Nicola S. Clayton, "How Intelligent Is a Cephalopod? Lessons from Comparative Cognition," *Biological Reviews* 96 (2021): 162–175, onlinelibrary.wiley.com/doi/10.1111/brv.12651; Alexandra K. Schnell and Nathan J. Emery, "Cephalopod Cognition," *Trends in Ecology & Evolution* 34 (2019): 812–825, cell.com/trends/ecology-evolution/fulltext/S0169-5347(18)30267-2; and Jennifer A. Mather, "Octopus Consciousness: The Role of Perceptual Richness," *NeuroSci* 2 (2021): 276–290, mdpi.com/2673-4087/2/3/20.

[364] Cephalopods exhibit rapid adaptive camouflage through the coordinated control of chromatophores, iridophores, and muscular papillae that allow instant changes in skin color, texture, and pattern. These neurally controlled organs enable species such as *Octopus vulgaris* and *Sepia officinalis* to achieve near-perfect background matching within milliseconds, effectively evading predators and ambushing prey. See Roger T. Hanlon, "Cephalopod Dynamic

Camouflage," *Current Biology* 17 (2007): R400–R404, cell.com/current-biology/fulltext/S0960-9822(07)01138-4.

[365] The nervous system of *Octopus vulgaris* contains roughly 500 million neurons in total—about six times that of a mouse. Of these, approximately 200 million are concentrated in the paired optic lobes and central brain (the supra- and subesophageal masses). At the same time, an estimated 300 million are distributed through the eight arms, primarily within the axial nerve cords and associated ganglia. This organization underscores the distributed, semi-autonomous control characteristic of cephalopod motor systems, with the arms collectively housing more neurons than the central brain itself. See Rachel M. Albertin and Paul S. Katz, "Cephalopod Genomes: Insights into the Molecular Basis of Cephalopod Brain Evolution," *Current Biology* (2024), pmc.ncbi.nlm.nih.gov/articles/PMC10792511/; Tomoyuki Shiratori et al., "Neuronal Organization and Arm Coordination in Cephalopods," *Frontiers in Neuroanatomy* 14 (2021): 565109, pmc.ncbi.nlm.nih.gov/articles/PMC7884766/; and Cassady S. Olson et al., "Neuronal Segmentation in Cephalopod Arms," *bioRxiv* (2024), pmc.ncbi.nlm.nih.gov/articles/PMC11160704/.

[366] Cephalopods exhibit exceptional visual specialization, including the ability to detect polarized light and incredibly high contrast levels, though they are essentially monochromatic. Studies confirm that coleoid cephalopods possess photoreceptor microvilli aligned to polarization planes, granting polarization sensitivity and contrast enhancement capabilities unparalleled among marine invertebrates. Despite lacking multiple photoreceptor types for color vision, they achieve fine control over the chromatophore, iridophore, and leucophore layers to dynamically modulate skin coloration and texture. See Dan-E. Nilsson, Sönke Johnsen, and Eric Warrant, "Cephalopod versus Vertebrate Eyes," *Current Biology* 33 (2023): R1067–R1105, pmc.ncbi.nlm.nih.gov/articles/PIIS0960982223009880/; and Paul S. Katz and Deirdre C. Lyons, "Cephalopod Vision: How to Build a

Better Eye," *Current Biology* 33 (2023): R16–R39, pmc.ncbi.nlm.nih.gov/articles/PIIS0960982222018450/.

[367] Cephalopods exhibit direct neural connectivity between visual and dermal systems, enabling near-instantaneous adaptive coloration without centralized processing. The optic lobes—comprising roughly three-quarters of the total brain volume—receive retinal input and project to chromatophore lobes, where motor neurons directly innervate chromatophore muscles to produce real-time environmental matching. This distributed architecture integrates sensory and motor control across the arms and skin, with up to two-thirds of the neurons residing in the peripheral nervous system. See Tessa G. Montague, *"Neural Control of Cephalopod Camouflage,"* Current *Biology* 33 (2023): R1067–R1105, cell.com/current-biology/fulltext/S0960-9822(23)01182-X; and Binyamin Hochner, *"An Embodied View of Octopus Neurobiology,"* Current Biology 22 (2012): R887–R892, cell.com/current-biology/fulltext/S0960-9822(12)01064-0.

[368] The complex coloration system of coleoid cephalopods involves multilayered optical and neural integration across three principal skin cell types—chromatophores, iridophores, and leucophores—each contributing distinct spectral and structural effects. Chromatophores are pigment sacs surrounded by radial muscle fibers, neurally activated to rapidly expand and contract, functioning as dynamic biological pixels under both central and peripheral control. Iridophores generate iridescent hues through multilayer reflectin platelets that create interference-based structural coloration, while leucophores act as broadband diffuse reflectors that adaptively mirror environmental light. This cooperative architecture, coupled with local neural ganglia capable of light-sensitive reflex responses via dermal opsins, allows cephalopod skin to perform distributed photoreception and pattern generation independent of the central brain. See Lydia M. Mäthger and Roger T. Hanlon, "Mechanisms and Behavioural Functions of Structural Coloration in Cephalopods," *Journal of the*

*Royal Society Interface* 5 (2008): S149–S163,
royalsocietypublishing.org/doi/10.1098/rsif.2008.0366.focus; M.
Desmond Ramirez and Todd H. Oakley, "Eye-Independent, Light-
Activated Chromatophore Expansion and Expression of
Phototransduction Genes in the Skin of *Octopus bimaculoides*," *Journal
of Experimental Biology* 218 (2015): 1513–1520,
pmc.ncbi.nlm.nih.gov/articles/PMC4448664/; Richard L. Sutherland et
al., "Cephalopod Coloration Model II: Multiple Layer Skin Effects,"
*Journal of the Optical Society of America A* 25 (2008): 2044–2055,
opg.optica.org/josaa/abstract.cfm?uri=josaa-25-8-2044; and
"Cephalopod Camouflage: Cells and Organs of the Skin," *Nature
Education Knowledge Project* (Nature Publishing Group, 2010),
nature.com/scitable/topicpage/cephalopod-camouflage-cells-and-
organs-of-the-144048968/.

[369] Cephalopod papillae are complex, muscle-controlled skin
projections that enable rapid, three-dimensional modulation of
texture for camouflage. Comparative morphological studies have
shown that papillae in octopuses and cuttlefish operate as muscular
hydrostats—structures extended and shaped by circular and
horizontal erector muscles, and retracted via opposing muscle
bundles. These papillae can expand or retract voluntarily, mimicking
natural textures such as coral or rock through independent neural
control of each unit. Their mechanism parallels mammalian arrector
pili reflexes but is actively controlled, allowing for instantaneous
texture changes integrated with chromatophore-based coloration.
See Justine J. Allen et al., "Comparative Morphology of Changeable
Skin Papillae in *Octopus* and *Cuttlefish*," *Journal of Morphology* 275
(2014): 371–390, onlinelibrary.wiley.com/doi/10.1002/jmor.20221; and
David G. O'Brien et al., "The Neural Basis of Dynamic Skin Patterning
in Cephalopods," *Current Biology* 33, no. 15 (2023): 3238–3247,
cell.com/current-biology/fulltext/S0960-9822(23)01182-X.

[370] Cephalopods maintain circulation through *three* distinct hearts—one systemic and two branchial—that coordinate oxygen transport via a closed, high-pressure system. The systemic heart distributes oxygenated blood throughout the body, while the paired branchial hearts pump deoxygenated blood through the gills for reoxygenation. Their blood is blue because it contains *hemocyanin*, a copper-based oxygen carrier rather than iron-based hemoglobin. This protein is highly effective in cold, low-oxygen environments. It remains stable under varying thermal and pH conditions, as shown by positive selection for temperature-adaptive modifications in octopod hemocyanin. However, their copper dependency also makes cephalopods vulnerable to environmental copper exposure, which can disrupt oxygen transport. See R. T. Hanlon and J. B. Messenger, "Cephalopod Circulatory and Respiratory Physiology," *Biological Bulletin* 227 (2014): 263–276, sciencedirect.com/science/article/pii/S105046481530005X; and Michael Oellermann et al., "Positive Selection in Octopus Haemocyanin Indicates Functional Links to Temperature Adaptation," *BMC Evolutionary Biology* 15 (2015): 133, pmc.ncbi.nlm.nih.gov/articles/PMC4491423/.

[371] Light is a form of *electromagnetic radiation*—a self-propagating wave produced by oscillating electric and magnetic fields oriented at right angles to each other. Its dual nature, expressed through *wave–particle duality*, is described by Maxwell's equations and confirmed by quantum mechanics. The position of any radiation within the electromagnetic spectrum is determined by its frequency and wavelength, ranging from long radio waves through microwaves, infrared, visible light, ultraviolet, X-rays, and gamma rays. The visible band humans perceive spans roughly 400–700 nanometers, representing only a small fraction of the universe's electromagnetic output. See Glenn Stark et al., "Light," *Encyclopaedia Britannica* (2025), britannica.com/science/light; Department of Physics, University of Maryland, "Properties of Light,"

science.umd.edu/faculty/wilkinson/bsci338/ACL7_light.pdf; and Oxford Instruments, "What Is Light: An Overview of the Properties of Light," andor.oxinst.com/learning/view/article/what-is-light.

[372] Polarization occurs when the electric field vectors of light waves are oriented along a single plane rather than distributed randomly in multiple directions, as in unpolarized light. This can result from reflection, scattering, or transmission through anisotropic materials, producing linearly (plane) polarized light in which oscillations are restricted to one orientation. See Thomas W. Cronin and Justin Marshall, *Patterns and Properties of Polarized Light in Air and Water*, *Philosophical Transactions of the Royal Society B* 366 (2011): 619–626, pmc.ncbi.nlm.nih.gov/articles/PMC3049010/.

[373] Cephalopods such as squids, octopuses, and cuttlefishes not only detect the angle of polarized light but can actively control the polarization of light reflected from their skin through specialized structures known as iridophores. This capability may serve both camouflage and intraspecies communication purposes. Behavioral and neurobiological research confirms that cephalopods exhibit polarization sensitivity, despite being essentially colorblind to wavelength, and rely instead on polarization contrast for visual discrimination. See Nadav Shashar and Thomas W. Cronin, "Do Cephalopods Communicate Using Polarized Light?" *Journal of Experimental Biology* 212 (2009): 2133–2140; and Judit R. Pungor and Cristopher M. Niell, "The Neural Basis of Visual Processing and Behavior in Cephalopods," *Current Biology* 33 (2023): R1106–R1118, doi:10.1016/j.cub.2023.08.093.

[374] Human color vision depends on specialized retinal photoreceptors called *rods* and *cones*, which convert incoming light into electrical signals transmitted to the brain. Rods detect light intensity and contrast, enabling low-light (scotopic) vision. In contrast, cones detect color and fine detail under bright conditions through three

wavelength-sensitive subtypes—short (blue), medium (green), and long (red). Both rods and cones contain the protein *opsin*, which initiates the phototransduction cascade that converts photons into neural impulses for visual perception. See Cleveland Clinic, "Photoreceptors (Rods and Cones): Anatomy & Function," my.clevelandclinic.org/health/body/photoreceptors-rods-and-cones; and Robert G. Weiner and T. Lamb, "Photoreceptors at a Glance," *Journal of Cell Science* 128 (2015): 4039–4045, pmc.ncbi.nlm.nih.gov/articles/PMC4712787/.

[375] Cephalopods perceive and manipulate polarized light through specialized retinal and dermal adaptations. Rhabdomeric photoreceptors in their retinas contain orthogonally arranged microvilli, enabling detection of specific polarization angles rather than color hue. This "orthogonal vision" allows perception of contrast, alignment, and surface properties invisible to humans. Simultaneously, iridophores in their skin produce and reflect linearly polarized light that can pass through pigmented chromatophores, creating dynamic, potentially concealed communication channels between individuals. Behavioral and anatomical evidence suggests that species such as Sepia officinalis and Loligo pealeii utilize polarization for both enhanced object detection and intraspecific signaling, which is undetectable to predators. See Nadav Shashar, Phillip S. Rutledge, and Thomas W. Cronin, "Polarization Vision in Cuttlefish—A Concealed Communication Channel?" *Journal of Experimental Biology* 199 (1996): 2077–2084; Lydia M. Mäthger and Roger T. Hanlon, "Anatomical Basis for Camouflaged Polarized Light Communication in Squid," *Biology Letters* 2 (2006): 494–496; and Nadav Shashar et al., "Polarization Vision in Octopus," *Current Biology* 20 (2010): 1–4.

[376] The expression *"Fermi Paradox"* is historically inaccurate. Enrico Fermi never published on the subject of extraterrestrial life and is known only to have asked *"Where is everybody?"* during a 1950 lunch at

Los Alamos, questioning the feasibility of interstellar travel rather than denying the possibility of intelligent life. The phrase itself first appeared in print in 1977 in connection with Michael Hart's 1975 argument, "they are not here; therefore they do not exist," later expanded by Frank Tipler. As Robert H. Gray demonstrates, the so-called paradox misattributes later reasoning to Fermi and is not a genuine paradox at all. See Robert H. Gray, *The Fermi Paradox Is Neither Fermi's Nor a Paradox*, *Astrobiology* 15 (2015): 195–199, arxiv.org/pdf/1605.09187.

## CHAPTER TEN

[377] Galileo Galilei's telescopic confirmation of heliocentrism in *Sidereus Nuncius* (1610) displaced Earth from the cosmic center and inaugurated the modern scientific method; Ignaz Semmelweis's mid-19th-century insistence on antiseptic handwashing prefigured germ theory and revolutionized medical hygiene; Alfred Wegener's 1912 theory of continental drift proposed that continents move across the planet's surface, anticipating plate tectonics; and Werner Heisenberg's 1925 formulation of matrix mechanics introduced the uncertainty principle and transformed physics at the quantum scale. Each exemplifies the intellectual upheaval invoked here—the courage to confront entrenched paradigms with empirical evidence. See S. L. Robbins, "Ignaz Semmelweis and the Birth of Infection Control," *Proceedings (Baylor University Medical Center)* 21, no. 4 (2008): 379–382, pmc.ncbi.nlm.nih.gov/articles/PMC2564400; H. R. Matthews, "Ignaz Semmelweis—The Handwasher's Legacy," *Cureus* 16 (2024): e68350, pmc.ncbi.nlm.nih.gov/articles/PMC11442886; "January 6, 1912: Alfred Wegener Presents His Theory of Continental Drift," *APS News* (Jan 2019), aps.org/apsnews/2019/01/alfred-wegener-theory-continental-drift; and "June/July 1925: Werner Heisenberg Pioneers Quantum Mechanics," *APS News* (July 2025), aps.org/apsnews/2025/07/werner-heisenberg-pioneers-quantum-mechanics.

[378] The role of oxygen as both the metabolic cornerstone and evolutionary prerequisite for complex multicellular life has long been regarded as axiomatic. Aerobic respiration, which relies on oxygen as the terminal electron acceptor, yields energy orders of magnitude higher than anaerobic pathways and is considered integral to the evolution of multicellularity and biological complexity. As Victor J. Thannickal notes, "multiple lines of evidence from evolutionary biology, geochemistry, and systems biology build a compelling case for a central role of $O_2$ in the evolution of complex multicellular life on Earth," linking atmospheric oxygenation events with the rise of metazoans. See Victor J. Thannickal, "Oxygen in the Evolution of Complex Life and the Price We Pay," *American Journal of Respiratory Cell and Molecular Biology* 40 (2009): 507–510, pmc.ncbi.nlm.nih.gov/articles/PMC2720141.

[379] Henneguya *salminicola* represents the first known multicellular organism to have entirely lost its mitochondrial genome and capacity for aerobic respiration. Genomic analysis confirms that the parasite lacks all genes associated with oxidative phosphorylation, indicating a complete transition to an anaerobic metabolism. See Dorothée Huchon et al., "A cnidarian parasite of salmon (*Myxozoa: Henneguya*) lacks a mitochondrial genome," *Proceedings of the National Academy of Sciences* 117 (2020): 5358–5363, pnas.org/doi/10.1073/pnas.1909907117.

[380] As Neme and Yanai observe, evolution frequently advances by reduction and simplification rather than through additive complexity. Such *regressive evolution* reflects the pragmatic nature of natural selection, which favors efficiency, adaptability, and function over form. See Rafael Neme and Itai Yanai, "Regressive evolution as an engine of biological innovation," *Biomimetics* 8 (2023): 362, pmc.ncbi.nlm.nih.gov/articles/PMC10452652.

[381] For detailed documentation of the Aguadilla (2013), USS *Omaha* (2019), and Nimitz (2004) encounters—each demonstrating

transmedium or non-ballistic flight characteristics—see prior discussion and sources in Chapters Five through Seven. These include the Scientific Coalition for UAP Studies' *Aguadilla Puerto Rico UAP Report: A Detailed Analysis* (2015), explorescu.org/post/2013-aguadilla-puerto-rico-uap-incident-report-a-detailed-analysis; and related analyses of anomalous motion and sensor data in Joseph, R. et al., *Unifying Scalar Fields and Field Repulsion: A New Approach to Understanding Unidentified Anomalous Phenomena and Trans-Medium Travel* (2024), researchgate.net/publication/389503379_Unifying_Scalar_Fields_and _Field_Repulsion_A_New_Approach_to_Understanding_Unidentifie d_Anomalous_Phenomena_and_Trans-Medium_Travel; and Joseph, R. et al., *Unidentified Anomalous Phenomena, Extraterrestrial Life, Plasmoids, Shape Shifters, Replicons, Thunderstorms, Lightning, Hallucinations, Aircraft Disasters, Ocean Sightings* (2024), researchgate.net/publication/383034675_Unidentified_Anomalous_P henomena_Extraterrestrial_Life_Plasmoids_Shape_Shifters_Replico ns_Thunderstorms_Lightning_Hallucinations_Aircraft_Disasters_Oc ean_Sightings.

[382] Cephalopods employ a wide range of complex avoidance and concealment strategies, including dynamic skin patterning, ink release, spatial learning, and context-dependent evasion behaviors. These responses demonstrate sophisticated neural control and adaptive decision-making in response to predation and environmental stimuli. See Graziano Fiorito et al., *Cephalopod Behavior: From Neural Control to Social Complexity, Frontiers in Marine Science* 9 (2022): 909192, frontiersin.org/journals/marine-science/articles/10.3389/fmars.2022.909192/full.

[383] Evidence for the colossal squid (*Mesonychoteuthis hamiltoni*) derives primarily from recovered beaks found in sperm whale stomachs, indicating predator–prey interactions and revealing its remarkable scale. Adults are estimated to reach lengths up to 46 feet (14 m),

including tentacles and masses approaching 1,000 pounds (≈ 450 kg). Human encounters have been limited to captured or deceased specimens; no confirmed in situ observation of a living adult existed until early 2025. See Stephen Dowling, *"Colossal Squid: The Eerie Ambassador from the Abyss,"* BBC Future (Jan 31, 2025), bbc.com/future/article/20250130-colossal-squid-the-eerie-ambassador-from-the-abyss; and Oceana, *"Colossal Squid (Mesonychoteuthis hamiltoni),"* oceana.org/marine-life/colossal-squid.

[384] The Antarctic squid *Gonatus antarcticus* was documented alive for the first time in December 2024, when researchers aboard the Schmidt Ocean Institute's *R.V. Falkor (too)* captured video of the species at a depth of about 7,000 feet in the Weddell Sea. Prior to this observation, *G. antarcticus* was known only from carcasses recovered in fishing nets and from beaks found in the stomachs of predatory marine animals. See Melissa Hobson, *"We've Never Seen This Rare Squid Alive in the Wild—Until Now,"* National Geographic (10 June 2025), nationalgeographic.com/animals/article/antarctic-squid-filmed-alive-first-video.

[385] The first complete sequencing of an octopus genome (*Octopus bimaculoides*) revealed extraordinary molecular and structural complexity, including massive genome rearrangements, an unprecedented expansion of protocadherin genes (critical for neural connectivity), and elevated transposon activity. These findings demonstrate that cephalopods possess unique genetic architectures not shared by other animals, thereby reshaping our understanding of invertebrate evolution and neurobiology. See Caroline B. Albertin et al., *"The Octopus Genome and the Evolution of Cephalopod Neural and Morphological Novelties,"* Nature 524 (2015): 220–224, nature.com/articles/nature14668.

[386] Cephalopods possess unparalleled control over their appearance, rapidly manipulating color, shape, and texture to evade predators or

deceive prey. Using coordinated networks of chromatophores, papillae, and iridophores, octopuses and cuttlefish can transform their bodies from smooth to rugged, mimic algae or rocks, and perform dynamic threat displays designed to startle or confuse. These behaviors—ranging from masquerade and mimicry to deliberate postural deception—are among the most complex adaptive camouflage strategies in the animal kingdom. See Drew Harvell, *"Shape-Shifting Cephalopods," American Scientist* 104, no. 5 (2016): 312–319, americanscientist.org/article/shape-shifting-cephalopods; and Sofia Quaglia, *"These Are the Weird and Wonderful Reasons Octopuses Change Shape and Color," National Geographic* (16 May 2024), nationalgeographic.com/animals/article/octopuses-squid-cuttlefish-cephalopod-camouflage-color-shape-changing.

[387] The Portuguese man o' war (*Physalia physalis*) is a colonial siphonophore composed of multiple highly specialized zooids that cannot survive independently yet function together as a single integrated organism. Some siphonophores also produce detachable dispersive units known as eudoxids, which separate from the parent colony and drift as individual-like stages, illustrating a modular system built of specialized, semi-autonomous parts capable of partial independence and self-renewal within the life cycle. See *Portuguese Man-of-War, National Geographic*, nationalgeographic.com/animals/invertebrates/facts/portuguese-man-of-war; and Maciej K. Mańko, Catriona Munro, and Lucas Leclère, "The Evolution of an Individual-Like Dispersive Stage in Colonial Siphonophores," *Current Biology* 35 (2025): 4946–4958, sciencedirect.com/science/article/pii/S0960982225011686.

[388] Numerous species across the animal kingdom exhibit regenerative capabilities, though the degree varies greatly by phylum. Among the most remarkable are echinoderms such as sea stars, whose members can regenerate entire body structures—including arms, nerve cords, and digestive tissues—from partial fragments. Experimental studies

on *Patiria miniata* have demonstrated complete whole-body regeneration following bisection, with the reformation of all lost organs and tissues over approximately two weeks. These findings align with broader metazoan patterns showing that echinoderms, cnidarians, platyhelminths, and annelids possess deeply conserved molecular mechanisms for regeneration. See Gregory A. Cary et al., "Analysis of Sea Star Larval Regeneration Reveals Conserved Processes of Whole-Body Regeneration Across the Metazoa," *BMC Biology* 17 (2019): 1–19, bmcbiol.biomedcentral.com/articles/10.1186/s12915-019-0633-9.

[389] Hydra has long served as a model organism for studying regeneration because of its exceptional ability to reform an entire organism from small tissue fragments or even dissociated cells. Classical and modern studies confirm that *Hydra* can regenerate complete, functional individuals within days, provided that a portion of the head region—containing as few as 5–15 epithelial cells with organizer potential—remains intact to guide patterning and morphogenesis. The head organizer functions as a nexus for reestablishing polarity and tissue differentiation, directing other cells to self-organize into a coherent structure. See Toshitaka Fujisawa, "Hydra Regeneration and Epitheliopeptides," *Developmental Dynamics* 226 (2003): 182–189, anatomypubs.onlinelibrary.wiley.com/doi/10.1002/dvdy.10221; and Brigitte Galliot and Eva-Maria Schmid, "Cellular, Molecular and Evolutionary Mechanisms of Hydra Regeneration," *Genes* 11, no. 6 (2020): 701, pmc.ncbi.nlm.nih.gov/articles/PMC7116057/.

[390] Structural anomalies in cephalopod appendages have been documented, including the first recorded case of double tentacle bifurcation in the squid *Moroteuthis ingens* and additional bifurcation and regenerative malformations in octopuses. These findings confirm that limb-division mutations, while rare, occur within cephalopod lineages and reflect the group's extraordinary morphological

plasticity. See T. J. Kubodera et al., "Case of an Octopus (*Callistoctopus ornatus*) with a Malformed Arm," *Animals* 15, no. 10 (2025): 1034, pmc.ncbi.nlm.nih.gov/articles/PMC11987900/; and Ángel F. González and Ángel Guerra, "First Observation of Double Tentacle Bifurcation in Cephalopods," *Marine Biodiversity Records* (2008): 1–6, researchgate.net/publication/231792034_First_observation_of_doubl e_tentacle_bifurcation_in_cephalopods.

[391] Extensive RNA editing in coleoid cephalopods enables rapid and precise alterations to protein synthesis without changes to the underlying DNA sequence. This A-to-I (adenosine-to-inosine) editing, catalyzed by ADAR enzymes, occurs at tens of thousands of sites in octopus and squid nervous systems, far exceeding the levels seen in most animals. These edits provide a mechanism for adaptive flexibility and environmental responsiveness, allowing cephalopods to fine-tune protein function dynamically. See J. J. C. Rosenthal et al., "Extensive Recoding of the Neural Proteome in Cephalopods by RNA Editing," *Annual Review of Animal Biosciences* 11 (2023): 11–32, pmc.ncbi.nlm.nih.gov/articles/PMC10467349/; and Tina Hesman Saey, "Octopuses and Squid Are Masters of RNA Editing While Leaving DNA Intact," *Science News* (May 6, 2023), sciencenews.org/article/octopus-squid-rna-editing-dna-cephalopods.

[392] Experimental and theoretical evidence support the involvement of quantum processes in several biological phenomena. Quantum coherence has been observed in photosynthetic exciton transfer, where pigment–protein complexes sustain wave-like energy propagation across bacteriochlorophyll molecules (Engel et al., *Nature* 446, 782–786, 2007). Spin-dependent magnetoreception in birds is likely to involve the quantum entanglement of radical pairs sensitive to weak magnetic fields (Cintolesi et al., *Chem. Phys.* 294, 385–399, 2003). Enzymatic catalysis has shown evidence of quantum tunneling in proton and electron transfer, while coherence and tunneling effects are proposed in olfaction and other sensory mechanisms. Together,

these findings demonstrate that quantum mechanics plays a measurable role in biological function across diverse systems. See Jennifer C. Brookes, "Quantum Effects in Biology: Golden Rule in Enzymes, Olfaction, Photosynthesis and Magnetodetection," *Proceedings of the Royal Society A* 473 (2017): 20160822, royalsocietypublishing.org/doi/10.1098/rspa.2016.0822; and Thorsten Ritz et al., "Quantum Effects in Biology: Bird Navigation," in *Quantum Effects in Biology*, eds. Mohseni et al. (Cambridge University Press, 2014), pp. 262–275.

[393] Optical projection tomography (OPT) is a three-dimensional mesoscopic imaging technique that generates volumetric reconstructions of translucent biological samples using scattered or transmitted light. It is conceptually analogous to X-ray computed tomography but operates with visible illumination, capturing multiple projection images from different angles and reconstructing the 3D structure via inverse Radon transformation. This method is widely used in developmental biology and biomedical imaging to visualize fixed and live samples in the millimeter to centimeter range. See Connor Darling et al., *"Optical Projection Tomography Implemented for Accessibility and Low Cost (OPTImAL),"* *Philosophical Transactions of the Royal Society A* 382 (2024): 20230101, royalsocietypublishing.org/doi/10.1098/rsta.2023.0101.

[394] Numerous studies confirm that cnidarians possess specialized mechanosensory and electroresponsive cells that regulate movement, orientation, and environmental awareness. These include hair-cell–like mechanoreceptors and cnidocytes, which are capable of generating electrical signals upon stimulation, thereby coordinating swimming and righting behaviors. Such systems demonstrate sensitivity to both mechanical and electrical cues, which cnidarians use to navigate flow conditions and orient within their surroundings. See Hannah Ozment et al., *"The cnidarian hair cell and the origin of*

*animal mechanosensation," eLife* 10 (2021): e74336,
pmc.ncbi.nlm.nih.gov/articles/PMC8846589.

[395] Documented reports of "missing time" and temporal disorientation
have appeared repeatedly in government and technical analyses of
UAP encounters. The 1961 Betty and Barney Hill case, detailed in a
Frontier Analysis Technical Service Report (UT025, 2003), notes a
two-hour lapse in recollection following the observation of an aerial
object and subsequent physiological effects. Similar temporal
anomalies were referenced in *UFO Encounter II*, a case compiled by the
UFO Subcommittee of the American Institute of Aeronautics and
Astronautics and preserved in CIA archives. See P. A. Budinger,
*Analysis of the Dress Worn by Betty Hill During the September 19, 1961
Abduction in New Hampshire* (Frontier Analysis Ltd., 2003),
documents.theblackvault.com/budinger/UT025.pdf; and *UFO
Encounter II, Sample Case Selected by the UFO Subcommittee of the AIAA*
(CIA, declassified), cia.gov/readingroom/docs/CIA-
RDP81R00560R000100010010-0.pdf.

[396] Cnidarian nematocysts contain one of the most chemically diverse
and potent arrays of toxins in the natural world, including
neurotoxins, pore-forming cytolysins, and enzymatic peptides
capable of disrupting ion channels and neuromuscular function.
These compounds are typically delivered by direct contact injection
but can also diffuse through surrounding water under specific
conditions, resulting in localized paralysis or sensory disruption.
Deep-sea variants exhibit particularly complex venom architectures,
shaped by the extreme environmental pressures they face. See Jason
Macrander and Marymegan Daly, *"Evolution of Venom Complexity in the
Cnidaria: The Toxinological and Ecological Underpinnings of Chemical
Diversity," Toxins* 7, no. 6 (2015): 2251–2271,
pmc.ncbi.nlm.nih.gov/articles/PMC4488701; and David J. Moran et al.,
*"Convergent evolution of complex venom arsenals in Cnidaria and Bilateria,"*

*Molecular Biology and Evolution* 32, no. 3 (2015): 740–753,
academic.oup.com/mbe/article/32/3/740/979655.

[397] *Irukandji syndrome* is a severe systemic reaction caused primarily by
the sting of the cubozoan jellyfish *Carukia barnesi* and related species.
Victims experience intense pain, hypertension, and the characteristic
sense of "impending doom," which is diagnostic of the condition. The
venom acts as a neural sodium-channel activator, causing massive
catecholamine release and *heightened neuronal excitability that can
progress to seizures, cardiac dysfunction, and pulmonary edema.* Broader
cnidarian venoms similarly disrupt neuronal communication through
ion-channel modulation and neurotransmitter imbalance, accounting
for their potent neurotoxic and convulsive effects. See Erwin L. Kong
and Thomas M. Nappe, *Irukandji Syndrome,* StatPearls (Treasure
Island, FL: StatPearls Publishing, 2025),
ncbi.nlm.nih.gov/books/NBK562264; and Russell Staggs and Jeffrey L.
Pay, *Cnidaria Toxicity,* StatPearls (Treasure Island, FL: StatPearls
Publishing, 2025), ncbi.nlm.nih.gov/books/NBK538170.

[398] Surveys and clinical analyses of individuals reporting encounters
with unidentified aerial phenomena describe recurring perceptual
and neurological effects, including *telepathic communication, intrusive
thoughts, perceived mental manipulation,* and post-encounter
electromagnetic sensitivity. Respondents in Thomas E. Reed et al.'s
large-scale study consistently reported these cognitive and sensory
anomalies, often accompanied by lasting aversion to electrical fields
or devices. A 2025 clinical review further supports these accounts,
noting that electromagnetic hypersensitivity, transient cortical
disruption, and altered states of consciousness have been
documented following close-encounter episodes. See Thomas E. Reed
et al., *A Study on Reported Contact with Non-Human Intelligence Associated
with Unidentified Aerial Phenomena,*
researchgate.net/publication/326151576_A_Study_on_Reported_Cont
act_with_Non-

Human_Intelligence_Associated_with_Unidentified_Aerial_Phenom
ena; and Luis Rafael Moscote-Salazar et al., *Neurological Effects of
Encounters with Unidentified Aerial Phenomena, Matrix Science Medica* 9
(2025): 63–66,
journals.lww.com/mtsm/fulltext/2025/07000/neurological_effects_of
_encounters_with.1.aspx.

[399] The vestibular system, located within the inner ear, is responsible
for maintaining equilibrium and spatial orientation through the
integration of sensory input from the semicircular canals and
otolithic organs with visual and proprioceptive cues. When these
signals fall out of synchrony—such as during mismatched visual and
motion perception—disorientation, vertigo, nausea, and dizziness
result, reflecting a breakdown in the body's sense of balance and
embodied awareness. See Timothy L. Thompson and Ronald Amedee,
*Vertigo: A Review of Common Peripheral and Central Vestibular Disorders,
The Ochsner Journal* 9, no. 1 (2009): 20–26,
pmc.ncbi.nlm.nih.gov/articles/PMC3096243/.

## CHAPTER ELEVEN

[400] Official defense and congressional records describe recurring
luminous phenomena in verified Unidentified Anomalous
Phenomena (UAP) encounters, including both radar- and sensor-
confirmed events. The *Human-Related and Radar-Verified UAP Report*
(DoD, 2024) documents multiple cases—such as metallic orbs,
triangular formations, and "jellyfish"-type objects—that exhibit
pulsing, flashing, or color-shifting luminescence visible in optical and
FLIR spectra. U.S. House testimony corroborates these findings,
summarizing defense imagery and pilot observations of self-luminous
orbs and structured light arrays responding to observer proximity.
Independent analyses further note that orb- and rod-shaped UAPs
frequently exhibit patterned, fluctuating fluorescence across visible
and infrared wavelengths. In contrast, educational summaries of

visual phenomena emphasize the recurring features of pulsation, multicolor variation, and atmospheric distortion associated with UAP light emissions. Collectively, these reports confirm that luminous or patterned light displays are among the most consistently observed characteristics in verified UAP encounters. See Department of Defense, *Human-Related and Radar-Verified UAP Report* (March 2024), media.defense.gov/2024/Mar/08/2003409233/-1/-1/0/DOPSR-CLEARED-508-COMPLIANT-HRRV1-08-MAR-2024-FINAL.PDF; "UFO Lights: Lighting Explained," PacLights Learning Center, paclights.com/explore/ufo-lights-lighting-explained/; Andrew D. Morgan and Brian Tyson, *Unidentified Anomalous Phenomena (UAP): Orb and Rod Object Form and Location Centric Abundance* (2024), researchgate.net/publication/385961112_Unidentified_Anomalous_P henomena_UAP_Orb_and_Rod_Object_Form_and_Location_Centric_ Abundance; and *U.S. House Committee on Oversight and Accountability Hearing on UAP Program Disclosures* (HHRG-118-GO12-20241113-SD003), congress.gov/118/meeting/house/117721/documents/HHRG-118-GO12-20241113-SD003.pdf.

[401] Bioluminescence has evolved independently in numerous terrestrial and aquatic lineages, appearing at least forty times across the tree of life. It is not limited to the ocean. Among land-dwelling organisms, it occurs most prominently in arthropods such as fireflies, click beetles, and railroad worms, in which distinct luciferase enzymes produce yellow-green or red light via oxygen-dependent reactions. Beyond insects, luminous systems have been identified in several annelid families, including scale worms (Polynoidae), which are capable of emitting green light from elytral photocytes. Additionally, multiple fungal lineages, such as Mycena and Armillaria, exhibit luciferin–luciferase chemistry that generates continuous green luminescence. Even mollusks participate in this evolutionary pattern: the limpet-like freshwater snail *Latia neritoides* produces a bright green mucus glow from a luciferin–luciferase system unique among known organisms. Together, these examples show that light

production is a convergent adaptation found across land and sea, spanning arthropods, annelids, fungi, and mollusks. See Ryo Futahashi et al., "Independent Evolution of Bioluminescence in Beetles Through Lateral Gene Transfer of Luciferase Genes," *Molecular Biology and Evolution* 41, no. 1 (2024), academic.oup.com/mbe/article/41/1/msad287/7510602; Gabriela V. Moraes et al., "Bioluminescence in Polynoid Scale Worms (Annelida: Polynoidae)," *Frontiers in Marine Science* 8 (2021), frontiersin.org/journals/marine-science/articles/10.3389/fmars.2021.643197/full; Camila M. Oliveira et al., "Diversity and Bioluminescence in Marine and Terrestrial Ecosystems," *Diversity* 16, no. 9 (2024), mdpi.com/1424-2818/16/9/539; and Yoshihiro Ohmiya et al., "Bioluminescence in the Limpet-Like Snail, *Latia neritoides*," researchgate.net/publication/250463771_Bioluminescence_in_the_Limpet-Like_Snail_Latia_neritoides.

[402] Quantitative surveys of deep-sea fauna show that bioluminescence is an exceptionally widespread adaptation, with estimates indicating that between 80 percent and 95 percent of marine organisms below 200 meters depth emit light. This prevalence has been documented across a wide range of taxa—including fish, squid, crustaceans, echinoderms, and gelatinous zooplankton—demonstrating that light production is a dominant ecological trait in the mesopelagic and hadal zones. The mechanism is chemically conserved: oxidation of luciferin by luciferase, often in the presence of coelenterazine or other imidazopyrazinone compounds, provides illumination for communication, predation, and defense. The distribution of this trait extends through multiple phyla, encompassing echinoderms, tunicates, crustaceans, clams, sea slugs, and sea snails, many of which either synthesize or acquire their luminescent substrates through feeding relationships. See Edith A. Widder, "Gleaning the Gleam: A Deep-Sea Webcam Sheds Light on Bioluminescent Ocean Life," *Scientific American* (2024),

scientificamerican.com/article/edith-widder-bioluminescence/; NOAA Ocean Exploration, "What Is Bioluminescence?" oceanexplorer.noaa.gov/ocean-fact/bioluminescence/; and Jeremy Mirza and Yuichi Oba, "Semi-Intrinsic Luminescence in Marine Organisms," in *Bioluminescence – Technology and Biology*, IntechOpen (2021), intechopen.com/chapters/78074.

[403] Long-term quantitative surveys in Monterey Bay confirm that bioluminescence is nearly universal among cnidarians. Analysis of 350,000 in situ observations collected over 17 years by remotely operated vehicles revealed that 97–99.7 percent of cnidarian taxa—including hydromedusae, siphonophores, and scyphozoans—are capable of producing their own light. These findings, drawn from a dataset encompassing 13 major marine taxa and depths to nearly 4,000 meters, demonstrate that bioluminescence is not only an ecological constant throughout the deep-water column but is especially dominant within Cnidaria, where almost all observed species exhibit intrinsic luminescent capability. See Séverine Martini and Steven H. D. Haddock, "Quantification of Bioluminescence from the Surface to the Deep Sea Demonstrates Its Predominance as an Ecological Trait," *Scientific Reports* 7 (2017): 45750, pmc.ncbi.nlm.nih.gov/articles/PMC5379559/; and Monterey Bay Aquarium Research Institute (MBARI), "New Study Shows That Three Quarters of Deep-Sea Animals Make Their Own Light," mbari.org/news/new-study-shows-that-three-quarters-of-deep-sea-animals-make-their-own-light/.

[404] *Vampyroteuthis infernalis*, commonly known as the vampire squid, is a deep-sea cephalopod uniquely classified as the sole extant member of its order, family, genus, and species—Order *Vampyromorphida*, Family *Vampyroteuthidae*, Genus *Vampyroteuthis*, and Species *V. infernalis*. Originally described by Carl Chun in 1903, it occupies an evolutionary position between squids and octopuses, retaining traits of both groups yet belonging to neither. Its gelatinous

body, cloak-like webbing, and retractile filaments reflect this lineage, distinguishing it as a phylogenetic relict of an ancient octopodiform branch. The entire body is covered in bioluminescent organs called photophores, which vary in size and structure and are capable of producing controlled flashes or prolonged glows. These photophores serve as a defensive adaptation: when threatened, the animal can emit bursts of light from its arm tips or release a cloud of bioluminescent mucus to confuse or distract predators in the deep sea's darkness. See *Vampire Squid* entry, *Wikipedia*, en.wikipedia.org/wiki/Vampire_squid.

[405] Stroboscopic or flickering light has been shown to measurably distort temporal perception, altering how the brain encodes duration and continuity. Controlled studies demonstrate that rhythmic visual stimulation entrains neural oscillations in the alpha range (8–12 Hz), leading to temporal compression, dilation, and discontinuities in perceived flow. Neurophysiological data indicate that pulsed light modifies cortical timing mechanisms and subjective experience of duration, sometimes producing sensations comparable to those reported in slow-motion or altered-state conditions. Together, these findings confirm that flickering light can disrupt the neural calibration of perceived time. See David M. Eagleman, "Human Time Perception and Its Illusions," *Current Opinion in Neurobiology* 18, no. 2 (2008): 131–136, pmc.ncbi.nlm.nih.gov/articles/PMC2866156; and Ram Kumar Pari, "Neural Dynamics of Stroboscopic Stimulation at Different Stimulation Frequencies," *bioRxiv* (2021), biorxiv.org/content/10.1101/2021.06.26.450044v1.full.

[406] Observational and laboratory studies confirm that *Vampyroteuthis infernalis* displays a distinctive defensive posture in which it inverts its arms and webbing over its body—often described as the "pineapple" or "umbrella" pose—drawing its glowing arm tips into a circular pattern around its head. This configuration conceals vital structures and transforms the animal's profile, likely reducing recognition by predators. The retracted, photophore-bearing arm tips

emit localized flashes, creating a luminous ring that produces the appearance of a reorganized, shifting light array. Such dynamic visual reconfiguration parallels many descriptions of patterned luminosity in verified UAP encounters, where light forms rearrange rapidly while maintaining apparent structural coherence. In addition, when disturbed, *V. infernalis* can release a dense, mucus-like cloud, infused with bioluminescent particles, that persists in the surrounding water. This glowing secretion adheres to approaching organisms, simultaneously disorienting and marking them while the squid escapes. See *Vampire Squid* entry, *Wikipedia*, en.wikipedia.org/wiki/Vampire_squid; and Hendrik J. T. Hoving and Bruce H. Robison, "Vampire Squid: Detritivores in the Oxygen Minimum Zone," *Proceedings of the Royal Society B* 279 (2012): 4559–4567, pmc.ncbi.nlm.nih.gov/articles/PMC3479720.

[407] The Hawaiian bobtail squid (*Euprymna scolopes*) maintains a highly specialized symbiosis with the luminous bacterium *Aliivibrio fischeri* (formerly *Vibrio fischeri*), which resides in a bilobed light organ within the squid's mantle cavity. The squid supplies its symbionts with sugars and amino acids derived from host metabolism. At the same time, the bacteria produce bioluminescence precisely matched to the intensity and spectral quality of downwelling moonlight—a camouflage system known as counter-illumination. This bacterial glow eliminates the squid's silhouette when viewed from below, rendering it nearly invisible to predators. Anatomical studies reveal that the light organ comprises a lens for focusing light, reflectors composed of crystalline tissue layers, and adjustable filters that balance bacterial luminescence with ambient light, all controlled by the squid's neural and ink sac structures, which act as an iris. The system exemplifies one of the most intricate known examples of mutualistic coevolution between animal and microbe. See Margaret J. McFall-Ngai, "Divining the Essence of Symbiosis: Insights from the Squid-Vibrio Model," *Annual Review of Microbiology* 75 (2021): 695–711, pmc.ncbi.nlm.nih.gov/articles/PMC8440403; and B. W. Jones and M.

K. Nishiguchi, "Counterillumination in the Hawaiian Bobtail Squid, *Euprymna scolopes* Berry (Mollusca: Cephalopoda)," *Marine Biology* 144 (2004): 1151–1155, researchgate.net/publication/225738384_Counterillumination_in_the _Hawaiian_bobtail_squid_Euprymna_scolopes_Berry_Mollusca_Cep halopoda.

[408] The deep-sea crown jelly *Atolla wyvillei*, sometimes called the "alarm jellyfish," is known for its distinctive bioluminescent defense display. When threatened, the animal emits a rapid series of concentric blue flashes that ripple around its bell in a circular pattern, resembling an underwater strobe. This luminescent ring is produced by light-emitting organs along the bell margin and is believed to serve as an anti-predator mechanism. Observations by NOAA's *Okeanos Explorer* and other research expeditions show that this "burglar alarm" display may attract larger predators toward the attacker, increasing the jelly's chance of escape. The genus *Atolla* is among the most widespread in the deep ocean, inhabiting the mesopelagic and bathypelagic zones worldwide, with A. wyvillei among the most frequently observed. See Monterey Bay Aquarium Research Institute (MBARI), "Scientists Discover a New Species of Deep-Sea Crown Jelly in Monterey Bay," mbari.org/news/scientists-discover-a-new-species-of-deep-sea-crown-jelly-in-monterey-bay/; and Sarah Keartes, "If There Was a Costume Contest in the Deep Sea, This Crazy UFO Jelly Would Win," *Earth Touch News* (2016), earthtouchnews.com/wtf/wtf/if-there-was-a-costume-contest-in-the-deep-sea-this-crazy-ufo-jelly-would-win/.

[409] Human brain rhythms operate within well-defined frequency bands, with theta oscillations typically falling between 4 and 7 Hz and alpha oscillations between 8 and 12 Hz. Research on neural entrainment has shown that periodic external stimulation—including rhythmic light pulses—can synchronize, or "entrain," ongoing cortical oscillations when delivered at matching frequencies. Defense

Department research on visual and auditory entrainment notes that flicker frequencies in the alpha and theta ranges can modulate arousal, attention, and perceptual processing, while psychophysical work on peak alpha frequency demonstrates that alpha-band rhythms act as an internal temporal "clock" for parsing visual events. Modifying or driving oscillatory activity at these frequencies alters temporal integration windows, perceptual segmentation, and the subjective flow of time itself. If an external luminous stimulus were capable of matching or entraining these intrinsic rhythms, subtle distortions in perceived duration or continuity could result. See Army Research Laboratory, *Human Perception of Temporal Target Discrimination* (1999), apps.dtic.mil/sti/tr/pdf/ADA365328.pdf; and Julie Freschl et al., "The Development of Peak Alpha Frequency and Its Role in Visual Temporal Processing: A Meta-Analysis," *Developmental Cognitive Neuroscience* 57 (2022): 101146, sciencedirect.com/science/article/pii/S1878929322000895.

[410] Coral bioluminescence appears to be among the oldest known light-producing adaptations in the animal kingdom. Recent phylogenetic analyses of anthozoans indicate that bioluminescence originated in a common octocoral ancestor during the Cambrian period, approximately 540 million years ago, making corals the earliest confirmed bioluminescent lineage. Corals also possess fluorescent proteins capable of absorbing harmful ultraviolet and high-energy blue wavelengths and re-emitting them as lower-energy hues, including pink, violet, and purple. This spectral shift serves a photoprotective role. Experimental work demonstrates that coral skeletons and tissues significantly reduce UV transmittance, dissipating absorbed radiation through fluorescence and thereby shielding overlying tissues from photonic damage. See *Ancient Glow: Scientists Discover 540 Million-Year-Old Bioluminescent Corals* (2024), scitechdaily.com/ancient-glow-scientists-discover-540-million-year-old-bioluminescent-corals/; Ruth Reef, Paulina Kaniewska, and Ove Hoegh-Guldberg, "Coral Skeletons Defend Against Ultraviolet

Radiation," *PLoS ONE* 4, no. 11 (2009): e7995,
researchgate.net/publication/40039497_Coral_Skeletons_Defend_aga
inst_Ultraviolet_Radiation; and Defense Documentation Center,
*Ultraviolet Radiation in the Marine Environment* (1965),
apps.dtic.mil/sti/tr/pdf/AD0630903.pdf.

[411] Across diverse bioluminescent organisms, one of the most
common light-producing pathways is the luciferin–luciferase
reaction, in which an enzyme known as luciferase catalyzes the
oxidation of a small-molecule substrate called luciferin. In this
reaction, luciferase binds luciferin in the presence of oxygen (and, in
many systems, ATP and magnesium), forms an activated
intermediate, and drives an oxidative chemistry that yields an
excited-state oxyluciferin. As this product relaxes to its ground state,
it releases a photon, producing visible light without the need for
external illumination. Comparative biochemical work identifies this
luciferase-catalyzed oxidation of luciferin as the fundamental
mechanism underlying many independently evolved bioluminescent
systems. See Spencer T. Adams Jr. and Stephen C. Miller, "Enzymatic
Promiscuity and the Evolution of Bioluminescence," *FEBS Journal* 287,
no. 7 (2020): 1369–1380, pmc.ncbi.nlm.nih.gov/articles/PMC7217382;
and "Everything About Luciferin and Luciferase," GoldBio,
goldbio.com/blogs/articles/everything-about-luciferin-luciferase.

[412] In cnidarian bioluminescence systems, such as those of *Aequorea*
jellyfish, calcium ions act as the immediate trigger for light emission
through activation of calcium-binding photoproteins like aequorin.
When $Ca^{2+}$ binds to the photoprotein, it initiates an intramolecular
oxidation of the luciferin coelenterazine, producing excited-state
coelenteramide and releasing blue light. Intracellular control over
calcium concentrations enables precise modulation of both timing
and intensity of emission. Magnesium, a critical stabilizing cofactor in
enzymatic pathways, contributes to ATP stability, intracellular pH
buffering, and cellular structural integrity, thereby supporting the

biochemical conditions necessary for controlled light production. See Yoshihiro Ohmiya and Takashi Hirano, "Shining the Light: The Mechanism of the Bioluminescence Reaction of Calcium-Binding Photoproteins," *Chemistry & Biology* 3, no. 5 (1996): 337–347, sciencedirect.com/science/article/pii/S1074552196901167; A. Milosavljevic et al., "The Emerging Use of Bioluminescence in Medical Research," *Biocybernetics and Biomedical Engineering* 39, no. 3 (2019): 652–665, sciencedirect.com/science/article/abs/pii/S0753332218302920; and Robert Whittam and Diana A. Wheeler, "Regulation of Intracellular Magnesium," *Proceedings of the Royal Society B* 258, no. 1352 (1994): 15–22, pmc.ncbi.nlm.nih.gov/articles/PMC4455825.

[413] The presence of biogenic magnetite in diverse organisms is well established. Kirschvink and Gould's foundational review documents the natural precipitation of magnetite ($Fe_3O_4$) in species spanning multiple phyla, including magnetotactic bacteria, chitons, honeybees, homing pigeons, sharks, and dolphins, demonstrating that many animals biomineralize magnetite crystals and incorporate them into tissues associated with magnetoreception. The paper further notes that magnetite occurs in both single-domain and superparamagnetic forms in these organisms and that its widespread occurrence "suggests that organic magnetite may be a common biological component." See Joseph L. Kirschvink and James L. Gould, "Biogenic Magnetite as a Basis for Magnetic Field Detection in Animals," *BioSystems* 13 (1981): 181–201, pubmed.ncbi.nlm.nih.gov/7213948/.

[414] Recent psychological research confirms that UAP encounters frequently produce not only surprise but marked emotional unease. De la Torre's study of 245 participants—including 93 direct witnesses found that UAP events often generate shock, disbelief, and "unsettled feelings," with many witnesses reporting the experience as life-changing. These reactions arise from the cognitive dissonance produced when anomalous phenomena defy existing perceptual and

conceptual frameworks, pushing attention, interpretation, and meaning-making processes beyond familiar limits. See Gabriel G. De la Torre, "Psychological Aspects in Unidentified Anomalous Phenomena (UAP) Witnesses," *International Journal of Astrobiology* 23 (2024): e4, 1–7, cambridge.org/core/journals/international-journal-of-astrobiology/article/psychological-aspects-in-unidentified-anomalous-phenomena-uap-witnesses/7D23058E10309384DDAA8E15607485CC.

[415] Multiple technical analyses confirm that the most anomalous aspect of reported UAP behavior is the ability to perform abrupt directional shifts, sustain extreme accelerations, and traverse distances with apparent instantaneous or near-instantaneous motion. Loeb and Kirkpatrick note observational cases in which objects appear to "instantaneously accelerate, change direction, or disappear," challenging conventional aerodynamic and inertial limits. Knuth, Powell, and Reali quantify these maneuvers across several well-documented encounters, including the 2004 Nimitz events, demonstrating accelerations ranging from tens to thousands of g and highlighting trajectory changes so rapid that they defy known propulsion or control surfaces. Their analysis further shows that these behaviors were corroborated by multiple sensing modalities, including visual tracking by pilots, radar returns, and infrared video, establishing that these were not one-off anomalies but rather repeated observations captured by sophisticated instrumentation. See Abraham Loeb and Sean M. Kirkpatrick, "Physical Constraints on Unidentified Aerial Phenomena," *submitted manuscript* (2023), lweb.cfa.harvard.edu/~loeb/LK1.pdf; and Kevin H. Knuth, Robert M. Powell, and Peter A. Reali, "Estimating Flight Characteristics of Anomalous Unidentified Aerial Vehicles," *Entropy* 21 (2019): 939, pmc.ncbi.nlm.nih.gov/articles/PMC7514271/.

[416] Human tolerance to acceleration imposes strict limits on any living occupant of a high-speed craft. Aeromedical guidance

consistently shows that even with training, G-suits, and anti-G straining maneuvers, most fighter pilots lose consciousness at around 9 g, and negative G-tolerance is far lower—typically only 2.5 to 3 g before "red-out" or capillary injury occurs. Positive G-force drives blood toward the lower extremities, reducing cerebral perfusion and causing symptoms such as gray-out, tunnel vision, blackout, and ultimately, gravity-induced loss of consciousness (G-LOC). Negative G-force forces blood toward the head, increasing intracranial and ocular pressure and rapidly overwhelming vascular tolerance. Both forms of acceleration stress impose hard biological limits on survivable motion, regardless of an aircraft's capability. See Ensign Eric Page, "Pulling Gs: The Pilot's Body Sets the Limit," *Proceedings* 134, no. 9 (2008), usni.org/magazines/proceedings/2008/september/pulling-gs-pilots-body-sets-limit; and Federal Aviation Administration, *Acceleration in Aviation: G-Force* (Aerospace Medical Education Division, 2021), faa.gov/pilots/safety/pilotsafetybrochures/media/acceleration.pdf.

[417] Radar operators aboard the USS *Princeton* and subsequent technical analysis confirm that during the 2004 Nimitz encounter, unidentified objects repeatedly dropped from approximately 28,000 feet to sea level in roughly 0.78 seconds. Knuth, Powell, and Reali's reconstruction of Senior Chief Kevin Day's radar data shows altitude changes of approximately 8,530 meters occurring in under one second, yielding estimated accelerations between 5,370 g and 5,950 g—orders of magnitude beyond the structural limits of any known airframe and far exceeding human or biological tolerance. Even the lowest calculated values imply inertial loads that no living organism or conventional engineered material could survive. See Kevin H. Knuth, Robert M. Powell, and Peter A. Reali, "Estimating Flight Characteristics of Anomalous Unidentified Aerial Vehicles in the 2004 Nimitz Encounter," *Proceedings* 33 (2019): 26, researchgate.net/publication/337985034_Estimating_Flight_Characte

ristics_of_Anomalous_Unidentified_Aerial_Vehicles_in_the_2004_Ni mitz_Encounter.

[418] The 1994 Northridge earthquake struck the Los Angeles region at 4:30 a.m. on January 17, 1994, with a moment magnitude of 6.7. Its epicenter was located roughly twenty miles west-northwest of downtown Los Angeles and only a few miles from West Los Angeles. Caltech's Southern California Earthquake Data Center notes that the blind-thrust rupture produced some of the strongest ground motions ever instrumentally recorded in a North American urban area, causing widespread damage to freeways, apartments, and commercial structures throughout the San Fernando Valley and Santa Monica regions—conditions consistent with the author's firsthand experience of the event. See Southern California Earthquake Data Center, "Northridge Earthquake," scedc.caltech.edu/earthquake/northridge1994.html.

[419] Studies of UAP reporting consistently show that witnesses do not conform to any single demographic, ideological, or psychological type. Ostdiek emphasizes that UAP encounters are "credibly reported by vast diversities of individuals across the public, private, and civil sectors... in countries around the globe," cutting across background, belief, and social position. What unites these reports is not a common profile but a shared phenomenological residue: experiences that carry a "consistent quality of differentness" and often leave witnesses grappling with a lingering cognitive or emotional dissonance as they attempt to integrate the event into existing frameworks of meaning. See Nate Ostdiek, *Strategic Construction of Anomalies: A Quantum International Relations Framework for UAP* (2025), digitalcommons.unomaha.edu/cgi/viewcontent.cgi?article=139.

[420] Technical analysis of the 2004 USS *Nimitz* encounter confirms that pilots and AEGIS radar operators observed an object descend from roughly 28,000 feet to near sea level in under one second, without

sonic boom, thermal signature, or visible turbulence. The Scientific Coalition for UAP Studies reconstructed the radar-tracked altitude shift as an approximately 28,000-foot drop occurring within 0.78 seconds, producing estimated accelerations exceeding 5000 g—far beyond survivable limits for any biological organism or known aerospace vehicle. The same investigation notes Commander David Fravor's account that the object, after hovering, abruptly accelerated out of view "in less than a second," leaving no shockwave, plume, or atmospheric disturbance. See Robert M. Powell et al., "A Forensic Analysis of Navy Carrier Strike Group Eleven's Encounter with an Anomalous Aerial Vehicle," Scientific Coalition for UAP Studies (2020), explorescu.org/post/2004-uss-nimitz-strike-navy-group-incident-report.

[421] Examination of major, independently documented UAP incidents across decades shows that credible reports consistently arise from trained observers using multiple sensing modalities. The SCU analysis notes earlier cases, such as Navy Lt. Graham Bethune's 1951 encounter—an event involving radar and visual confirmation years before modern sensor systems—and includes Japan Air Lines Captain Kenju Terauchi's 1986 JAL 1628 incident in Alaska among the historically significant, well-documented sightings analyzed in the extended study. The same report provides detailed reconstructions of the 2004 USS *Nimitz* encounters, incorporating radar data from Senior Chief Kevin Day, pilot testimony from Commander David Fravor, and ATFLIR video analysis. Taken together, these cases illustrate consistent patterns in eyewitness testimony and sensor-tracked behavior across disparate eras, locations, and witness populations. See Kevin H. Knuth, Robert M. Powell, and Peter A. Reali, "Estimating Flight Characteristics of Anomalous Unidentified Aerial Vehicles in the 2004 Nimitz Encounter," *Entropy* 21 (2019): 939, pmc.ncbi.nlm.nih.gov/articles/PMC7514271/.

[422] Combat Air Patrol (CAP) missions rely on a pre-briefed rendezvous location—known as a CAP point—that serves as a silent fallback position for aircraft to regroup without radio coordination. As documented in the SCU analysis of the 2004 USS *Nimitz* encounters, CAP points are established in advance, are not transmitted over open channels, and are known only to the participating aircrew and controllers. Following Commander Fravor's initial visual engagement, the object he pursued accelerated away "like it was shot out of a rifle," and AEGIS radar operators aboard the USS *Princeton* reported multiple similar contacts in the area. Senior Chief Kevin Day further stated that shortly after Fravor disengaged, the primary object appeared "exactly at the CAP point," a location approximately sixty miles from the pilots' position—an event strongly suggesting intentional navigation rather than coincidence. See Robert M. Powell et al., "A Forensic Analysis of Navy Carrier Strike Group Eleven's Encounter with an Anomalous Aerial Vehicle," Scientific Coalition for UAP Studies (2020), explorescu.org/post/2004-uss-nimitz-strike-navy-group-incident-report.

## CHAPTER TWELVE

[423] Historical shifts in scientific understanding have repeatedly required abandoning long-held explanatory models when empirical evidence proved them untenable. The transition from geocentrism to heliocentrism illustrates this process: ancient and medieval cosmologies positioned Earth at the center of the universe until observational inconsistencies and later Copernican analysis overturned the model. Likewise, nineteenth-century public-health discourse increasingly challenged miasma theory—the belief that diseases such as cholera spread through "bad air"—as evidence mounted for waterborne transmission, culminating in the acceptance of germ theory. See Matthew Williams, "What Is the Geocentric Model of the Universe?" *Universe Today* (2016), universetoday.com/articles/geocentric-model; and Stephen Halliday,

"Death and miasma in Victorian London: an obstinate belief," *BMJ* 323 (2001): 1469–1471, pmc.ncbi.nlm.nih.gov/articles/PMC1121911/

[424] The scene described in this section derives from Schindler's List (1993), Steven Spielberg's dramatization of Oskar Schindler's efforts to save Jewish prisoners during the Holocaust and his complex relationship with SS officer Amon Göth. The dialogue quoted— Schindler's challenge to Göth's conception of power, Göth's responses, and the later moment of self-pardon—appears as part of the film's depiction of Göth's brutality and Schindler's attempt to redirect it. While the film interprets these historical figures through a narrative lens, its portrayal of Göth aligns with documented accounts of his command of the Plaszów concentration camp, his arbitrary killings, and his eventual conviction for war crimes. See Schindler's List, Britannica, britannica.com/topic/Schindlers-List; and "Amon Göth," Britannica, britannica.com/biography/Amon-Goth.

[425] The role of quantum processes in biological function is well documented across multiple domains of life. Research demonstrates that quantum coherence facilitates highly efficient exciton transport during photosynthesis, that quantum tunneling contributes to enzymatic reaction rates, that avian magnetoreception relies on radical-pair dynamics within cryptochrome proteins, that olfactory receptors may distinguish odorants through vibrational energy signatures, and that proton tunneling and other quantum effects influence DNA stability and repair. See Brookes, *Proceedings of the Royal Society A*, royalsocietypublishing.org/doi/10.1098/rspa.2016.0822; and Marais et al., *Journal of the Royal Society Interface*, pmc.ncbi.nlm.nih.gov/articles/PMC6283985/.

[426] The description of photosynthetic exciton dynamics and the role of quantum superposition in energy transfer is supported by research into coherence-driven transport in photosynthetic complexes.

Quantum superposition refers to a particle's ability to exist in multiple states or locations simultaneously until a measurement collapses it into a single outcome. Evidence suggests that excitonic energy in photosynthetic systems explores multiple molecular pathways simultaneously before selecting the most efficient route, thereby contributing to the high efficiency of photosynthesis. See Graham R. Fleming and Akihito Ishizaki, "Theoretical studies of quantum coherence in photosynthetic light harvesting," *New Journal of Physics* 12, no. 055004 (2010), iopscience.iop.org/article/10.1088/1367-2630/12/5/055004; and Ilamaran Sivarajah, "The Quantum Mechanics of Photosynthesis," AZoQuantum (2022), azoquantum.com/Article.aspx?ArticleID=281.

[427] The estimate that Homo sapiens emerged roughly 200,000 years ago is supported by genetic, archaeological, and paleoanthropological evidence indicating a deep divergence of early modern human lineages in Africa during the late Middle Pleistocene. Mitochondrial DNA coalescence models and fossil calibrations consistently place the appearance of anatomically modern humans around this timeframe, with subsequent population expansions occurring later in eastern Africa. See Teresa Rito et al., "A dispersal of Homo sapiens from southern to eastern Africa immediately preceded the out-of-Africa migration," Scientific Reports 9, no. 4728 (2019), nature.com/articles/s41598-019-41176-3.

[428] The idea that a biological organism could "delaminate" from familiar spacetime geometry is speculative. Still, the analogy draws on established features of quantum systems in which states evolve through superposition rather than along a single classical trajectory. As studies of the quantum–classical boundary show, quantum states can occupy multiple potential configurations simultaneously, diverging from classical path constraints until coherence is lost through decoherence processes. These behaviors provide conceptual, though not literal, precedents for describing nonclassical navigation

through spacetime. See E. Aldo Arroyo, "Exploring the Transition Between Quantum and Classical Mechanics," *arXiv:2405.18564* (2024), arxiv.org/pdf/2405.18564.

[429] Newton's first law establishes that objects resist changes to their state of motion unless acted upon by an external force, defining inertia as the inherent tendency of mass to oppose acceleration. This principle underlies everyday motion and frames how bodies move through gravitational and inertial fields, where rest and uniform motion reflect the structure of spacetime described by inertial frames in modern physics. The biological and mechanical examples in this passage — birds generating lift and thrust through wing morphology and muscle-driven flight mechanics, and aircraft using engines, lift, and aerodynamic control surfaces — are well documented in both biological and aeronautical literature, illustrating how living and engineered systems overcome resistive forces associated with motion. See Ahmed Ragab Mahmoud Salah, *Newton's First Law of Motion: Inertia and Applications in Mechanics* (2025), researchgate.net/publication/395710919_Newton%27s_First_Law_of_Motion_Inertia_and_Applications_in_Mechanics; Andrew A. Biewener, "Biomechanics of Avian Flight," *Current Biology* 32 (2022): R1042–R1172, sciencedirect.com/science/article/pii/S0960982222010843; Federal Aviation Administration, *Pilot's Handbook of Aeronautical Knowledge*, Chapter 5 (2023), faa.gov/sites/faa.gov/files/07_phak_ch5_0.pdf.

[430] Resonance describes the behavior of systems that naturally oscillate at particular frequencies and respond most strongly when driven at those same frequencies. When two systems share or approach a common resonant frequency, energy transfer between them becomes more efficient, producing interactions that feel smoother or unusually amplified, a principle evident across mechanical, electrical, and quantum domains. As widely noted in physics explainers, resonance can synchronize oscillations in

everything from vibrating glass to atomic transitions, and parallel resonance in electrical systems. This illustrates how matched frequencies can dramatically amplify or streamline interactions. See Ben Brubaker, "How the Physics of Resonance Shapes Reality," *Quanta Magazine* (2022), quantamagazine.org/how-the-physics-of-resonance-shapes-reality-20220126; Eaton, "What Is Harmonic Resonance and How Does It Work?" (2025), eaton.com/us/en-us/products/controls-drives-automation-sensors/harmonics/harmonics-faq-video-library/what-is-harmonic-resonance-and-how-does-it-work-.html.

[431] In classical mechanics, multi-particle systems exhibit chaotic, collision-driven motion governed by Newtonian interactions, with trajectories diverging sensitively as particles scatter, exchange momentum, and redistribute energy. By contrast, quantum-coherent states allow multiple constituents to evolve in a correlated fashion, maintaining phase relationships that cause them to behave collectively rather than independently. Theoretical treatments of coherent energy transfer demonstrate how quantum phases can synchronize excitation dynamics across coupled systems, producing unified, wave-like propagation rather than classical, uncorrelated motion. See David Tong, *Dynamics and Relativity*, Lecture Notes, University of Cambridge (Section 5, "Systems of Particles"), damtp.cam.ac.uk/user/tong/relativity/five.pdf; Seogjoo Jang et al., "Theory of Coherent Resonance Energy Transfer," *Journal of Chemical Physics* 129 (2008): 101104, researchgate.net/publication/23557862_Theory_of_coherent_resonan ce_energy_transfer.

[432] Scientific consensus recognizes five major mass extinction events in Earth's history, each marked by the rapid loss of a substantial percentage of species across land and sea. These include the Ordovician–Silurian extinction (~444 million years ago), Late Devonian extinction (~372 million years ago), Permian–Triassic extinction (~252 million years ago), Triassic–Jurassic extinction (~201

million years ago), and the Cretaceous–Paleogene extinction event (~66 million years ago), all of which are well-documented in paleontological and geological records. Contemporary research further indicates that Earth is now entering, or may already be in, a sixth major extinction event—the Holocene extinction—driven primarily by human activity, habitat destruction, pollution, climate change, and accelerating biodiversity loss. Current extinction rates have risen far above background levels, and numerous assessments warn that a significant proportion of global species are at risk unless immediate conservation action is taken. See *Major Mass Extinctions*, Encyclopaedia Britannica, britannica.com/list/major-mass-extinctions; "Mass Extinctions," EBSCO Research Starters (2025), ebsco.com/research-starters/geology/mass-extinctions.

[433] The deep ocean is widely recognized as one of Earth's most stable and extinction-resilient habitats, buffered from many surface-level catastrophes that drive mass die-offs in terrestrial and shallow-marine environments. Fossil evidence from multiple extinction intervals shows that deep-sea ecosystems often persist through upheavals that devastate surface biota, and that benthic communities can remain comparatively intact even when upper-ocean productivity collapses. At the same time, while much of the deep ocean depends on surface-derived organic matter through the biological pump, research indicates that not all deep-ocean zones follow this rule uniformly, and some ecosystems – including vent- and seep-based chemosynthetic communities – demonstrate partial decoupling from surface food webs. Recent syntheses of deep-sea ecology likewise emphasize that key processes governing deep-ocean stability, carbon transport, and biological resilience remain incompletely understood, leaving open the possibility that not every deep-ocean organism or habitat is constrained by surface productivity dynamics. See Heather Birch et al., "Ecosystem Function After the K/Pg Extinction: Decoupling of Marine Carbon Pump and Diversity," *Proceedings of the Royal Society B* 288 (2021): 20210863,

royalsocietypublishing.org/doi/10.1098/rspb.2021.0863; Nikk Ogasa, "In the Wake of History's Deadliest Mass Extinction, Ocean Life May Have Flourished," *Science News* (2023), sciencenews.org/article/historys-deadliest-mass-extinction-ocean-life; *Deep Sea Synthesis Report* (2022), stateoftheocean.org/wp-content/uploads/2022/11/DeepSea-Synthesis-31oct-high.pdf.

[434] The two-dimensional "Flatland" analogy originates from Edwin A. Abbott's 1884 novella *Flatland: A Romance of Many Dimensions*, which has long been used to illustrate how beings confined to lower-dimensional spaces would perceive the intrusion of a higher-dimensional object only as shifting cross-sections. A sphere passing through a plane, for instance, would appear to a flat observer not as a sphere but as a series of changing shapes—just as a three-dimensional finger would manifest in Flatland as a dot, then a widening oval, then a dot again. This framework has become a standard pedagogical tool for conveying higher-dimensional geometry and the limits of perception. See Colin C. Adams, "A Forgivably Flat Classic," *American Scientist* 98, no. 6 (2010), americanscientist.org/article/a-forgivably-flat-classic.

[435] Convergent evolution refers to the process by which distinct evolutionary lineages independently acquire similar traits in response to comparable selective pressures, rather than shared ancestry. Classic examples include the evolution of echolocation in bats and toothed whales, limbless body forms in burrowing reptiles, and antifreeze proteins in Arctic and Antarctic fishes—all cases in which unrelated organisms arrive at comparable solutions to similar environmental challenges. As recent genomic studies show, such convergence can arise at multiple biological levels, from repeated changes in the same genes or pathways to the emergence of parallel phenotypes via distinct molecular routes, underscoring how natural selection, constraint, and limited genetic pathways can produce similar adaptations across distant taxa. See Timothy B. Sackton and

Nathan Clark, "Convergent Evolution in the Genomics Era: New Insights and Directions," *Philosophical Transactions of the Royal Society B* 374 (2019): 20190102, royalsocietypublishing.org/doi/10.1098/rstb.2019.0102.

## CHAPTER THIRTEEN

[436] The deep evolutionary roots of both cnidarians and cephalopods are well established in the paleontological record. Phylogenetic and fossil evidence place cnidarian lineages among the earliest metazoans, with medusozoan and hydroidolin groups arising by the late Precambrian–early Cambrian interval, on the order of hundreds of millions of years ago, and consistent with estimates around 580 million years for their first appearance in the fossil record. Cephalopods, by contrast, emerge later in the early Paleozoic, with shell-bearing forms appearing in the early Cambrian and diversifying into nautiloid and coleoids by roughly 500 million years ago. See A. Bentlage et al., "Tackling the phylogenetic conundrum of Hydroidolina (Cnidaria: Medusozoa: Hydrozoa)," *PeerJ* 9 (2021): e12104, peerj.com/articles/12104/; "Octopuses, Squids, and Relatives," *Smithsonian Ocean Portal*, ocean.si.edu/ocean-life/invertebrates/octopuses-squids-and-relatives.

[437] Cnidarians are unusual among metazoans in that some lineages possess introns within their mitochondrial DNA, a genome that is typically compact, intron-poor, and highly streamlined. In the sea anemone *Metridium senile*, for example, the mitochondrial COI gene contains a group I intron that encodes a homing endonuclease, enabling precise self-splicing and targeted insertion into specific mitochondrial sites. Even more striking is the ND5 intron of the same species, which carries entire protein-coding genes nested within it—an exceptionally rare genomic architecture among animals and strong

evidence of the distinctive structural complexity of cnidarian mitochondrial genomes. See C. B. Bridge et al., "Mitochondrial genome structure in cnidaria," *Proceedings of the Royal Society B* 263 (1996): 157–163, pmc.ncbi.nlm.nih.gov/articles/PMC1460033/; N. Beagley et al., "Two mitochondrial group I introns in a metazoan, the sea anemone *Metridium senile*: one intron contains an open reading frame for a putative homing endonuclease," *Molecular Biology and Evolution* 13 (1996): 137–146, pubmed.ncbi.nlm.nih.gov/8643626/.

[438] Modern cephalopod genomic studies reveal that their genomes are unusually rearranged and structurally complex compared to other animal lineages, with extensive shuffling and large sets of gene clusters that appear unique to coleoids and are associated with their distinctive neural and behavioral traits. These lineages also possess an exceptional abundance of A-to-I RNA editing sites—numbering in the tens of thousands—far exceeding those documented in any other known organism. Editing is especially enriched in neural tissues, where it contributes to functional diversity and rapid physiological adaptability. See Joshua Rosenthal et al., "The pace of transcriptomic innovation: A-to-I RNA editing in vertebrates and cephalopods," *Nature Communications* 13 (2022): 2220, pmc.ncbi.nlm.nih.gov/articles/PMC9068888/; Eli Eisenberg et al., "RNA editing in cephalopods," *Nature Communications* 13 (2022): 2221, nature.com/articles/s41467-022-29694-7.

[439] The origins of mitochondria and chloroplasts are firmly established in evolutionary biology. Both organelles descend from free-living bacteria that entered into a stable symbiotic relationship with early host cells, eventually becoming permanent, heritable components of the eukaryotic lineage. This endosymbiotic merger is believed to have given rise to the first eukaryotes, thereby setting the stage for all complex, multicellular life on Earth. See John F. Allen, "Why chloroplasts and mitochondria retain their own genomes and genetic systems: Colocation for redox regulation of gene expression,"

*Proceedings of the National Academy of Sciences* 112 (2015): 10231–10238,
pmc.ncbi.nlm.nih.gov/articles/PMC4547249/.


440 Arthur C. Clarke's well-known formulation, "Any sufficiently
advanced technology is indistinguishable from magic," originates
from his third law, first articulated in a revision of his essay "Profiles
of the Future" and later collected as part of the widely cited set now
known as Clarke's Three Laws. The statement reflects Clarke's view
that technological sophistication, when far enough beyond the
observer's frame of reference, becomes perceptually indistinguishable
from the supernatural. See "Clarke's Three Laws," *EBSCO Research
Starters* (2024), ebsco.com/research-starters/science/clarkes-three-
laws.


441 Darwin's work on the Galápagos finches demonstrates how beak
morphology diverged across islands in direct response to local
ecological pressures, particularly food availability, making them a
classic example of natural selection in action. Modern analyses
continue to confirm the adaptive significance and evolutionary
trajectories of these beak variations across distinct habitats. See Peter
R. Grant and B. Rosemary Grant, "From microcosm to macrocosm:
adaptive radiation of Darwin's finches," *Evolutionary Journal of the
Linnean Society* 3 (2024): kzae006,
academic.oup.com/evolinnean/article/3/1/kzae006/7686731; and Peter
T. Boag and Peter R. Grant, "Evolution of Darwin's finches caused by a
rare climatic event," *Proceedings of the Royal Society B* 251 (1993): 111–117,
pmc.ncbi.nlm.nih.gov/articles/PMC2830240/.


442 Cnidarians display both epigenetic and developmental plasticity
that supports the idea of "genomic strangeness." Studies of DNA
methylation in corals and other cnidarians show that CpG
methylation is not static but dynamically deployed across the
genome, particularly in association with transposons and stress
responses, indicating a flexible epigenetic system that can modulate

genome defense and gene regulation over time. In parallel, several cnidarian species are capable of ontogeny reversal, or reverse development, in which later life stages reactivate genetic programs characteristic of earlier forms and revert to more juvenile morphologies. This back-transformation is achieved through processes such as transdifferentiation, programmed cell death, and proliferation of specific cell populations, allowing cnidarians to revert to earlier developmental stages under adverse conditions. See Hua Ying et al., "The role of DNA methylation in genome defense in Cnidaria and other invertebrates," *Molecular Biology and Evolution* 40 (2023): msac018, pmc.ncbi.nlm.nih.gov/articles/PMC8857917/; Stefano Piraino et al., "Reverse development in Cnidaria," *Canadian Journal of Zoology* 82 (2004): 1748–1754, cdnsciencepub.com/doi/10.1139/z04-174.

[443] In general relativity, any concentration of mass–energy produces curvature in spacetime. In principle, even the intense, localized energy densities associated with nuclear reactors and nuclear reactions can generate extremely small gravitational and temporal effects. This possibility has motivated both experimental searches for tiny time distortions near operating reactors and theoretical work on gravitational wave generation from nuclear processes, all of which treat nuclear systems as sources of minute but real space-time perturbations. See Victor Tangermann, "Scientists Are Investigating If Time Warps Near a Nuclear Reactor," *Futurism* (2021), futurism.com/scientists-investigating-time-warps-nuclear-reactor; Giorgio Fontana and Robert M. L. Baker Jr., "Generation of Gravitational Waves with Nuclear Reactions," *AIP Conference Proceedings* 813 (2006): 1352–1358, pubs.aip.org/aip/acp/article-abstract/813/1/1249/813905.

[444] Recent U.S. government assessments confirm that a substantial number of UAP reports remain unresolved despite standardized intake, technical review, and multi-agency analysis. While many cases are ultimately attributed to prosaic objects, a significant body of

reports continues to lack sufficient data for definitive explanation, demonstrating that well-documented yet unexplained encounters persist within the official record. See *Fiscal Year 2024 Consolidated Annual Report on Unidentified Anomalous Phenomena*, All-domain Anomaly Resolution Office (AARO), National Archives Catalog ID 493470289, catalog.archives.gov/id/493470289.

## CHAPTER FOURTEEN

[445] The origin of life remains one of the most actively debated questions in modern science, with significant discussion surrounding the mechanisms of abiogenesis and the extent to which its processes can be reconstructed from available evidence. Contemporary reviews highlight both the progress and the persistent uncertainties in origin-of-life research, noting the conceptual divides between historical and ahistorical facets of the problem as well as the competing frameworks—such as "replication first," "metabolism first," and hybrid models—that seek to explain life's earliest emergence. See Addy Pross and Robert Pascal, *The origin of life: what we know, what we can know and what we will never know,* Open Biology 3 (2013): 120190, pmc.ncbi.nlm.nih.gov/articles/PMC3718341/; and Jack Lasky, *Abiogenesis,* EBSCO Research Starters (2024), ebsco.com/research-starters/biology/abiogenesis.

[446] Euthycarcinoids are an extinct arthropod lineage that lived from the Cambrian through the Triassic periods and are widely regarded as strong candidates for the earliest animals to transition onto land. Fossil evidence suggests that these organisms inhabited warm, tidal pools, likely spawning in these environments to protect their eggs from marine predators and accelerate larval development. Researchers have proposed that neoteny—the selective retention of juvenile traits—may have driven earlier sexual maturity in response to the harsh, fluctuating conditions of tidal zones. Morphological comparisons also reveal a notable resemblance between

euthycarcinoids and the juvenile stages of fuxianhuiid arthropods, leading to speculation about a possible ancestral connection. Trace fossils and depositional contexts further support the hypothesis that euthycarcinoids alternated between aquatic and subaerial environments rather than living entirely on land. See Simon Braddy, *The First Animals on Land Evolved in Warm Tidal Nursery Pools 500 Million Years Ago,* Sci.News (2024), sci.news/paleontology/euthycarcinoids-13132.html; S. J. Braddy, *Euthycarcinoid ecology and evolution, Neues Jahrbuch für Geologie und Paläontologie* (2024): doi 10.1127/njgpa/2024/1199, biostor.org/reference/166000; and Gregory D. Edgecombe, David A. Legg, and Xianguang Hou, *Ancient Land Arthropods in the Cambrian, Science* 272 (1996): 746–749, science.org/doi/pdf/10.1126/science.272.5262.746.

[447] The Cambrian explosion refers to the rapid diversification of animal life during the early Cambrian, a period marked by the emergence of numerous major body plans, the evolution of hard skeletons, and a dramatic expansion of ecological strategies. Although the exact tempo of this diversification remains debated, most estimates place the principal interval of evolutionary innovation within a window of roughly 20–30 million years. During this period, animals evolved new sensory systems, limb morphologies, feeding structures, and widespread biomineralized exoskeletons, leaving behind a rich fossil record that anchors modern interpretations of early metazoan evolution. See *Cambrian explosion,* Encyclopaedia Britannica (2024), britannica.com/science/Cambrian-explosion; and Derek E.G. Briggs, *The Cambrian explosion, Current Biology* 25 (2015): R845–R875, sciencedirect.com/science/article/pii/S0960982215004984.

[448] Fossil and molecular evidence indicate that both cnidarians and early cephalopods trace their origins to the Cambrian, with soft-bodied stem-group cephalopods such as *Nectocaris pteryx* extending the cephalopod lineage deeper into the Middle Cambrian and

molecular phylogenies placing significant cnidarian diversification near or before this same interval. See Martin R. Smith and Jean-Bernard Caron, *Primitive soft-bodied cephalopods from the Cambrian,* Nature 465 (2010): 469–472, nature.com/articles/nature09068; and Eunji Park et al., *Estimation of divergence times in cnidarian evolution based on mitochondrial protein-coding genes and the fossil record, Molecular Phylogenetics and Evolution* 62 (2012): 329–345, sciencedirect.com/science/article/abs/pii/S1055790311004374.

[449] Cambrian ecosystems saw a marked intensification of bioturbation, as burrowing organisms began reworking seafloor sediments to depths not previously observed in earlier intervals. This sediment mixing enhanced oxygen penetration and facilitated the recycling and mobilization of nutrients, contributing to the development of increasingly complex benthic food webs. Although recent analyses have refined earlier assumptions about the extent of sediment oxygenation driven solely by bioturbation, the ecological significance of these early burrowers remains well supported within the broader context of Cambrian environmental change. See A.R. Cribb et al., *Ediacaran–Cambrian bioturbation did not extensively oxygenate sediments in shallow marine settings, Geobiology* 21 (2023): 847–864, onlinelibrary.wiley.com/doi/10.1111/gbi.12550; and Lei Jiang et al., *Pulses of atmosphere oxygenation during the Cambrian radiation of animals, Earth and Planetary Science Letters* 590 (2022): 117565, sciencedirect.com/science/article/abs/pii/S0012821X22002011.

[450] Sharks first appear in the fossil record approximately 450 million years ago, with early chondrichthyan remains indicating the group's deep Paleozoic origins. Their long evolutionary tenure is often attributed to a highly efficient, conservatively maintained body plan, with many lineages exhibiting substantial morphological continuity over vast spans of geological time. Studies of shark diversity and macroevolution emphasize both their persistence and the relative stability of their core anatomical features, even as ecological roles and

species richness have shifted over hundreds of millions of years. See Alan Pradel et al., *The evolutionary history of sharks*, PLOS Biology 19 (2021): e3001108, journals.plos.org/plosbiology/article?id=10.1371/journal.pbio.3001108 ; and Zuzana Križnar and Luka Gale, *Diversity and evolutionary patterns of Paleozoic chondrichthyans*, Diversity 16 (2024): 147, mdpi.com/1424-2818/16/3/147.

[451] Gestalt psychology describes a set of perceptual principles—including symmetry, continuity, similarity, and proximity—that shape how humans organize and interpret sensory information. These principles function as unconscious rules that allow us to navigate complex visual and spatial environments by grouping elements into coherent wholes rather than isolated parts. See Stephen E. Palmer, *The role of symmetry in perceptual organization*, Proceedings of the National Academy of Sciences 109 (2012): 11369–11376, pmc.ncbi.nlm.nih.gov/articles/PMC3482144/.

[452] Research in cognitive and evolutionary psychology demonstrates that human perception and decision-making depend heavily on heuristics—rapid, efficient mental shortcuts shaped by natural selection to improve survival rather than optimize objective accuracy. These heuristics enable swift responses under threat but can also produce predictable perceptual and cognitive errors. See Patrick R. Steffen, Dawson Hedges, and Rebekka Matheson, *The Brain Is Adaptive Not Triune: How the Brain Responds to Threat, Challenge, and Change*, Frontiers in Psychiatry 13 (2022): 802606, pmc.ncbi.nlm.nih.gov/articles/PMC9010774/.

[453] Pareidolia is widely recognized in psychology as the human tendency to perceive faces or other familiar patterns in inanimate objects or random visual stimuli. It is considered a normal cognitive function rooted in the brain's pattern-recognition systems, emerging from the evolutionary advantage of rapidly identifying socially or

biologically relevant shapes. Common examples include perceiving faces in clouds, bark, or reflective surfaces—illustrations frequently noted in the psychological literature. See Janine Ungvarsky, *Pareidolia,* EBSCO Research Starters (2023), ebsco.com/research-starters/health-and-medicine/pareidolia.

[454] Research in cognitive psychology consistently shows that eyewitness accounts judged most credible often contain minor discrepancies, especially in peripheral details. Such variation reflects natural differences in perception, attention, and memory encoding rather than deception or fabrication. When multiple observers independently agree on the central features of an event but diverge slightly on secondary attributes—such as colors, angles, or brief impressions—this pattern is considered a hallmark of authentic collective recall. See Andrew J. Russ, Melanie Sauerland, Charlotte E. Lee, and Markus Bindemann, "Individual Differences in Eyewitness Accuracy Across Multiple Lineups of Faces," *Cognitive Research: Principles and Implications* 3 (2018): 30, cognitiveresearchjournal.springeropen.com/articles/10.1186/s41235-018-0121-8.

[455] The reliability of reported events increases substantially when independent modes of detection corroborate the same occurrence, a principle widely recognized in research methodology. In observational sciences, the convergence of multiple data streams—whether from separate observers, distinct instruments, or differing methodological approaches—is understood to strengthen the validity and credibility of findings. This same logic applies when radar, infrared, or other sensors independently match eyewitness observations, producing a form of instrumental triangulation that provides compelling evidence of an underlying real event. See Anita Bans-Akutey and Benjamin Makimilua Tiimub, *Triangulation in Research, Academia Letters* (2021): Article 3392, academia.edu/51125516/Triangulation_in_Research.

[456] The difference in arrival times between light and sound is a foundational observation in physics: electromagnetic waves propagate dramatically faster than acoustic waves, with light traveling at approximately 299,792,458 m/s in a vacuum, while even the fastest theoretical speed of sound in any material is orders of magnitude slower. This contrast underlies everyday experiences such as seeing lightning before hearing thunder and is widely used in physics education as an introductory demonstration of wave-speed differences. See *Speed of Light,* EBSCO Research Starters (2024), ebsco.com/research-starters/physics/speed-light; and Leah Crane, "Physicists have discovered the ultimate speed limit of sound," *New Scientist* (2020), newscientist.com/article/2256743-physicists-have-discovered-the-ultimate-speed-limit-of-sound/.

[457] Quantum entanglement refers to the nonclassical correlation between two particles whose shared quantum state causes measurement outcomes on one to be instantly reflected in the other, regardless of distance. Although these correlations appear to imply instantaneous influence, they cannot be used for communication because the outcome of any individual measurement is inherently random—a constraint formalized in the no-communication theorem. More broadly, quantum systems exist in superpositions of all possible states until measured, at which point the wavefunction collapses into a specific outcome from the observer's perspective. See Mark M. Wilde, *The No-Communication Theorem and Quantum Physics,* arxiv.org/abs/2409.11067; Whitney Clavin, *Untangling Quantum Entanglement, Caltech Magazine* (2019), magazine.caltech.edu/post/untangling-entanglement; and *What Is Quantum Superposition? Caltech Science Exchange* (2023), scienceexchange.caltech.edu/topics/quantum-science-explained/quantum-superposition.

[458] LIGO—the Laser Interferometer Gravitational-Wave Observatory— operates twin 4-km interferometers in Hanford, Washington, and

Livingston, Louisiana, designed to detect gravitational waves produced by extreme astrophysical events such as black hole mergers, neutron star collisions, and certain types of supernovae. These waves produce space-time strains on the order of $10^{-21}$, compressing and stretching distances by amounts far smaller than a proton's width. LIGO measures these distortions by splitting a laser beam into two perpendicular arms, reflecting the beams off suspended mirrors, and recombining them to detect interference shifts caused by minute changes in arm length. See *FAQ*, LIGO Laboratory (2024), ligo.caltech.edu/page/faq; *The Laser Interferometer Gravitational-Wave Observatory and the First Direct Observation of Gravitational Waves*, The Royal Swedish Academy of Sciences (2017), nobelprize.org/uploads/2018/06/advanced-physicsprize2017-1.pdf; and *LLO Laser SOP Template (LIGO-M1400242-V1)*, LIGO Document Control Center, dcc-llo.ligo.org.

[459] The common statement that physical objects are "mostly empty space" is an oversimplification. In quantum mechanics, the region surrounding an atomic nucleus is filled by the electron cloud—a probabilistic distribution described by wavefunctions, indicating where electrons are most likely to be found. Rather than empty volume, this space is occupied by continuously distributed electron probability densities and quantum fluctuations, which determine the atom's interactions and apparent solidity. See Matthew Williams, "What Is the Electron Cloud Model?" *Universe Today* (2016), universetoday.com/articles/electron-cloud-model.

[460] The atomic nucleus is extraordinarily small compared to the overall size of the atom, with standard estimates placing nuclear dimensions at roughly 100,000 times smaller than the atom itself. This extreme scale difference is frequently illustrated by analogies comparing the atom to a large stadium and the nucleus to a marble or peanut at its center—emphasizing that nearly all of the atom's volume is occupied by the surrounding electron cloud. See Ali Sundermier,

"The particle physics of you," *Symmetry Magazine* (2015), symmetrymagazine.org/article/the-particle-physics-of-you; and *The atomic nucleus*, Radioactivity.eu.com (2024), radioactivity.eu.com/categories/phenomenon/atomic_nucleus.

[461] The apparent solidity of everyday objects arises not from physical contact between atoms but from quantum-mechanical interactions, chiefly the electrostatic repulsion between negatively charged electron clouds and the Pauli exclusion principle, which prevents identical fermions—such as electrons—from occupying the same quantum state. Together, these effects produce the resistance we interpret as firmness when macroscopic bodies press against one another; in essence, solidity emerges from interacting fields rather than from atoms as hard, indivisible particles. See Masato Itani, Shunji Muto, and Yuichi Motoyama, "Quantum mechanical origin of internal resistance existing inside metallic hydrides," *Journal of Condensed Matter Physics* 7 (2016): 084104, pmc.ncbi.nlm.nih.gov/articles/PMC6386640/; and Christopher L. Cook et al., "Atomic-scale origins of solid stiffness revealed by in situ electron microscopy," *Scientific Reports* 8 (2018): 25493, nature.com/articles/s41598-018-25493-7.

[462] Electrons are not classical point-like marbles but quantum wavefunctions—spread-out probability amplitudes whose squared modulus gives the likelihood of finding the electron in a given region of space. This framework reflects the inherently delocalized and probabilistic nature of subatomic particles prior to measurement. Moreover, the apparent "emptiness" within atoms is permeated by quantum fluctuations and virtual-particle activity, as described in quantum field theory, revealing a dynamic vacuum rather than true emptiness. See Shan Gao, "Interpreting the wave function—what are electrons? and how do they move?," *Discusiones Filosóficas* 14, no. 22 (2013): 13–23, scielo.org.co/scielo.php?script=sci_arttext&pid=S0124-61272013000100002.

[463] Paleoanthropological research places the earliest hominins—those more closely related to humans than to chimpanzees—at roughly six to seven million years ago. Fossils such as *Sahelanthropus tchadensis* (approximately 6–7 million years old) and Orrorin tugenensis (around 6 million years old) mark the earliest known points in the human lineage following its divergence from the chimpanzee lineage. See Jenny Wong and Lisa Hendry, "The origin of our species," Natural History Museum, nhm.ac.uk/discover/the-origin-of-our-species.html.

CHAPTER FIFTEEN

[464] The Orchestrated Objective Reduction (Orch-OR) model, developed by Sir Roger Penrose and Stuart Hameroff, proposes that consciousness arises not solely from computational processes but from the orchestrated collapse of quantum superpositions occurring within cytoskeletal microtubules, with these objective reductions linked to the fundamental geometry of space-time. The theory argues that such collapses reflect non-computable processes embedded in quantum gravity and may underlie the structure of conscious awareness itself. See Stuart Hameroff and Roger Penrose, "Consciousness in the universe: A review of the 'Orch OR' theory," *Physics of Life Reviews* 11 (2014): 39–78, sciencedirect.com/science/article/pii/S1571064513001188.

[465] The concept of quantum decision trees has been explored as a framework for modeling how quantum-coherent systems may evolve toward preferred outcomes without classical interference. These models treat branching possibilities as superposed states that can be guided or "steered" along probabilistic paths before measurement collapse. See Subhash Kak, "On Quantum Decision Trees" (2017), arxiv.org/pdf/1703.03693; and Songfeng Lu and Samuel L. Braunstein, "Quantum decision tree classifier," *Quantum Information Processing* 13 (2014): 757–770, link.springer.com/article/10.1007/s11128-013-0687-5.

[466] Wheeler's delayed-choice experiment and its modern realizations show that the measurement context—chosen only after a photon has passed an initial beam splitter—determines whether the photon displays wave-like or particle-like behavior. In the delayed-choice formulation, a single photon is sent toward an interferometer, and the decision to observe interference (wave behavior) or which-path information (particle behavior) is postponed until after the photon has already entered the apparatus. Experimental implementations confirm the theoretical prediction: the photon's detected behavior matches the delayed measurement setting, leading to the appearance of retroactive influence. See Marijn Waaijer and Jan van Neerven, "Delayed choice experiments: an analysis in forward time," *Quantum Studies: Mathematics and Foundations* 11 (2024): 391–408, link.springer.com/article/10.1007/s40509-024-00328-5; and Changyu Huang, Yong-Chang Huang, and Yi-You Nie, "New quantum physics, solving puzzles of Wheeler's delayed choice and a particle's passing N slits simultaneously and quantum oscillator in experiments," *Scientific Reports* 12 (2022): 14410, pmc.ncbi.nlm.nih.gov/articles/PMC9402567/.

[467] Quantum computing uses qubits, which can exist in a superposition of 0 and 1, allowing quantum processors to represent and manipulate many possible states simultaneously. When combined with entanglement and quantum interference, this enables quantum systems to evaluate multiple computational pathways in parallel before measurement collapses the state to a definite outcome. See *What Is Quantum-Centric Supercomputing?* IBM, ibm.com/think/topics/quantum-centric-supercomputing; and *Quantum Computing Explained,* National Institute of Standards and Technology (NIST), nist.gov/quantum-information-science/quantum-computing-explained.

[468] In fungi, the familiar mushroom is only the reproductive structure; the primary organism is the mycelium, an extensive, interconnected

network of hyphae that permeates soil or decaying substrate. This mycelial matrix constitutes the true biological individual, supporting growth, nutrient exchange, and communication across large spatial scales. At the same time, the visible fruiting bodies represent only brief, emergent structures for spore dispersal. See Mercedes A. Segura Campos, *The Mycelium as a Network, Microbiology Spectrum* 5 (2017): FUNK-0033-2017, journals.asm.org/doi/10.1128/microbiolspec.funk-0033-2017.

[469] Models of distributed artificial intelligence describe systems in which individual computational nodes operate semi-independently while contributing to an emergent, network-level intelligence. In such architectures, awareness or decision-making is not confined to a single location but arises from the coordinated behavior of many interconnected elements, analogous to a cloud-based or spatially distributed cognitive framework. See João Leite and Luís Moniz Pereira, "Distributed AI and the Future of Intelligence," *AI and Ethics* 3 (2023): 747–759, sciencedirect.com/science/article/pii/S266730532300056X.

[470] Human memory is widely understood to arise from activity-dependent changes at synapses, with learning and recall involving biochemical and structural modifications to neuronal circuits. Long-term memory formation is associated with synaptic remodeling, changes in gene expression, and the consolidation of neural activity patterns over time. However, the precise mechanisms remain an active area of research. See Eric R. Kandel, "The molecular biology of memory: cAMP, PKA, CRE, CREB-1, CREB-2, and CPEB," *Cold Spring Harbor Perspectives in Biology* 4 (2012): a021758, pmc.ncbi.nlm.nih.gov/articles/PMC4484970/.

[471] The holographic memory model posits that neural information may be encoded not as localized traces but as distributed interference patterns across neuronal microcircuits. This framework draws on

theoretical work proposing that memory emerges through wave-like propagation and pattern interference, resulting in non-local, holographically distributed storage rather than discrete synaptic repositories. See Alexey D. Redozubov, "Holographic Memory: A Novel Model of Information Processing by Neuronal Microcircuits," in *The Physics of the Mind and Brain Disorders*, eds. Ioan Opris and Mikhail F. Casanova (Springer, 2017), 271–295, link.springer.com/chapter/10.1007/978-3-319-29674-6_13.

[472] The creation of a hologram involves splitting a coherent laser beam into an object beam and a reference beam, directing one toward the object and the other straight to the recording medium; the interference pattern created where these beams meet encodes both the amplitude and phase of the scattered light, allowing the full three-dimensional structure of the object to be stored on a two-dimensional surface. Each portion of the resulting hologram contains the entire image, with resolution decreasing as the fragment shrinks, reflecting the inherent redundancy and non-local information storage characteristic of holography. See *Holography*, EBSCO Research Starters, 2022, at ebsco.com/research-starters/history/holography and Tracy V. Wilson, "How Holograms Work," *HowStuffWorks*, updated 2023, at science.howstuffworks.com/hologram.htm

[473] The term "probability gradient" refers to how a probability distribution changes as its underlying variables shift, describing the sensitivity of an outcome's likelihood to slight variations in conditions. This concept appears in mathematical treatments of the evolution of probability measures and gradient-flow dynamics. See *Gradient Flows of Probability Measures*, in Handbook of Differential Equations: Evolutionary Equations, Vol. 1 (2007), sciencedirect.com/science/chapter/handbook/abs/pii/S1874571707800 041?via%3Dihub; and Xiaohui Chen and Hans-Georg Müller, "Wasserstein gradients for the temporal evolution of probability distributions," Electronic Journal of Statistics 15 (2021): 4061–4084,

projecteuclid.org/journals/electronic-journal-of-statistics/volume-15/issue-2/Wasserstein-gradients-for-the-temporal-evolution-of-probability-distributions/10.1214/21-EJS1883.full.

[474] The Magic Castle is a historic private clubhouse for magicians located in Hollywood, California, housed in a 1909 château-style mansion perched above Franklin Avenue and serving as the home of the Academy of Magical Arts. It is widely regarded as a legendary, members-only venue known for its intimate performance spaces, strict dress code, and distinctive atmosphere, which have made it an iconic center of stage magic and illusion in Los Angeles. See *Academy History*, The Academy of Magical Arts, magiccastle.com/history/; and *Magic Castle*, Wikipedia, en.wikipedia.org/wiki/Magic_Castle.

[475] Three-card monte is a long-standing confidence game in which victims are deceived into betting on the location of a "money card" among three cards, with the outcome predetermined through sleight of hand and the coordinated actions of shills. Historically recognized as a classic short con, the game has been documented for centuries and is designed not to entertain but to defraud, with the mark having no genuine chance of success at any point. See *Three-card monte*, Wikipedia, en.wikipedia.org/wiki/Three-card_monte.

## CHAPTER SIXTEEN

[476] Contemporary reporting on the Cisco Grove incident identifies Donald Shrum as a young missile technician employed by Aerojet who worked on Polaris and Titan systems, and who embarked on the September 1964 bow-hunting trip with two colleagues before becoming separated in the Tahoe National Forest. Multiple accounts also note that Shrum initially withheld his identity for professional reasons, using pseudonyms for decades and permitting his real name to be associated with the case only in the 2000s, a delay attributed to concerns about ridicule and its potential impact on his aerospace

career. See "The Cisco Grove UFO Incident: Don Shrum's 1964 Alien Encounter in Tahoe National Forest," tvi.show/nexus/the-cisco-grove-ufo-incident-don-shrums-1964-alien-encounter-in-tahoe-national-forest; Vicky Verma, "Man Fought With Three Alien Robots Whole Night in California's Forest in 1964," howandwhys.com/cisco-grove-alien-encounter/; Marcus Lowth, "The 1964 Cisco Grove Encounter," ufoinsight.com/aliens/encounters/cisco-grove-encounter.

[477] The primary accounts of the Cisco Grove incident describe Shrum's overnight ordeal as beginning with a silent light moving through the trees and revealing a large cylindrical craft, followed by the emergence of smaller saucer-like craft and multiple entities—including two humanoid figures in metallic suits and a robot-like figure—that approached his tree repeatedly. These reports document the entities shaking the tree, emitting a visible vapor that caused Shrum to lose consciousness on multiple occasions, and attempting to climb toward him. They also record Shrum's defensive actions, including firing arrows at the figures and dropping burning materials such as clothing and matches, as well as the prolonged, hours-long standoff that continued until the beings withdrew shortly before dawn. See "The Cisco Grove UFO Incident: Don Shrum's 1964 Alien Encounter in Tahoe National Forest," tvi.show/nexus/the-cisco-grove-ufo-incident-don-shrums-1964-alien-encounter-in-tahoe-national-forest; Vicky Verma, "Man Fought With Three Alien Robots Whole Night in California's Forest in 1964," howandwhys.com/cisco-grove-alien-encounter/; Marcus Lowth, "The 1964 Cisco Grove Encounter," ufoinsight.com/aliens/encounters/cisco-grove-encounter.

[478] Several accounts of the Cisco Grove incident note that one of Shrum's hunting companions, Vincent Alvarez, independently observed the unusual light in the forest that same night, later describing a glowing craft rising rapidly into the sky. Reporting also states that Alvarez provided a written affidavit regarding what he saw,

and that the group found physical traces the following day—such as arrows and charred clothing—consistent with elements of Shrum's account. See Vicky Verma, "Man Fought With Three Alien Robots Whole Night in California's Forest in 1964," howandwhys.com/cisco-grove-alien-encounter/; Marcus Lowth, "The 1964 Cisco Grove Encounter," ufoinsight.com/aliens/encounters/cisco-grove-encounter.

479 Multiple summaries of the Cisco Grove case report that the incident prompted an official response from the U.S. Air Force, with investigators from Wright-Patterson Air Force Base—identified as Captain McLoy (or McClory) and Sergeant Barnes—dispatched cross-country to interview Shrum and review the evidence. These sources describe a priority notification referencing a "UFO landing with entities" and note that the investigators questioned Shrum extensively, examined his recovered arrows, and treated the case as one of the small fraction of reports that merited further attention under Project Blue Book-era protocols. See "The Cisco Grove UFO Incident: Don Shrum's 1964 Alien Encounter in Tahoe National Forest," tvi.show/nexus/the-cisco-grove-ufo-incident-don-shrums-1964-alien-encounter-in-tahoe-national-forest; Marcus Lowth, "The 1964 Cisco Grove Encounter," ufoinsight.com/aliens/encounters/cisco-grove-encounter.

480 Discussions of the Cisco Grove incident in contemporary analyses and community forums outline several competing interpretations. Some contributors argue that Shrum's symptoms and perceptions could reflect a hallucinogenic or otherwise altered physiological state, while others propose the event may have involved a covert military activity or chemical test misinterpreted under stress. A third position maintains that the encounter proceeded as Shrum described, involving nonhuman beings and advanced craft. See "Cisco Grove / Donald Shrum (AKA 'Schrum') Incident (California; 1964)," forums.forteana.org/index.php?threads/cisco-grove-donald-shrum-

aka-schrum-incident-california-1964.66278/; "S4 E26: Aliens in the Forest – The Cisco Grove Incident," thatwouldberadpodcast.com/s4-e26-aliens-in-the-forest-the-cisco-grove-incident/.

[481] Historical documentation shows that during the period in which the Cisco Grove incident occurred, U.S. intelligence agencies were actively carrying out covert experimentation with psychoactive substances and other chemical agents, frequently without the knowledge or consent of the people involved. The CIA's MK-Ultra program, undertaken from 1953 to 1964, included the surreptitious administration of LSD and related compounds to unwitting civilians, incarcerated people, hospital patients, and others, and operated outside standard ethical constraints. Broader Cold War–era programs also involved secret testing of biological and chemical agents in populated areas, exposing large numbers of people without informed consent and under conditions later deemed unethical. See "MK-ULTRA," britannica.com/topic/MKULTRA; Leonard A. Cole, "Open-Air Biowarfare Testing and the Evolution of Values," Health Security 14 (2016): 315–322, pmc.ncbi.nlm.nih.gov/articles/PMC5041545/.

[482] Clinical descriptions of LSD and related psychedelics emphasize that these substances disrupt normal sensory processing and cognition, often causing hallucinations, distorted perception of space and time, impaired coordination, and rapid shifts in thought and mood. Such states frequently involve confusion, disorientation, and difficulty maintaining coherent focus or organized behavior, with experiences that are typically unstable and variable as brain networks become desynchronized. Psychedelic states are further characterized by transient, evolving imagery rather than sustained, consistent perceptions or goal-directed actions. See "Hallucinogens: LSD, Peyote, Psilocybin, PCP & Other Psychedelic Drugs," my.clevelandclinic.org/health/articles/6734-hallucinogens-lsd-peyote-psilocybin-and-pcp; "How Psychedelics Affect the Brain," americanbrainfoundation.org/how-psychedelics-affect-the-brain/.

[483] Reporting on the 1994 Ariel School incident in Ruwa, Zimbabwe, records that 62 schoolchildren were present on the grounds during the event, with a subset of them claiming to have directly witnessed the landed craft and beings. Accounts from those direct witnesses have remained notably consistent in interviews conducted many years later, contributing to the case's enduring status as an unresolved and widely discussed encounter. See Sean Christie, "Remembering Zimbabwe's Great Alien Invasion," *Mail & Guardian* (2014), mg.co.za/article/2014-09-04-remembering-zimbabwes-great-alien-invasion/.

[484] Geographic and municipal data identify Ruwa as a small agricultural community situated about 25 kilometers from Harare, Zimbabwe's capital, with a population size and settlement structure reflecting a semi-rural character despite its proximity to the urban center. Contemporary reporting on the Ariel School case further notes that the school was a relatively expensive private institution whose student body in the 1990s primarily consisted of children from affluent families in Harare who were sent there for a higher-end educational environment. See "Ruwa Local Board," ucaz.org.zw/members/ruwa-local-board/; Brian Dunning, "The 1994 Ruwa Zimbabwe Alien Encounter," Skeptoid #760 (2020), skeptoid.com/episodes/760.

[485] Reporting on the Ariel School incident notes that several sightings of unusual aerial phenomena were made in the region in the days leading up to the event. Contemporary coverage describes multiple witnesses observing bright objects or "balls of fire" moving silently across the sky, with some interpreting them as meteors or comets while others viewed them as unidentified craft. Additional accounts from local residents and journalists indicate similar observations in areas such as Hwange, Bulawayo, and Kariba during the same period, contributing to a brief regional cluster of sightings before the schoolyard encounter. See Sean Christie, "Remembering Zimbabwe's

Great Alien Invasion," *Mail & Guardian* (2014), mg.co.za/article/2014-09-04-remembering-zimbabwes-great-alien-invasion/; "The Day the Aliens Landed," *The Herald* (2014), heraldonline.co.zw/the-day-the-aliens-landed/.

[486] Accounts of the Ariel School investigation identify two principal researchers who documented the children's testimony soon after the incident. Dr. John Mack, a psychiatrist and professor at Harvard Medical School, conducted in-depth interviews with the witnesses during his visit to Ruwa. Cynthia Hind, serving as the African representative for MUFON and editor of *UFO Afrinews*, also investigated the case, interviewing pupils and school staff and compiling early reports on the event. See Brian Dunning, "The 1994 Ruwa Zimbabwe Alien Encounter," Skeptoid #760 (2020), skeptoid.com/episodes/760; Sean Christie, "Remembering Zimbabwe's Great Alien Invasion," *Mail & Guardian* (2014), mg.co.za/article/2014-09-04-remembering-zimbabwes-great-alien-invasion/.

[487] Critical evaluations of the Ariel School investigation note that John Mack interviewed the children in small groups and that his setup allowed some pupils to watch portions of other interviews, a circumstance described as a methodological weakness because it increases the risk of inadvertent influence or convergence of recollections. Analyses also point out that Cynthia Hind asked the children to draw what they had witnessed and later selected specific drawings to share publicly. This process may have introduced selection bias by emphasizing illustrations that aligned with a UFO-centered interpretation while omitting others. See Brian Dunning, "The 1994 Ruwa Zimbabwe Alien Encounter," Skeptoid #760 (2020), skeptoid.com/episodes/760; James Felton, "The Ariel School Phenomenon: What Really Happened When 68 Children Witnessed a UFO?" IFLScience (2022), iflscience.com/the-ariel-school-

phenomenon-what-really-happened-when-68-children-witnessed-a-
ufo-63873.

[488] Launch records confirm that a Zenit-2 rocket carrying the *Cosmos 2290* (Orlets-2) payload was launched from the Baikonur Cosmodrome on August 26, 1994, approximately three weeks before the Ariel School incident. While some commentators have suggested a connection between this launch and a wave of aerial sightings reported in Zimbabwe, contemporary analyses indicate that the widespread fireball observed across southern Africa on September 14, 1994, was caused by the atmospheric re-entry of the Zenit-2 second-stage booster, catalogued by NORAD as SL-16 R/B (ID 23219). This re-entry event is distinct from the launch itself and does not involve debris falling directly from Baikonur. Although the re-entry accounts for certain regional sky sightings in the days immediately preceding the Ariel School encounter, it does not establish a direct causal link to the events reported at the school. See "Cosmos 2290 | Zenit-2," *Next Spaceflight*, nextspaceflight.com/launches/details/1301/; Brian Dunning, "The 1994 Ruwa Zimbabwe Alien Encounter," *Skeptoid* #760 (2020), skeptoid.com/episodes/760; James Felton, "The Ariel School Phenomenon: What Really Happened When 68 Children Witnessed a UFO?" *IFLScience* (2022), iflscience.com/the-ariel-school-phenomenon-what-really-happened-when-68-children-witnessed-a-ufo-63873.

[489] Contemporary reporting and later analyses of the Ariel School incident describe the children witnessing one or more silver, disc-like craft descending toward a wooded area adjacent to the playground, with several pupils stating that figures dressed in black with large, prominent eyes emerged and moved toward them. Accounts gathered by John Mack include multiple children who claim the beings communicated telepathically, conveying concerns about environmental degradation and warning of dire consequences if humanity continued on a destructive path; one child recalled being

explicitly told that the Earth would be destroyed if people did not change their behavior. News coverage and interviews conducted soon after the event also note that the witnesses consistently rejected the idea that the objects were airplanes. See Sean Christie, "Remembering Zimbabwe's Great Alien Invasion," *Mail & Guardian* (2014), mg.co.za/article/2014-09-04-remembering-zimbabwes-great-alien-invasion/; "The Day the Aliens Landed," *The Herald* (2014), heraldonline.co.zw/the-day-the-aliens-landed/; James Felton, "The Ariel School Phenomenon: What Really Happened When 68 Children Witnessed a UFO?" IFLScience (2022), iflscience.com/the-ariel-school-phenomenon-what-really-happened-when-68-children-witnessed-a-ufo-63873.

[490] Some analyses of the Ariel School incident note that not all of the children interpreted the figures as extraterrestrial. A number of students drew on regional folklore to explain what they saw, identifying the beings as tokoloshes or related spirit figures from Shona and Ndebele tradition, reflecting local cultural frameworks for interpreting unusual or frightening encounters. See Robbie Mitchell, "The Ariel UFO Incident: When Aliens Visited a School in Zimbabwe," historicmysteries.com/unexplained-mysteries/ariel-ufo-children/28719/.

[491] Later commentary on the Ariel School case includes the claim, presented in the 2023 Netflix documentary *Encounters*, that a former student named Dallyn stated he initiated the incident as a prank by telling other children that a distant "shiny rock" was a UFO. In the documentary, he characterized the ensuing reactions as an unexpected instance of mass hysteria. His account stands in contrast to an earlier on-camera interview from roughly 15 years prior, in which he described seeing a craft with lights that "flash[ed] a different color in the sky." See "Ariel School UFO Incident," *Wikipedia*, en.wikipedia.org/wiki/Ariel_School_UFO_incident; Dallyn interview, youtube.com/watch?v=AQqzhzABfaY.

[492] Accounts of the Delphos Ring incident describe 16-year-old Ron
Johnson witnessing a glowing, dome- or mushroom-shaped object
hovering silently near a tree on his family's farm in November 1971,
with his parents also reporting that they saw the object as it ascended
and departed. Subsequent descriptions emphasize the physical trace
left behind: an approximately eight-foot circular ring of pale,
desiccated soil that appeared to glow and was markedly drier and
lighter in color than the surrounding ground. Reporting further notes
that local law enforcement, including Sheriff Ralph Enlow, visited the
site, photographed the ring, and confirmed its unusual appearance,
including faint luminescence and soil conditions distinct from those
of the rest of the property. See Jennifer Jones, "The Delphos Ring:
Kansas' Mysterious UFO Incident,"
thedeadhistory.com/2025/06/18/the-delphos-ring/; Gerrard Kaonga,
"'Eye-opening' Scorch Marks on Kansas Farm Show Where Family
Watched 'UFO' Land," unilad.com/community/ufo-scorch-marks-
evidence-ground-farm-kansas-glowing-958087-20231010.

[493] Laboratory analyses of soil samples taken from the Delphos ring
indicated that the affected material had become hydrophobic,
resisting water absorption even when mechanically agitated.
Chemical testing further identified alterations in composition,
including the presence of an extracted water-soluble compound not
found in control soil and structural changes consistent with chemical
modification rather than natural variation. See Erol A. Faruk, "A New
Appraisal of the Data of the Delphos CE2 1971 Case,"
explorescu.org/post/a-new-appraisal-of-the-data-of-the-delphos-ce2-
1971-case.

[494] Certain natural biological processes can produce circular soil
formations that may superficially resemble the Delphos ring. Studies
of fungal ecology have documented "fairy ring"–forming species
whose underground mycelial growth generates sharply defined
circular patches, sometimes accompanied by shifts in soil hydration

or nutrient availability. Research also shows that bacterial colonies can self-organize into concentric ring patterns under environmental stress, and that some fungal species are capable of true bioluminescence. See "Riding the Wave: Response of Bacterial and Fungal Microbiota Associated with the Spread of the Fairy Ring Fungus *Calocybe gambosa*," ScienceDirect, sciencedirect.com/science/article/abs/pii/S0929139321000846; Clare Watson, "Scientists Spot Eerily Sophisticated Patterns in 'Simple' Bacteria Colonies," ScienceAlert, sciencealert.com/scientists-spot-eerily-sophisticated-patterns-in-simple-bacteria-colonies; Anderson G. Oliveira et al., "Circadian Control Sheds Light on Fungal Bioluminescence," *Current Biology*, sciencedirect.com/science/article/pii/S0960982215001608.

[495] Analyses of UAP history frequently note the persistence of a theory within UFO literature that non-Earth-based intelligences show particular interest in human nuclear technology. Reporting and research surveys connect this idea to repeated clusters of sightings near nuclear test sites and weapons facilities, and to the long-standing belief within UAP discourse that nuclear detonations and weapons programs may trigger outside observation. Contemporary articles and reviews of historical incidents further document the extensive cataloging of UAP–nuclear correlations by researchers and commentators. See Eric Lagatta, "Did aliens spy on our nuclear tests? Study finds signs of UFOs near US sites in 1950s," *USA Today* (2025), usatoday.com/story/news/nation/2025/10/28/uap-nuclear-testing-study/86941783007/; Tim Ventura, "The UAP Nuclear Correlation," altpropulsion.com/the-uap-nuclear-correlation/.

[496] The teaching from Case 29 of the *Mumonkan* (Gateless Gate) is a classic Zen koan in which two monks debate whether it is the wind or the flag that moves, prompting the Sixth Patriarch Huineng to reply that it is the mind that moves. Traditional commentaries preserve this phrasing and identify it as one of the most widely cited

illustrations of Zen insight into perception and consciousness. See "Not the Wind; Not the Flag," zentexts.org/blog/2020-07-05-not-wind-not-flag/; "Mumonkan Case 29," moonwaterdojo.com/blog/mumonkan-case-29.

## CHAPTER SEVENTEEN

[497] Convergent evolution is widely documented as a pattern in which distantly related lineages independently arrive at similar traits or functional solutions in response to comparable ecological pressures, illustrating that evolution can repeatedly reach the same outcomes across disparate groups. This includes well-studied examples such as the independent evolution of streamlined bodies in marine vertebrates, camera-type eyes in vertebrates and cephalopods, and numerous other morphological and behavioral parallels. See C. Tristan Stayton, "What does convergent evolution mean? The interpretation of convergence and its implications in the search for limits to evolution," *Interface Focus* 5 (2015): 20150039, pmc.ncbi.nlm.nih.gov/articles/PMC4633856; Holly Chetan-Welsh and Lisa Hendry, "Convergent evolution explained with 13 examples," Natural History Museum, nhm.ac.uk/discover/convergent-evolution.html.

[498] The Smithsonian Human Origins Program places the appearance of anatomically modern *Homo sapiens* at roughly 300,000 years ago, based on fossil evidence from African sites such as Jebel Irhoud. This establishes that an event occurring around 500,000 years ago would precede the emergence of modern humans by approximately 200,000 years. See "Homo sapiens," Smithsonian National Museum of Natural History, humanorigins.si.edu/evidence/human-fossils/species/homo-sapiens.

[499] Fossil evidence places the earliest known hominins at approximately 7 to 6 million years ago, based on *Sahelanthropus*

*tchadensis*, one of the oldest species in the human lineage. This establishes the minimum range for the origin of hominids and supports the statement that hominids have existed for roughly 6 to 7 million years. See "Sahelanthropus tchadensis," Smithsonian National Museum of Natural History, humanorigins.si.edu/evidence/human-fossils/species/sahelanthropus-tchadensis.

[500] Geological data from the U.S. National Park Service places the start of the Cambrian Period at approximately 541 million years ago, marking the beginning of a significant diversification of complex life. This aligns with the widely used estimate that the Cambrian began around 540 million years ago. See "Cambrian Period—541 to 485.4 MYA," National Park Service, nps.gov/articles/000/cambrian-period.htm.

[501] Quantum biology research demonstrates that quantum effects underlie several biological processes in both plants and animals, including photon absorption in vision, exciton transport in photosynthesis, and radical-pair spin dynamics involved in magnetoreception. These examples illustrate that quantum phenomena such as quantized energy states, coherence-assisted transport, and spin-dependent reactions play functional roles across diverse living systems. See Pedro H. Alvarez et al., "Quantum phenomena in biological systems," *Frontiers in Quantum Science and Technology* (2024), frontiersin.org/articles/10.3389/frqst.2024.1466906/full.

[502] Quantum proton tunneling has been identified as a viable mechanism for DNA base-pair tautomerisation, with theoretical models showing that protons can traverse the hydrogen-bond barrier between nucleotide bases on timescales far shorter than biological processes allow—down to femtoseconds, or less than a quadrillionth of a second. This tunneling-driven transfer produces transient

tautomeric forms that can survive long enough to generate replication mismatches, thereby contributing to point mutation rates. Recent open-quantum-systems modeling of the G–C base pair demonstrates that quantum tunneling increases proton-transfer rates by several orders of magnitude compared to classical predictions and yields a tautomeric occupation probability large enough to influence mutation frequency. These findings are consistent with broader explanations of proton tunneling as a rapid process occurring within ~$10^{-15}$ seconds and frequent enough to affect mutational outcomes. See Louie Slocombe, Marco Sacchi, and Jim Al-Khalili, "An open quantum systems approach to proton tunnelling in DNA," *Communications Physics* 5 (2022): 109, nature.com/articles/s42005-022-00881-8; Jenny Morber, "'Spooky' quantum biology might cause your DNA to mutate," *Big Think*, bigthink.com/hard-science/quantum-biology-mutation.

[503] The sickle-cell allele is a classic example of heterozygote advantage: individuals inheriting two copies develop sickle cell disease, which carries severe physiological consequences, while carriers with only one copy exhibit significant resistance to *Plasmodium falciparum* malaria. This balanced polymorphism has been extensively documented, with epidemiological and molecular data showing that the heterozygous genotype confers protection in malaria-endemic regions, while the homozygous genotype causes the full clinical disorder. See Philip E. Thuma et al., "The prevalence of sickle cell trait and glucose-6-phosphate dehydrogenase deficiency in a malaria-endemic region of Zambia," *American Journal of Tropical Medicine and Hygiene* 98 (2018): 1424–1427, pmc.ncbi.nlm.nih.gov/articles/PMC6171532.

[504] Early genomic research held that only a small fraction—roughly 1–2%—of the human genome encoded functional proteins, leading the remaining noncoding regions to be labeled "junk DNA." Subsequent large-scale functional genomics work, including ENCODE-era

analyses and later critiques, has demonstrated that many noncoding elements participate in regulatory activity, chromatin organization, transcription factor binding, and other processes that influence gene expression. Modern reviews emphasize that while not all noncoding DNA is adaptive, substantial portions once classified as "junk" play key biochemical and regulatory roles, revealing a far more complex and functionally integrated genome than early models proposed. See Nelson J.R. Fagundes et al., "What We Talk About When We Talk About 'Junk DNA,'" *Genome Biology and Evolution* 14 (2022): evac055, pmc.ncbi.nlm.nih.gov/articles/PMC9086759; A.N. Wittkopp et al., "Short tandem repeats bind transcription factors to tune eukaryotic gene expression," *Science* 380 (2023): eadd1250, science.org/doi/10.1126/science.add1250.

## EPILOGUE

[505] Research in cognitive and integrative neuroscience supports the view that selective attention evolved as an adaptive mechanism for filtering sensory information in the service of survival-driven behavior. Comparative and evolutionary analyses show that attentional systems help organisms prioritize behaviorally relevant stimuli and suppress distractions, functions that would have been essential for early humans engaged in tasks such as foraging, hunting, or predator avoidance. These mechanisms are understood not merely as cognitive refinements but as deeply rooted ecological adaptations found widely across taxa, emphasizing the role of focused attention in navigating complex environments. See Tidhar Lev-Ari, Hadar Beeri, and Yoram Gutfreund, "The Ecological View of Selective Attention," *Frontiers in Integrative Neuroscience* 16 (2022): 856207, frontiersin.org/articles/10.3389/fnint.2022.856207/full; Wouter Kruijne et al., "Tracking the allocation of attention using brief, rare stimuli," *Frontiers in Neuroscience* 9 (2015): 55, pmc.ncbi.nlm.nih.gov/articles/PMC4364301.

506 Both physics and perceptual neuroscience provide empirical evidence that slight differences in an observer's position, motion, or orientation can produce measurable changes in what is perceived. In physics, relativistic frame dependence shows that temporal and spatial measurements vary systematically with the motion and reference frame of the observer, establishing that observation is not absolute but depends on relative speed and position. In perceptual science, sensory judgments—including time perception—are demonstrably altered by variations in sensory input and viewpoint, with experiments showing that even minor changes in stimulus dynamics can shift observers' internal estimates of events. Together, these domains confirm that observer-dependent variation is a real and quantifiable feature of physical measurement and perceptual processing. See Gergely Szirkó, "Why Is Time Frame-Dependent in Relativity?" philsci-archive.pitt.edu/2462; Misha B. Ahrens and Maneesh Sahani, "Observers exploit stochastic models of sensory change to help judge the passage of time," *Current Biology* 21 (2011): 200–206, cell.com/current-biology/fulltext/S0960-9822(10)01711-2.

507 Plato introduces the allegory of the cave in *Republic* Book VII, describing prisoners confined in a dark cavern who see only shadows cast by objects passing before a fire behind them, mistaking these shadows for reality. The text further explains that only upon being released and ascending toward the light can they perceive the proper relation between shadow and form. However, this transition is initially painful and disorienting. See Plato, *Republic*, Book VII, trans. Oleg Bychkov, "The Allegory of the Cave," web.sbu.edu/theology/bychkov/plato%20republic%207.pdf; Daniel R. DeNicola, "Plato's Cave and the Stubborn Persistence of Ignorance," *MIT Press Reader*, thereader.mitpress.mit.edu/platos-cave-and-the-stubborn-persistence-of-ignorance

# ABOUT THE AUTHOR

Keaton Ryon writes across fiction and nonfiction, exploring the borderlands where science, mystery, and imagination meet. Keaton's work follows hidden patterns, unanswered questions, and the deeper stories woven into the natural world and human experience.

Learn more at **keatonryon.com**.
Continue the journey of *Abyssal Sapience* at **abyssalsapience.com**.